$$-\lambda \int e^{-\lambda t} t \ dt$$

$$t = v \qquad du = e^{-\lambda t}$$
$$dv = 1 \qquad v = e^{-\lambda t}$$

$$v \ \ + t e^{-\lambda t} + \lambda \int e^{-\lambda t} \ dt$$

$$t e^{-\lambda t} + \lambda e^{-\lambda t} \Big|_0^x$$

$$x e^{-\lambda x} + \lambda e^{-\lambda x} - \lambda$$

ELEMENTARY
NUMERICAL
ANALYSIS

ELEMENTARY NUMERICAL ANALYSIS

KENDALL ATKINSON
University of Iowa

JOHN WILEY & SONS
New York
Chichester
Brisbane
Toronto
Singapore

Library of Congress Cataloging in Publication Data:

Atkinson, Kendall E.
　Elementary numerical analysis.

　Bibliography: p. 407
　Includes index.
　1. Numerical analysis.　I. Title.

QA297.A83 1985　　　519.4　　　84-11974
ISBN 0-471-89733-7

Printed in the United States of America

10　9

To my daughters
Elizabeth and Kathryn

PREFACE

This book gives an introduction to numerical analysis, and it is intended for use by undergraduates in the sciences, mathematics, and engineering. The main prerequisite for using the book is a one-year course in the calculus of functions of one variable; some knowledge of computer programming is also needed. With this background, the book can be used for a sophomore-level course in numerical analysis. The last two chapters of the textbook are on numerical methods for linear algebra and ordinary differential equations. A background in these subjects would be helpful, but these chapters include the necessary theoretical material.

Students taking a course in numerical analysis do so for a variety of reasons. Some will need it in other subjects, in research, or in their careers. Others will be taking it to broaden their knowledge of computing. When I teach this course, I have several objectives for the students. First, they should obtain an intuitive and working understanding of some numerical methods for the basic problems of numerical analysis (as specified by the chapter headings). Second, they should gain some appreciation of the concept of error and of the need to analyze and predict it. And third, they should develop some experience in the implementation of numerical methods using a computer. This should include an appreciation of computer arithmetic and its effects.

The book covers most of the standard topics in a numerical analysis course, and it also explores some of the main underlying themes of the subject. Among these are the approximation of problems by simpler problems, the construction of algorithms, iteration methods, error analysis, stability, asymptotic error formulas, and the effects of machine arithmetic. Because of the level of the course, emphasis has been placed on obtaining an intuitive understanding of both the problem at hand and the numerical methods being used to solve it. The examples have been carefully chosen to develop this understanding, not just to illustrate an algorithm. Proofs are included only where they are sufficiently simple and where they add to an intuitive understanding of the result.

For the introduction to computer programming, the preferred language in the world of scientific computing is Fortran. In this text I have used Fortran 77, the new Fortran standard. It permits much better programming practices, and the programs written in it are much easier to understand than those written in Fortran 66. I have found that students experienced in Pascal can learn Fortran 77 very rapidly during the course.

The Fortran programs are included for several reasons. First, they illustrate the construction of algorithms. Second, they save the students from having to write as many programs, allowing them to spend more time on experimentally learning about a numerical method. After all, the main focus of the course should be numerical analysis, not learning how to program. Third, the programs provide examples of the language Fortran 77 and of good programming practices using it. Of course, the students should write some programs of their own. Some of these can be simple modifications of the included programs, for example, modifying the Simpson integration code to one for the trapezoidal rule. Other programs should be more substantial and original. The Fortran 77 programs of this text are available on a floppy disk, included with the instructor's manual for the text.

There are exercises at the end of each section in the book. These are of several types. Some exercises provide additional illustrations of the theoretical results given in the section, and many of these can be done with either a hand calculator or with a simple computer program. Other exercises are to further explore the theoretical material of the section, perhaps developing some additional theoretical results. In some sections, exercises are given that require more substantial programs; many of these exercises can be done in conjunction with package programs such as those discussed in Appendix C.

In teaching a one-semester course from this textbook, I cover the material in the order given here. The material can be taught in some other order, but I suggest that Chapters 1 through 3 be covered first. Following that, Chapter 8 on linear algebra can be included at any point. The material on polynomial interpolation in Chapter 5 will be needed before covering Chapters 6, 7, and 9. The textbook contains more than enough material for a one-semester course, and the instructor has considerable leeway in what to leave out.

I would like to thank my colleagues Dan Anderson, Herb Hethcote, and Keith Stroyan of the University of Iowa for their comments on the text. I also thank the reviewers of the manuscript for their suggestions, which were very helpful in preparing the final version of the book. I would like to thank Lois Friday for having done an excellent job of typing the book. Several classes of students have used preliminary versions of this text, and I thank them for their forbearance and suggestions. The staff of John Wiley have been very supportive and helpful in this project, and the text is much better as a result of their efforts, for which

I am grateful. Finally, I would like to thank my wife, Alice, for her patience and support, something that is much needed in a project such as this.

Kendall E. Atkinson
Iowa City, Iowa
April 1984

CONTENTS

ELEMENTARY
NUMERICAL
ANALYSIS

ONE

TAYLOR POLYNOMIALS

Numerical analysis uses results and methods from many other areas of mathematics, particularly calculus and linear algebra. In this chapter we consider one of the most useful tools from calculus, Taylor's theorem. This will be needed for both the development and understanding of most of the numerical methods taken up in this text.

The first section introduces Taylor polynomials as a way to evaluate other functions approximately; and the second section gives a precise formula, Taylor's theorem, for the error in these polynomial approximations. The final section discusses the evaluation of polynomials.

Other material from calculus is given in the appendixes. Appendix A contains a complete review of mean-value theorems and Appendix B reviews other results from calculus, algebra, geometry, and trigonometry.

1.1 THE TAYLOR POLYNOMIAL

Most functions $f(x)$ that occur in mathematics cannot be evaluated exactly in any simple way. For example, consider evaluating $f(x) = \cos(x)$, e^x, or $\sqrt{x}$, without using a calculator or computer. To evaluate such

1

expressions, we use functions $\hat{f}(x)$ which are almost equal to $f(x)$ and are easier to evaluate. The most common class of approximating functions $\hat{f}(x)$ are the polynomials. They are easy to work with and they are usually an efficient means of approximating $f(x)$. Among polynomials, the most widely used is the Taylor polynomial. There are other more efficient approximating polynomials, and we study some of them in Chapter 6. But the Taylor polynomial is comparatively easy to construct, and it is often a first step in obtaining more efficient approximations. The Taylor polynomial is also important in several other areas of mathematics.

Let $f(x)$ denote a given function, for example, e^x or $\log(x)$. The Taylor polynomial is constructed to mimic the behavior of $f(x)$ at some point $x = a$. As a result, it will be nearly equal to $f(x)$ at points x near to a. To be more specific, find a linear polynomial $p_1(x)$ for which

$$p_1(a) = f(a) \qquad\qquad (1.1)$$
$$p_1'(a) = f'(a)$$

Then it is easy to verify that the polynomial is uniquely given by

$$p_1(x) = f(a) + (x - a)f'(a) \qquad\qquad (1.2)$$

The graph of $y = p_1(x)$ is tangent to that of $y = f(x)$ at $x = a$.

Example

Let $f(x) = e^x$ and $a = 0$. Then

$$p_1(x) = 1 + x$$

The graphs of f and p_1 are given in Figure 1.1. Note that $p_1(x)$ is approximately e^x when x is near 0.

To continue the construction process, consider finding a quadratic polynomial $p_2(x)$ that approximates $f(x)$ at $x = a$. Since there are three coefficients in the formula of a quadratic polynomial,

$$p_2(x) = b_0 + b_1 x + b_2 x^2$$

it is natural to impose three conditions on $p_2(x)$ in order to determine

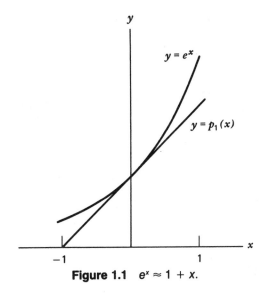

Figure 1.1 $e^x \approx 1 + x$.

them. To better mimic the behavior of $f(x)$ at $x = a$, we require

$$
\begin{aligned}
p_2(a) &= f(a) \\
p_2'(a) &= f'(a) \\
p_2''(a) &= f''(a)
\end{aligned}
\tag{1.3}
$$

It can be checked that these are satisfied by the formula

$$
p_2(x) = f(a) + (x - a)f'(a) + \frac{1}{2}(x - a)^2 f''(a)
\tag{1.4}
$$

Example

Continuing the previous example of $f(x) = e^x$, we have

$$
p_2(x) = 1 + x + \frac{1}{2}x^2
$$

We can continue this process of mimicking the behavior of $f(x)$ at $x = a$. Let $p_n(x)$ be a polynomial of degree n, and require it to satisfy

$$
p_n^{(j)}(a) = f^{(j)}(a), \qquad j = 0, 1, \ldots, n
\tag{1.5}
$$

where $f^{(j)}(x)$ is the order j derivative of $f(x)$. Then

$$p_n(x) = f(a) + (x - a)f'(a)$$

$$+ \frac{(x - a)^2}{2!} f''(a) + \ldots + \frac{(x - a)^n}{n!} f^{(n)}(a) \quad (1.6)$$

$$= \sum_{j=0}^{n} \frac{(x - a)^j}{j!} f^{(j)}(a)$$

In the formula,

$$j! = \begin{cases} 1 & , \quad j = 0 \\ j(j - 1) \ldots (2)(1), & \quad j = 1, 2, 3, 4, \ldots \end{cases}$$

and is called "j factorial."

Example

Again let $f(x) = e^x$ and $a = 0$. Then

$$f^{(j)}(x) = e^x, \qquad f^{(j)}(0) = 1, \qquad \text{for all} \quad j \geq 0$$

Thus

$$p_n(x) = 1 + x + \frac{1}{2!} x^2 + \ldots + \frac{1}{n!} x^n = \sum_{j=0}^{n} \frac{x^j}{j!} \quad (1.7)$$

Table 1.1 contains values of $p_1(x)$, $p_2(x)$, $p_3(x)$, and e^x at various values of x in $[-1, 1]$. For a fixed x, the accuracy improves as the degree n increases. And for a polynomial of fixed degree, the accuracy decreases as x moves away from $a = 0$.

Table 1.1 Taylor Approximations to e^x

x	$p_1(x)$	$p_2(x)$	$p_3(x)$	e^x
-1.0	0	0.500	0.33333	0.36788
-0.5	0.5	0.625	0.60417	0.60653
-0.1	0.9	0.905	0.90483	0.90484
0	1.0	1.000	1.00000	1.00000
0.1	1.1	1.105	1.10517	1.10517
0.5	1.5	1.625	1.64583	1.64872
1.0	2.0	2.500	2.66667	2.71828

Example

Let $f(x) = \log(x)$ and $a = 1$. Then $f(1) = \log(1) = 0$. By induction,

$$f^{(j)}(x) = (-1)^{j-1}(j-1)!\frac{1}{x^j}$$

$$f^{(j)}(1) = (-1)^{j-1}(j-1)!, \quad j \geq 1$$

If this is used in (1.6), the Taylor polynomial is given by

$$p_n(x) = (x-1) - \frac{1}{2}(x-1)^2$$

$$+ \frac{1}{3}(x-1)^3 - \ldots + (-1)^{n-1}\frac{1}{n}(x-1)^n \quad (1.8)$$

$$= \sum_{j=1}^{n} \frac{(-1)^{j-1}}{j}(x-1)^j$$

PROBLEMS

① Using (1.8), compare $\log(x)$ and its Taylor polynomials of degrees 1, 2, and 3, in the manner of Table 1.1. Do this on the interval $[\frac{1}{2}, \frac{3}{2}]$.

② Produce the linear and quadratic Taylor polynomials for the following cases.
    (a) $f(x) = \sqrt{x}, a = 1$
    (b) $f(x) = \sin(x), a = \pi/4$
    (c) $f(x) = e^{\cos(x)}, a = 0$

③ Produce a general formula for the degree n Taylor polynomials for the following functions, all using $a = 0$ as the point of approximation.
    (a) $1/(1-x)$
    (b) $\sin(x)$
    (c) $\sqrt{1+x}$

4. Compare $f(x) = \sin(x)$ with its Taylor polynomials of degrees 1, 3, and 5, on the interval $-\pi/2 \leq x \leq \pi/2$; $a = 0$. Produce a table in the manner of Table 1.1.

5. (a) Produce the Taylor polynomials of degrees 1, 2, 3, and 4 for $f(x) = e^{-x}$, with $a = 0$ the point of approximation.
    (b) Using the Taylor polynomials for e^t, substitute $t = -x$ to obtain polynomial approximations for e^{-x}. Compare with (a).

6. Repeat problem 5 with $f(x) = e^{x^2}$.

7. The quotient

$$g(x) = \frac{e^x - 1}{x}$$

is undefined for $x = 0$. Approximate e^x using Taylor polynomials of degrees 1, 2, and 3, in turn, to determine a natural definition of $g(0)$.

8. Repeat problem 7 with $g(x) = (1/x) \log (1 + x)$.

9. (a) As an alternative to the linear Taylor polynomial, construct a linear polynomial $q(x)$, satisfying

$$q(a) = f(a), \qquad q(b) = f(b)$$

for given points a and b.

  (b) Apply this to $f(x) = e^x$ with $a = 0$ and $b = 1$. For $0 \leq x \leq 1$, numerically compare $q(x)$ with the linear Taylor polynomial of this section.

10. For $f(x) = e^x$, construct a cubic polynomial $q(x)$ for which

$$q(0) = f(0), \qquad q(1) = f(1)$$
$$q'(0) = f'(0), \qquad q'(1) = f'(1)$$

Compare it to e^x and the Taylor polynomial $p_3(x)$ of (1.6).
Hint: Write $q(x) = b_0 + b_1 x + b_2 x^2 + b_3 x^3$. Determine b_0 and b_1 from the conditions at $x = 0$. Then obtain a linear system of two equations for the remaining coefficients b_2 and b_3.

1.2 THE ERROR IN TAYLOR'S POLYNOMIAL

To make practical use of the Taylor polynomial approximation to $f(x)$, we need to know its accuracy. The following theorem gives the main way of estimating this accuracy. We present it without proof, since it is given in most calculus texts.

Theorem 1.1 (Taylor's Theorem)

Assume that $f(x)$ has $n + 1$ continuous derivatives on an interval $\alpha \leq x \leq \beta$, and let the point a belong to that interval. For the Taylor polynomial $p_n(x)$ of (1.6), let $R_n(x) \equiv f(x) - p_n(x)$ denote the remainder in approximating $f(x)$ by $p_n(x)$. Then,

$$R_n(x) = \frac{(x - a)^{n+1}}{(n + 1)!} f^{(n+1)}(c_x), \qquad \alpha \leq x \leq \beta \qquad (1.9)$$

with c_x an unknown point between a and x.

Example

Let $f(x) = e^x$ and $a = 0$. The Taylor polynomial is given in (1.7). From the above theorem, the approximation error is given by

$$e^x - p_n(x) = \frac{x^{n+1}}{(n + 1)!} e^c, \qquad n \geq 0 \qquad (1.10)$$

with c between 0 and x.

As a special case, take $x = 1$. Then from (1.7),

$$e \approx p_n(1) = 1 + 1 + \frac{1}{2!} + \frac{1}{3!} + \ldots + \frac{1}{n!}$$

and from (1.10),

$$e - p_n(1) = R_n(1) = \frac{e^c}{(n + 1)!}, \qquad 0 < c < 1$$

From definitions of e given in most calculus texts, it is easy to obtain a bound of $e < 3$. Thus we can bound $R_n(1)$:

$$\frac{1}{(n + 1)!} \leq R_n(1) \leq \frac{e}{(n + 1)!} < \frac{3}{(n + 1)!}$$

This uses the inequality $e^0 \leq e^c \leq e^1$. As an actual numerical example, suppose we want to approximate e by $p_n(1)$ with

$$R_n(1) \leq 10^{-9}$$

Since we only know an upper bound for $R_n(1)$, we can obtain the desired error by making the upper bound satisfy

$$\frac{3}{(n + 1)!} \leq 10^{-9}$$

This is true when $n \geq 12$; thus $p_{12}(1)$ is a sufficiently accurate approximation to e.

The formulas (1.6) and (1.9) can be used to form approximations and remainder formulas for most of the standard functions encountered in undergraduate mathematics. For later reference, we give some of the more important ones.

$$e^x = 1 + x + \frac{x^2}{2!} + \ldots + \frac{x^n}{n!} + \frac{x^{n+1}}{(n+1)!} e^c \qquad (1.11)$$

$$\sin(x) = x - \frac{x^3}{3!} + \frac{x^5}{5!} - \ldots + (-1)^{n-1} \frac{x^{2n-1}}{(2n-1)!}$$
$$+ (-1)^n \frac{x^{2n+1}}{(2n+1)!} \cos(c) \qquad (1.12)$$

$$\cos(x) = 1 - \frac{x^2}{2!} + \frac{x^4}{4!} - \ldots + (-1)^n \frac{x^{2n}}{(2n)!}$$
$$+ (-1)^{n+1} \frac{x^{2n+2}}{(2n+2)!} \cos(c) \qquad (1.13)$$

$$\frac{1}{1-x} = 1 + x + x^2 + \ldots + x^n + \frac{x^{n+1}}{1-x}, \qquad x \neq 1 \qquad (1.14)$$

$$(1+x)^\alpha = 1 + \binom{\alpha}{1}x + \binom{\alpha}{2}x^2 + \ldots$$
$$+ \binom{\alpha}{n}x^n + \binom{\alpha}{n+1}x^{n+1}(1+c)^{\alpha-n-1} \qquad (1.15)$$

In this last formula, α is any real number. The coefficients $\binom{\alpha}{k}$ are called binomial coefficients and are defined by

$$\binom{\alpha}{k} = \frac{\alpha(\alpha-1)\ldots(\alpha-k+1)}{k!}, \qquad k = 1, 2, 3, \ldots$$

In all of the formulas, except (1.14), the point c is between 0 and x.

By rearranging the terms in (1.14), we obtain the sum of a finite geometric series or progression,

$$1 + x + x^2 + \ldots + x^n = \frac{1 - x^{n+1}}{1 - x}, \qquad x \neq 1 \qquad (1.16)$$

And by letting $n \to \infty$ in (1.14) when $|x| < 1$, we obtain the infinite *geometric series*

$$\frac{1}{1-x} = 1 + x + x^2 + x^3 + \ldots = \sum_{j=0}^{\infty} x^j, \qquad |x| < 1 \quad (1.17)$$

Example

Approximate $\cos(x)$ for $|x| \leq \pi/4$, with an error of no greater than 10^{-5}. Since the point c in the remainder of (1.13) is unknown, we consider the worst possible case and make it satisfy the desired error bound:

$$|R_{2n+1}(x)| \leq \frac{x^{2n+2}}{(2n+2)!} \leq 10^{-5}, \qquad \text{for } |x| \leq \pi/4$$

This uses $|\cos(c)| \leq 1$. For this inequality to be true, we must have

$$\frac{(\pi/4)^{2n+2}}{(2n+2)!} \leq 10^{-5}$$

which is satisfied when $n \geq 3$. The desired approximation is

$$\cos(x) \approx 1 - \frac{x^2}{2!} + \frac{x^4}{4!} - \frac{x^6}{6!}$$

Not all Taylor polynomials or remainder terms are created directly from (1.6) and (1.9). Rather, the above standard series are manipulated. For example, to obtain a series for $\log(1-t)$, integrate (1.14). More precisely,

$$\int_0^t \frac{dx}{1-x} = \int_0^t (1 + x + x^2 + \ldots + x^n)\, dx + \int_0^t \frac{x^{n+1}}{1-x}\, dx$$

$$-\log(1-t) = t + \frac{1}{2}t^2 + \frac{1}{3}t^3 + \ldots + \frac{1}{n+1}t^{n+1} + \int_0^t \frac{x^{n+1}}{1-x}\, dx$$

$$\log(1-t) = -\left(t + \frac{1}{2}t^2 + \ldots + \frac{1}{n+1}t^{n+1}\right) - \int_0^t \frac{x^{n+1}}{1-x}\, dx \quad (1.18)$$

This is valid for $t < 1$. The remainder term can be simplified by applying the integral mean value theorem (Theorem A.3 in Appendix A) to obtain

$$\int_0^t \frac{x^{n+1}}{1-x}\, dx = \frac{1}{1-c}\int_0^t x^{n+1}\, dx = \left(\frac{1}{1-c}\right)\frac{t^{n+2}}{n+2}$$

for some c between 0 and t.

PROBLEMS

①　Bound the error in using $p_3(x)$ to approximate e^x on $[-1,1]$, with $p_3(x)$ and its remainder given in (1.11). Compare to the results given in Table 1.1.

2.　Find the degree 2 Taylor polynomial for $f(x) = e^x \sin(x)$, about the point $a = 0$. Bound the error in this approximation when $-\pi/4 \leq x \leq \pi/4$.

③　(a)　Bound the error in the approximation

$$\sin(x) \approx x$$

for $-\pi/4 \leq x \leq \pi/4$.

(b)　Since this is a good approximation for small values of x, also consider the "percentage error"

$$\frac{\sin(x) - x}{\sin(x)} \doteq \frac{\sin(x) - x}{x}$$

Bound the absolute value of the latter quantity for $-\delta \leq x \leq \delta$. Pick δ to make the percentage error less than 1%.

④　How large should the degree $2n - 1$ be chosen in (1.12) to have

$$|\sin(x) - p_{2n-1}(x)| \leq 0.001$$

for all $-\pi/2 \leq x \leq \pi/2$. Check your result by evaluating the resulting $p_{2n+1}(x)$ at $x = \pi/2$.

⑤　Write out $p_4(x)$ for (1.15) with $\alpha = 1/2$. Bound the error if $0 \leq x \leq 1/2$.

6.　How large should n be chosen in (1.11) to have

$$|e^x - p_n(x)| \leq 10^{-5}, \qquad -1 \leq x \leq 1$$

7.　Verify (1.14).
　　Hint: Multiply both sides by $1 - x$ and simplify.

⑧ (a) Obtain a Taylor polynomial with remainder for $f(t) = 1/(1 + t^2)$, about $a = 0$.

  Hint: Substitute $x = -t^2$ into (1.14).

 (b) Obtain a Taylor polynomial with remainder for $g(x) = \tan^{-1}x$. Do this by integrating the result in (a) and using

$$\tan^{-1}(x) = \int_0^x \frac{dt}{1 + t^2}$$

9. (a) Using (1.18), obtain the Taylor polynomial with a remainder for $\log(1 + x)$.

 (b) Use (1.18) and (a) to obtain a Taylor polynomial with a remainder for

$$f(x) = \log\left[\frac{1 + x}{1 - x}\right], \qquad -1 < x < 1$$

⑩ Consider evaluating π by using

$$\pi = 4\tan^{-1}(1)$$

Using the results of problem 8, how many terms would be needed in the Taylor approximation $\tan^{-1}(x) \doteq p_{2n-1}(x)$ to calculate π with an accuracy of 10^{-10}? Is this a practical method of evaluating π?

1.3 POLYNOMIAL EVALUATION

The evaluation of a polynomial would appear to be a straightforward task. It is not, and to illustrate the possibilities, we consider the evaluation of

$$p(x) = 3 - 4x + 5x^2 - 6x^3 + 7x^4 - 8x^5$$

From a programmer's perspective, the simplest method of evaluation is to compute each term independently of the remaining terms. More precisely, the term cx^k is computed in a program by

$$c*x**k$$

This requires k multiplications with most Fortran compilers. With this approach, there will be

$$1 + 2 + 3 + 4 + 5 = 15 \text{ multiplications}$$

in the evaluation of $p(x)$.

The second method of evaluation is more efficient. We compute each power of x by multiplying x times the preceding power of x, as

$$x^3 = x(x^2), \qquad x^4 = x(x^3), \qquad x^5 = x(x^4) \tag{1.19}$$

Thus each term cx^k takes two multiplications for $k > 1$. The resulting evaluation of $p(x)$ uses

$$1 + 2 + 2 + 2 + 2 = 9 \text{ multiplications}$$

a considerable savings over the first method, especially with higher degree polynomials.

The third method is called *nested multiplication*. With it, we write and evaluate $p(x)$ in the form

$$p(x) = 3 + x(-4 + x(5 + x(-6 + x(7 - 8x))))$$

The number of multiplications is only 5, an additional saving over the second method. The nested multiplication method is the preferred evaluation procedure; and its advantage increases as the degree of the polynomial becomes larger.

Consider the general polynomial of degree n,

$$p(x) = a_0 + a_1 x + a_2 x^2 + \ldots + a_n x^n, \qquad a_n \neq 0 \tag{1.20}$$

If we use the second method, with the powers of x being computed as in (1.19), then the number of multiplications in the evaluation of $p(x)$ equals $2n - 1$. For the nested multiplication method, write and evaluate $p(x)$ in the form

$$p(x) = a_0 + x(a_1 + x(a_2 + \ldots + x(a_{n-1} + a_n x) \ldots) \tag{1.21}$$

This uses only n multiplications, a savings of about 50% over the second method. Both methods use n additions.

Example

Evaluate the Taylor polynomial $p_5(x)$ for $\log(x)$ about $a = 1$. A general formula is given in (1.18). From it,

$$p_5(x) = (x - 1) - \frac{1}{2}(x - 1)^2 + \frac{1}{3}(x - 1)^3 - \frac{1}{4}(x - 1)^4 + \frac{1}{5}(x - 1)^5$$

Let $z = (x - 1)$ and write

$$p_5(x) = z\left(1 + z\left(-\frac{1}{2} + z\left(\frac{1}{3} + z\left(-\frac{1}{4} + \frac{1}{5}z\right)\right)\right)\right)$$

In a computer program, you would store the coefficients in decimal form, to an accuracy consistent with the arithmetic of the machine.

We give a more formal algorithm for (1.21) because of its connection with some other topics. Suppose we want to evaluate $p(x)$ at some number z. Define a sequence of coefficients b_i as follows:

$$b_n = a_n$$
$$b_{n-1} = a_{n-1} + zb_n$$
$$b_{n-2} = a_{n-2} + zb_{n-1} \qquad (1.22)$$
$$\vdots$$
$$b_0 = a_0 + zb_1$$

Then

$$p(z) = b_0 \qquad (1.23)$$

The coefficients b_i are the successive computations within a matching pair of parentheses in (1.21). b_{n-1} is the innermost computation, and b_0 is the final one. When looked at in this manner, the nested multiplication method is called *Horner's method* or *synthetic division*, and it is given in many beginning algebra texts.

Using the coefficients of (1.22), define the polynomial

$$q(x) = b_1 + b_2x + b_3x^2 + \ldots + b_nx^{n-1} \qquad (1.24)$$

It can be shown that

$$p(x) = b_0 + (x - z)q(x) \qquad (1.25)$$

$q(x)$ is the quotient from dividing $p(x)$ by $x - z$, and b_0 is the remainder. The proof of (1.25) is left as problem 3 for the reader. This result is used in connection with polynomial rootfinding methods to reduce the degree of a polynomial when a root has been found.

An Example Program

We conclude the section by giving a Fortran function subprogram for evaluating a particular polynomial. The polynomial is chosen as an approximation to the function

$$S(x) = \frac{1}{x} \int_0^x \frac{\sin(t)\, dt}{t}, \qquad -1 \le x \le 1 \qquad (1.26)$$

Begin by using the standard series for $\sin(t)$, given by (1.12) with x replaced by t. Dividing by t and integrating over $[0, x]$, we get

$$S(x) = \frac{1}{x} \int_0^x \left[1 - \frac{t^2}{3!} + \frac{t^4}{5!} - \cdots + (-1)^{n-1} \frac{t^{2n-2}}{(2n-1)!} \right] dt + R_{2n-2}(x)$$

$$= 1 - \frac{x^2}{3!3} + \frac{x^4}{5!5} - \cdots + (-1)^{n-1} \frac{x^{2n-2}}{(2n-1)!(2n-1)}$$

$$+ R_{2n-2}(x) \quad (1.27)$$

$$R_{2n-2}(x) = \frac{1}{x} \int_0^x (-1)^n \frac{t^{2n}}{(2n+1)!} \cos(c_t)\, dt$$

The point c_t is between 0 and t. Since $|\cos(c_t)| \le 1$,

$$|R_{2n-2}(x)| \le \frac{1}{x} \int_0^x \frac{t^{2n}}{(2n+1)!}\, dt = \frac{x^{2n}}{(2n+1)!(2n+1)} \qquad (1.28)$$

We will choose the degree so that

$$|R_{2n-2}(x)| \le 5 \times 10^{-9} \qquad (1.29)$$

a fairly arbitrary choice. From (1.28), this is satisfied for $|x| \le 1$ if $2n + 1 \ge 11$. The polynomial we use is

$$p(x) = 1 - \frac{x^2}{3!3} + \frac{x^4}{5!5} - \frac{x^6}{7!7} + \frac{x^8}{9!9}, \qquad -1 \le x \le 1 \quad (1.30)$$

The following program uses nested multiplication to evaluate the degree

4 polynomial $g(u)$ obtained by letting $u = x^2$ in $p(x)$,

$$g(u) = 1 - \frac{u}{18} + \frac{u^2}{600} - \frac{u^3}{35280} + \frac{u^4}{3265920}$$

The function subprogram is written in Fortran 77, and it uses several features not available in earlier versions of Fortran. (1) The array A is allowed to have a zero subscript. (2) Named constants can be defined in a PARAMETER statement. This is useful in this case because the coefficients are as accurate as the arithmetic allows, no matter what the length of a single precision number. (3) In the DO loop the index variable is allowed to decrease.

```
      FUNCTION SF(X)
C
C     THIS EVALUATES AN APPROXIMATION TO
C
C
C              1       X  1
C     SF(X) = - INTEGRAL  - SIN(T)*DT
C              X       O  T
C
C     FOR -1 .LE. X .LE. 1.   THE APPROXIMATION IS THE
C     DEGREE 8 TAYLOR POLYNOMIAL FOR SF(X).  ITS
C     MAXIMUM ERROR IS 5.0E-9, PROVIDED THE COMPUTER
C     ARITHMETIC ALLOWS A RELATIVE ERROR OF THAT SIZE.
C
      DIMENSION A(0:4)
      PARAMETER(C1=-1.0/18.0,C2=1.0/600.0,
     *          C3=-1.0/35280.0,C4=1.0/3265920.0)
      DATA A/1.0,C1,C2,C3,C4/,
     *     NDEG/4/
C
C     INITIALIZE.
C
      U=X*X
      SF=A(NDEG)
C
C     DO NESTED MULTIPLICATION.
C
      DO 10 I=NDEG-1,0,-1
         SF=A(I)+U*SF
10       CONTINUE
      RETURN
      END
```

PROBLEMS

1. (a) Implement the function *SF* on your computer. Evaluate it at the points $x = 0, .1, .2, \ldots, 1.0$.

(b) Modify the program by decreasing the degree of the approximation from 8 to 6 in (1.30). This can be done in the program by simply changing the value of *NDEG* from 4 to 3. Evaluate this new function at the same points as in (a), and see whether there is any difference on your computer.

2. Prepare a function subprogram for evaluating

$$C(x) = \frac{1}{x} \int_0^x \frac{1 - \cos(t)}{t^2} \, dt, \qquad -1 \le x \le 1$$

with the same error tolerance as in (1.29) for $S(x)$. For comparison purposes, $C(1) = 0.486385376235$.

3. Show (1.25). Compute the quantity $b_0 + (x - z)q(x)$ by substituting (1.24), and collect together common powers of x. Then simplify those coefficients using (1.22). It may be easier to initially restrict the proof to a low degree, say $n = 3$.

4. Show $p'(z) = q(z)$ in (1.25).

5. Evaluate

$$p(x) = 1 - \frac{x^3}{3!} + \frac{x^6}{6!} - \frac{x^9}{9!} + \frac{x^{12}}{12!}$$

as efficiently as possible. How many multiplications are necessary? Assume all coefficients have been computed and stored for later use.

6. For $f(x) = e^x$, find a Taylor approximation that is in error by at most 10^{-7} on $[-1,1]$. Using this approximation, write a function subprogram to evaluate e^x. Compare it to the standard value of e^x obtained from the Fortran function EXP(x); calculate the difference between your approximation and EXP(x) at 21 evenly spaced points in $[-1, 1]$.

TWO

COMPUTER
REPRESENTATION
OF NUMBERS

One of the more important topics in numerical analysis is the study of the actual arithmetic operations that are used in calculations, most of which are done using computers. This chapter looks at the representation of numbers on computers and at the errors involved in this representation. Chapter Three discusses the errors in the basic arithmetic operations and combinations of them.

For representing numbers and doing calculations with them, most computers use the binary number system or some variant of it, rather than using the decimal system. The first section of this chapter gives an introduction to the binary number system and to the conversion between it and the decimal system. The second section gives the floating point representation of a number and discusses the error involved in using it.

2.1 THE BINARY NUMBER SYSTEM

In the decimal system, a number such as 342.105 means

$$3 \cdot 10^2 + 4 \cdot 10^1 + 2 \cdot 10^0 + 1 \cdot 10^{-1} + 0 \cdot 10^{-2} + 5 \cdot 10^{-3} \quad (2.1)$$

Numbers written in the decimal system are interpreted as a sum of multiples of integral powers of 10. There are 10 digits, denoted by 0, 1, ..., 9. We say 10 is the base of the decimal system.

The binary system represents all numbers as a sum of multiples of integral powers of 2. There are two digits, 0 and 1; and two is the base of the binary system. The digits 0 and 1 are called *bits*, which is short for *binary digits*. For example, the number 1101.11 in the binary system represents the number

$$1 \cdot 2^3 + 1 \cdot 2^2 + 0 \cdot 2^1 + 1 \cdot 2^0 + 1 \cdot 2^{-1} + 1 \cdot 2^{-2} = 13.75 \quad (2.2)$$

in the decimal system. For clarity when discussing a number with respect to different bases, we will enclose it in parentheses and give the base as a subscript. For example,

$$(1101.11)_2 = (13.75)_{10}$$

To convert a general binary number to its decimal equivalent, proceed as for (2.2).

Example

Consider $(111 \ldots 1)_2$ with n consecutive 1s to the left of the binary point. This has the decimal equivalent

$$2^{n-1} + 2^{n-2} + \ldots + 2^1 + 2^0 = 2^n - 1$$

using formula (1.16) for the sum of a geometric series. Thus,

$$(11 \ldots 11)_2 = (2^n - 1)_{10} \quad (2.3)$$

where the binary number has n digits.

Table 2.1 Binary Addition

+	1	1 0	1 1	1 0 0	1 0 1
1	1 0	1 1	1 0 0	1 0 1	1 1 0
1 0	1 1	1 0 0	1 0 1	1 1 0	1 1 1
1 1	1 0 0	1 0 1	1 1 0	1 1 1	1 0 0 0
1 0 0	1 0 1	1 1 0	1 1 1	1 0 0 0	1 0 0 1
1 0 1	1 1 0	1 1 1	1 0 0 0	1 0 0 1	1 0 1 0

Table 2.2 Binary Multiplication

×	1	1 0	1 1	1 0 0	1 0 1
1	1	1 0	1 1	1 0 0	1 0 1
1 0	1 0	1 0 0	1 1 0	1 0 0 0	1 0 1 0
1 1	1 1	1 1 0	1 0 0 1	1 1 0 0	1 1 1 1
1 0 0	1 0 0	1 0 0 0	1 1 0 0	1 0 0 0 0	1 0 1 0 0
1 0 1	1 0 1	1 0 1 0	1 1 1 1	1 0 1 0 0	1 1 0 0 1

The arithmetic operations are much the same in the binary system and the decimal system, one difference being that in binary fewer digits are allowed. As examples, consider the following addition and multiplication calculations:

$$
\begin{array}{r}
1\ 1\ 1\ 1\ 0 \\
+\quad 1\ 1\ 0\ 1 \\
\hline
1\ 0\ 1\ 0\ 1\ 1
\end{array}
\qquad
\begin{array}{r}
1\ 1\ 1 \\
\times\quad 1\ 1\ 0 \\
\hline
0\ 0\ 0 \\
1\ 1\ 1\quad \\
1\ 1\ 1\quad\quad \\
\hline
1\ 0\ 1\ 0\ 1\ 0
\end{array}
$$

Tables 2.1 and 2.2 contain addition and multiplication tables for the binary numbers corresponding to the decimal digits 1, 2, 3, 4, and 5.

Conversion from Decimal to Binary

We will give methods for converting decimal integers and decimal fractions to binary integers and binary fractions. These methods will be in a form convenient for hand computation. Different algorithms are required when doing such conversions within a binary computer, but we will not give them in this text.

Suppose that x is an integer written in decimal. We want to find coefficients $a_0, a_1, a_2, \ldots, a_n$, all 0 or 1, for which

$$a_n \cdot 2^n + a_{n-1} \cdot 2^{n-1} + \ldots + a_1 \cdot 2^1 + a_0 \cdot 2^0 = x \qquad (2.4)$$

The binary integer will be

$$(a_n a_{n-1} \cdots a_0)_2 = (x)_{10} \qquad (2.5)$$

To find the coefficients, begin by dividing x by 2, and denote the quotient by x_1. The remainder is a_0. Next divide x_1 by 2, and denote the quotient by x_2. The remainder is a_1. Continue this process, finding $a_2, a_3, a_4, \ldots,$ a_n in succession.

Example

The following shortened form of the above is convenient for hand computation. Convert $(19)_{10}$ to binary.

$$
\begin{array}{lll}
2\,\lfloor\,19 & = x & \\
\quad 2\,\lfloor\,9 & = x_1 & a_0 = 1 \\
\quad\quad 2\,\lfloor\,4 & = x_2 & a_1 = 1 \\
\quad\quad\quad 2\,\lfloor\,2 & = x_3 & a_2 = 0 \\
\quad\quad\quad\quad 2\,\lfloor\,1 & = x_4 & a_3 = 0 \\
\quad\quad\quad\quad\quad 0 & = x_5 & a_4 = 1
\end{array}
$$

Thus, $(19)_{10} = (10011)_2$.

Suppose now that x is a decimal fraction and that it is positive and less than 1.0. Then we want to find coefficients $a_1, a_2, a_3, \ldots,$ all 0 or 1, for which

$$a_1 \cdot 2^{-1} + a_2 \cdot 2^{-2} + a_3 \cdot 2^{-3} + \ldots = x \tag{2.6}$$

The binary fraction will be

$$(.\,a_1 a_2 a_3 \ldots)_2 = (x)_{10} \tag{2.7}$$

To find the coefficients, begin by denoting $x = x_1$. Multiply x_1 by 2, and denote $x_2 = \text{Frac}\,(2x_1)$, the fractional part of $2x_1$. The integer part $\text{Int}(2x_1)$ equals a_1. Repeat the process. Multiply x_2 by 2, letting $x_3 = \text{Frac}(2x_2)$ and $a_2 = \text{Int}(2x_2)$. Continue in the same manner, obtaining $a_3, a_4, \ldots$ in succession.

Example

Find the binary form of 5.703125. We break the numbers into an integer and a fraction part. As in the previous example, we find

$$(5)_{10} = (101)_2$$

For the fractional part $x_1 = x = 0.703125$, use the above algorithm.

$$
\begin{array}{lll}
2x_1 = 1.40625 & x_2 = 0.40625 & a_1 = 1 \\
2x_2 = 0.8125 & x_3 = 0.8125 & a_2 = 0 \\
2x_3 = 1.625 & x_4 = 0.625 & a_3 = 1 \\
2x_4 = 1.25 & x_5 = 0.25 & a_4 = 1 \\
2x_5 = 0.5 & x_6 = 0.5 & a_5 = 0 \\
2x_6 = 1.0 & x_6 = 0 & a_6 = 1
\end{array}
$$

Thus the binary equivalent is 0.101101; combining it with the earlier result, we get

$$(5.703125)_{10} = (101.101101)_2$$

Example

Convert the decimal fraction 0.1 to its binary equivalent. Using the above procedure, we obtain

$$(0.1)_{10} = (0.00011001100110 \ldots)_2 \tag{2.8}$$

an infinite repeating binary fraction. Numbers have a finite binary fractional form if and only if they are expressible as a sum of a finite number of negative powers of 2. The decimal number 0.1 is not expressible as such a finite sum, because 5 is not a factor of the base. This result has implications when working with finite decimal numbers on a binary computer, and we consider those implications later in the chapter.

Example

Convert the infinite repeating binary fraction

$$x = (0.1010101010101 \ldots)_2 \tag{2.9}$$

to its decimal equivalent. Converting to powers of 2 gives us

$$
\begin{aligned}
x &= 2^{-1} + 2^{-3} + 2^{-5} + 2^{-7} + \cdots \\
&= \frac{1}{2}\left[1 + \frac{1}{4} + \frac{1}{16} + \frac{1}{64} + \cdots \right]
\end{aligned}
$$

The quantity in brackets is an infinite geometric series. Using (1.17), we obtain

$$x = \frac{1}{2} \cdot \frac{1}{1 - \frac{1}{4}} = \frac{1}{2} \cdot \frac{4}{3} = \frac{2}{3}$$

Written as a decimal fraction,

$$x = (0.66666 \ldots)_{10}$$

Hexadecimal Numbers

The base of the hexadecimal system is 16, and the digits are usually denoted by

$$0, 1, \ldots, 9, A, B, C, D, E, F$$

Thus, $(F)_{16} = (15)_{10}$. As before, a number written in the hexadecimal system is a sum of multiples of integral powers of 16. For example,

$$(AC.17)_{16} = (10 \cdot 16^1 + 12 \cdot 16^0 + 1 \cdot 16^{-1} + 7 \cdot 16^{-2})_{10}$$
$$= (172.08984375)_{10}$$

Algorithms can be given to convert from decimal to hexadecimal, in analogy with those given above for binary numbers; see problem 7.

There is a close connection between the binary and hexadecimal systems, and it is easy to convert between them. To convert from hexadecimal to binary, replace each hex digit by its equivalent binary representation using four binary digits.

Example

Convert $(AC.17)_{16}$ to binary. It is given by

$$(10101100.00010111)_2$$

since

$$(A)_{16} = (1010)_2, \ldots, (7)_{16} = (0111)_2$$

To convert x from binary to hexadecimal, break x into packets of four bits, going both left and right from the binary point. Then convert those to their hex equivalent.

Example

Convert $x = (110101.11001)_2$ to hexadecimal. We write this as

$$(0011 \quad 0101.1100 \quad 1000)_2 = (35.C8)_{16}$$

Example

Convert the binary fraction in (2.9) to hexadecimal.

$$x = (0.1010 \quad 1010 \quad 1010 \quad 1010 \quad \dots)_2$$
$$= (. \, AAA \dots)_{16},$$

and by what was shown following (2.9), this is equivalent to the fraction $\frac{2}{3}$.

The hexadecimal system is used on IBM computers, and the binary system is used on most other lines of computers.

PROBLEMS

1. Convert the following binary numbers to decimal form.
   (a) 110011   (b) 101.1101   (c) 0.1000001
2. Expand Tables 1 and 2 to include one more column and row.
3. Convert the following decimal numbers to binary format.
   (a) 366   (b) 4.25   (c) 1/6   (d) π
4. Find the binary expressions for the decimal expressions 0.2, 0.4, 0.8, and 1.6.
   Hint: Use (2.8).
5. Find the rational number or fraction corresponding to the infinite repeating binary fraction

$$x = (0.110110110\dots)_2$$

   Hint: Write $x = 2^{-1} + 2^{-2} + 2^{-4} + 2^{-5} + 2^{-7} + 2^{-8} + \dots$
   $$= [2^{-1} + 2^{-4} + 2^{-7} + 2^{-10} + \dots]$$
   $$+ [2^{-2} + 2^{-5} + 2^{-8} + \dots]$$

Treat each series in brackets in a way similar to that used for (2.9).

6. Convert the following hexadecimal numbers to both decimal and binary.
   (a)   1F.C   (b)   FF. . . F repeated n times   (c)   11.1

7. Generalize the algorithms for decimal → binary conversion to decimal → hexadecimal conversion. For conversion of integers, divide by 16 rather than 2; and for decimal fractions less than one, multiply by 16 rather than 2. Convert the following decimal numbers to hexadecimal.
   (a)   161   (b)   0.3359375   (c)   0.1

8. Let x be written in binary form, $0 \leq x \leq 1$,

$$x = (. \, a_1 a_2 a_3 \dots)_2$$

Give a geometric meaning to the coefficients $a_1, a_2, a_3, \dots$ in terms of where x is located in the interval $[0,1]$.

2.2 FLOATING-POINT NUMBERS

To simplify the explanation of the floating-point representation of a number, first consider dealing with a nonzero decimal number x. It can be written in a unique way as

$$x = \sigma \cdot \bar{x} \cdot 10^e \qquad (2.10)$$

where $\sigma = +1$ or -1, e is an integer, and $0.1 \leq \bar{x} < 1$. These three quantities are called the sign, exponent, and mantissa, respectively, of the representation (2.10). As an example,

$$12.462 = (.12462) \cdot 10^2$$

with the sign $\sigma = +1$, the exponent $e = 2$, and the mantissa $\bar{x} = .12462$.

The decimal floating-point representation of a number x is basically that given in (2.10), with limitations on the number of digits in $\bar{x}$ and on the size of e. For example, suppose we limit the number of digits in $\bar{x}$ to four and the size of e to between -99 and $+99$. We would say that a computer with such a representation has a four decimal digit floating-point arithmetic. As a corollary to the limitation on the length of $\bar{x}$, no number can have more than its first four digits stored accurately. We will return to this topic later in the section.

Now consider a binary number x. In analogy with (2.10), we can write

$$x = \sigma \cdot \bar{x} \cdot 2^e \qquad (2.11)$$

where $\sigma = +1$ or -1, e is an integer, and $\bar{x}$ is a binary fraction satisfying

$$(.1)_2 \leq \bar{x} < 1 \qquad (2.12)$$

In decimal, $\frac{1}{2} \leq \bar{x} < 1$. For example, if $x = (1101.10111)_2$, then $\sigma = +1$, $e = 4 = (100)_2$, and $\bar{x} = (.110110111)_2$.

The floating-point representation of a binary number x consists of (2.11) with a restriction on the number of binary digits in $\bar{x}$ and on the size of e. We will give two examples of such representations for current computers. But keep in mind that the general ideas and framework are what is important, not the details of these particular machines.

Examples

1. CDC 6000 and 7000 series machines and the related CYBER machines use a 60 bit word to represent all floating point numbers. The floating-point word is described schematically in Figure 2.1. The mantissa $\bar{x}$ contains 48 binary digits or bits, and the exponent is limited to $-975 \leq e \leq 1071$. This unusual range for e is due to the design of the arithmetic operations on the CDC machines.

2. PRIME computers use a 32 bit word to represent floating-point numbers, in the form shown in Figure 2.2. The mantissa $\bar{x}$ contains 23 binary digits, and e is limited by $-128 \leq e \leq 127$. In fact, $e + 128$ is stored rather than e as a signed quantity, but that is not important from our perspective.

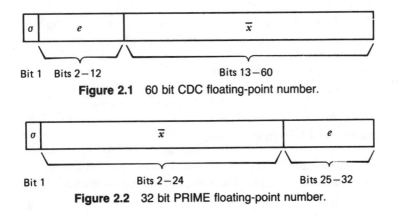

Bit 1   Bits 2–12   Bits 13–60

Figure 2.1   60 bit CDC floating-point number.

Bit 1   Bits 2–24   Bits 25–32

Figure 2.2   32 bit PRIME floating-point number.

Accuracy of Floating-Point Representation

Consider how accurately a number can be stored in the floating-point representation. To do this, we look for the largest integer M having the property that any integer x satisfying $0 \leq x \leq M$ can be stored or represented exactly in floating-point form. Since x is an integer, we will have to convert it to a binary fraction $\bar{x}$ with a positive exponent e in (2.11). If n is the number of binary digits in the mantissa, then it is fairly easy to convince oneself that all integers less than or equal to

$$(.11 \ldots 1)_2 \cdot 2^n$$

can be stored exactly, where this mantissa contains n binary digits, all 1. This is the integer composed of n consecutive 1s; and from (2.3), it equals $2^n - 1$. In fact, 2^n also stores exactly; but there are not enough digits in the mantissa to store $2^n + 1$, as this would require $n + 1$ binary digits in the mantissa. Thus, an n digit binary floating-point representation implies

$$M = 2^n \tag{2.13}$$

Any positive integer $\leq M$ can be represented exactly in this floating-point representation.

Examples

1. On CDC machines, $M = 2^{48} = 2.81 \cdot 10^{14}$. All integers with 14 or fewer decimal digits and some with 15 digits will store exactly in the CDC floating-point representation. We say that the CDC representation contains at least 14 decimal digits of accuracy.

2. On PRIME machines, $M = 2^{23} = 8388608$. Thus all six digit decimal integers will store exactly, as well as most seven digit integers. We say the PRIME arithmetic contains six to seven decimal digits of accuracy in its floating-point representation.

Rounding and Chopping

If the number x in (2.11) does not have a mantissa $\bar{x}$ that will fit in the available space of n binary bits, then it must be shortened. Currently, this is done in two different ways, depending on the computer. The simplest technique is to simply truncate or *chop* $\bar{x}$ to n binary digits, ignoring the remaining digits. The second method is to *round* $\bar{x}$ to n digits, based on the size of the part of $\bar{x}$ following digit n. More precisely, if digit $n + 1$ is zero, chop $\bar{x}$ to n digits; otherwise, chop $\bar{x}$ to n digits and add 1 to the last digit of the result. Regardless of whether chopping or rounding is being used, we will denote the machine floating-point version of a number x by $fl(x)$.

It can be shown that the error in representing x by $fl(x)$ satisfies

$$fl(x) = x \cdot (1 + \epsilon) \qquad (2.14)$$

Since ϵ is small, this says that $fl(x)$ is a slight perturbation of x. If chopping is used,

$$-2^{-n+1} \leq \epsilon \leq 0 \qquad (2.15)$$

and if rounding is used,

$$-2^{-n} \leq \epsilon \leq 2^{-n} \qquad (2.16)$$

The most important characteristics of chopping are (1) the worst possible error is twice as large as when rounding is used, and (2) the sign of the error $x - fl(x)$ is the same as the sign x. This last characteristic is the worst of the two. In many calculations it will lead to no possibility of cancellation of errors; examples are given in Chapter 3. With rounding, the worst possible error is only half as large as for chopping. More important, the error $x - fl(x)$ is negative for half of the cases and positive for the other half. This leads to better error propagation behavior in large calculations.

Examples

1. For CDC computers, chopping is used. Since there are $n = 48$ binary digits in the mantissa, we have

$$-2^{-47} \leq \epsilon \leq 0 \qquad (2.17)$$

for the formula (2.14)

2. For PRIME computers, both rounding and chopping are available. There are $n = 23$ digits in the mantissa; if rounding is used,

$$-2^{-23} \leq \epsilon \leq 2^{-23} \qquad (2.18)$$

in formula (2.14) for the error in $fl(x)$.

Double-Precision Floating-Point Numbers

What is described above is usually referred to as a single-precision number. In Fortran, it is said to be of REAL type. With many computations, the length of the mantissa does not contain enough digits of accuracy. As compensation, most computers also have another form of floating-point number with a longer mantissa; it is called a double-precision number.

Examples

1. With CDC computers, the double precision number occupies two words of memory, 120 bits in total length. The mantissa is 108 bits in length, far longer than is needed for most practical computations. Since single-precision numbers of CDC computers are usually adequate, double precision is implemented by software rather than hardware, making it much slower to work with.

2. PRIME computers have a 64 bit double-precision number. Figure 2.3 contains a schematic of it. Note that both the exponent and mantissa are larger. The exponent has the limits $-32768 \leq e \leq 32767$. The mantissa contains 47 bits; therefore, the number M in (2.13) is 1.41×10^{14}. This means that all 14 digit decimal integers can be stored exactly in double precision, and we say that PRIME double precision contains 14 digits of accuracy.

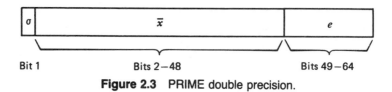

Bit 1   Bits 2–48   Bits 49–64

Figure 2.3   PRIME double precision.

Hexadecimal Floating-Point Numbers

Most IBM computers use hexadecimal arithmetic rather than binary arithmetic. For a general nonzero hexadecimal number x, we write

$$x = \sigma \cdot \bar{x} \cdot 16^e \tag{2.19}$$

with $\sigma = +1$, e an integer exponent, and

$$\frac{1}{16} = (0.1)_{16} \leq \bar{x} < 1 \tag{2.20}$$

A schematic of a single-precision hexadecimal number is shown in Figure 2.4. Such a number contains 32 bits, with the mantissa containing the first six hexadecimal digits of $\bar{x}$. Because of (2.20), the first hexadecimal digit of $\bar{x}$ could be as large as $(F)_{16} = (1111)_2$ or as small as $(1)_{16} = (0001)_2$. Consequently, $\bar{x}$ could contain as few as 21 binary digits or as many as 24 binary digits of accuracy, counting from the first nonzero binary digit in $\bar{x}$. The exponent e is limited by $-64 \le e \le 63$.

Generalizing (2.13), the largest integer M is given by

$$16^6 = 16777216$$

and thus all seven digit decimal integers can be stored exactly. We say there are seven digits of accuracy. Double-precision numbers have the same bounds on the exponents, but expand the mantissa to 14 hexadecimal digits. This leads to 16 decimal digits of accuracy in double precision.

The formula (2.14) for $fl(x)$ generalizes to base 16 and to other number bases. In general, let β be the base of the number system, and let n be the number of digits in the floating-point representation relative to this base. Then the floating-point representation of a number x satisfies

$$fl(x) = (1 + \epsilon)x \tag{2.21}$$

with ϵ satisfying

$$
\begin{aligned}
-\beta^{-n+1} &\le \epsilon \le 0 & \text{chopped representation} \\
-\tfrac{1}{2}\beta^{-n+1} &\le \epsilon \le \tfrac{1}{2}\beta^{-n+1} & \text{rounded representation}
\end{aligned}
\tag{2.22}
$$

For single precision on the IBM, $\beta = 16$ and $n = 6$, and

$$-16^{-5} \le \epsilon \le 0$$

because chopping is used. With double precision, $n = 14$, and this lower bound is replaced by -16^{-13}. The use of base 16 leads to much larger errors in rounding or chopping, as compared to the corresponding errors on a binary machine.

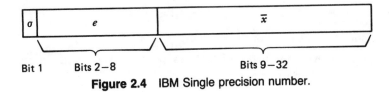

Bit 1   Bits 2–8   Bits 9–32

Figure 2.4   IBM Single precision number.

Programming Problems

Numbers that have finite decimal expansions may have infinite binary expansions. Consider the following computer code on a binary machine that uses chopping:

```
        X = 0.0
   10   X = X + 0.1
        PRINT *,X,SQRT(X)
        IF(X.NE.1.0) GO TO 10
```

This code forms an infinite loop. Clearly, the author intended it to stop when $X = 1.0$; but because 0.1 is not represented exactly in the computer, the number 1.0 is never attained exactly. A different kind of looping test should be used, one not subject to rounding or chopping errors.

PROBLEMS

1. Find the number M, the upper bound for integers representable in floating-point, for CDC double-precision floating-point numbers. How many decimal digits of accuracy can be stored in this double-precision form?

2. Repeat problem 1 for double-precision floating-point numbers on IBM computers.

3. Some microcomputers use binary floating-point numbers with 8 bits for the exponent and 32 bits for the mantissa. They also use rounding. Find the number M of (2.13), and use it to determine the accuracy of the representation. Also, find the bounds on ϵ in (2.14).

4. By considering the bounds on the exponent e in the floating-point representation, we can find the largest and smallest positive numbers expressible in the machine. Using the bounds given earlier for e on CDC machines, $-975 \leq e \leq 1071$, show that all nonzero numbers x in the machine satisfy

$$2^{-976} \leq |x| < 2^{1071}$$

Convert this to bounds involving powers of 10.

5. Repeat problem 4, for PRIME computers, using both single- and double-precision representations.

6. Repeat problem 4 for IBM computers.

⑦ In much current mathematical software, a quantity called the *unit round* is used to let the program know the precision of the floating-point representation on the computer on which the program is being run. It is defined as the smallest positive number u, representable on the machine, for which $1 + u > 1$ on the arithmetic of the machine. On CDC computers with single-precision arithmetic, show $u = 2^{-47}$.

8. Find the unit round of problem 7 for IBM machines in single- and double-precision arithmetic.

⑨ Find the unit round of problem 7 for PRIME machines, in single- and double-precision arithmetic, with rounding.

10. As an example that hexadecimal and binary arithmetic are not completely equivalent, let

$$x = 2^{-1} + 2^{-22} = (.800004)_{16}$$

Compute

$$u = x/8 \qquad z = 8 \cdot u$$

in their IBM single-precision representations, using IBM arithmetic with chopping. Show that $z \neq x$. How many binary bits of accuracy are lost in x? Would z be computed correctly on a binary computer?

⑪. Predict the output of the following section of code if it is run on a binary computer that uses chopping.

```
        I = 0
        X = 0.0
        H = .1
   10   I = I + 1
        X = X + H
        PRINT *,I,X
        IF(X .LT. 1.0) GO TO 10
```

Would the outcome be any different if the statement "X = X + H" was replaced by "X = I*H"?

12. Establish (2.14) for a binary computer with rounding, using the following argument.
    (a) Show that

$$-2^{e-n-1} \leq x - fl(x) \leq 2^{e-n-1}$$

(b) Show that

$$\frac{|x - fl(x)|}{|x|} \leq 2^{-n}$$

(c) Let

$$\frac{x - fl(x)}{x} = -\epsilon$$

and then solve for $fl(x)$. What are the bounds on ϵ?

THREE

ERROR

Much of numerical analysis is concerned with how to solve a problem numerically, that is, developing a sequence of calculations that will give a satisfactory answer. Part of this process is the consideration of the errors that arise in these calculations, whether from rounding errors in arithmetic operations or from some other source. Throughout this book, as we look at the numerical solution of various problems, we will simultaneously consider the errors involved in whatever computational procedure is being used.

In this chapter, we give a few general results on error. Section 3.1 contains definitions and some important examples. Section 3.2 discusses the propagation of error in arithmetic calculations. Section 3.3 looks at errors involved in some summation procedures, and it illustrates the significant advantage of rounding over truncation in arithmetic calculations.

3.1 ERRORS: DEFINITIONS, SOURCES, AND EXAMPLES

The error in a computed quantity is defined as

$$\text{Error} = \text{true value} - \text{approximate value}$$

The relative error is a measure of the error in relation to the size of the true value being sought:

$$\text{Relative error} = \frac{\text{error}}{\text{true value}}$$

To simplify the notation when working with these quantities, we will usually denote the true and approximate values of a quantity x by x_T and x_A, respectively. Then we write

$$\text{Error } (x_A) = x_T - x_A$$

$$\text{Rel } (x_A) = \frac{x_T - x_A}{x_T}$$

As an illustration, consider the well-known approximation

$$\pi \approx \frac{22}{7}$$

Here $x_T = \pi = 3.14159265\ldots$ and $x_A = \frac{22}{7} = 3.1428571\ldots$,

$$\text{Error } (22/7) = \pi - \frac{22}{7} \doteq -0.00126$$

$$\text{Rel } (22/7) = \frac{\pi - (22/7)}{\pi} \doteq -0.000402$$

Another example of error measurement is given by the Taylor remainder (1.9).

An idea related to relative error is that of *significant digits*. For a number x_A, the number of its leading digits that are correct relative to the corresponding digits in the true value x_T is called the number of significant digits in x_A. For a more precise definition, assuming the numbers are written in decimal, calculate the magnitude of the error, $|x_T - x_A|$. If this error is less than or equal to five units in the $(m + 1)$st digit of x_T, counting rightward from the first nonzero digit, then we say x_A has at least m significant digits of accuracy relative to x_T.

Example

(a) $x_A = 0.222$ has three digits of accuracy relative to $x_T = 2/9$.

(b) $x_A = 23.496$ has four digits of accuracy relative to $x_T = 23.494$.

(c) $x_A = 0.02138$ has just two digits of accuracy relative to $x_T = 0.02144$.

(d) $x_A = 22/7$ has three digits of accuracy relative to $x_T = \pi$.

It can be shown that if

$$\left| \frac{x_T - x_A}{x_T} \right| \leq 5 \cdot 10^{-m-1} \tag{3.1}$$

then x_A has m significant digits with respect to x_T. Most people find it easier to measure relative error than significant digits; in some textbooks, satisfaction of (3.1) is used as the definition of x_A having m significant digits of accuracy.

Sources of Error

Imagine solving a scientific–mathematical problem, and suppose this involves a computational procedure. Errors will usually be involved in this process, often of several different kinds. We will give a rough classification of the kinds of error that might occur.

(E1) Mathematical equations are used to represent physical reality, a process that is called mathematical modeling. This modeling introduces error into the real-world problem that you are trying to solve. For example, the simplest model for population growth is given by

$$N(t) = N_0 e^{kt} \tag{3.2}$$

where $N(t)$ equals the population at time t, and N_0 and k are positive constants. For some stages of growth of a population, when it has unlimited resources, this can be an accurate model. But more often, it will overestimate the actual population for large t.

The error in a mathematical model falls outside the scope of numerical analysis, but it is still an error with respect to the solution of the overall scientific problem of interest.

(E2) Blunders and mistakes are a second source of error, a familiar one to almost everyone. In the precomputer era, blunders generally consisted of isolated arithmetic errors, and elaborate check schemes were used to detect them. Today the mistakes are more likely to be programming errors. To detect such errors, it is important to have some way of checking the accuracy of the program output. When first running the program, use cases for which you know the correct answer. With a complex program, break it into smaller subprograms, each of which can be tested separately. And when you believe the entire program is correct and are running it for cases of interest, maintain a watchful eye as to whether the output is reasonable or not.

(E3) Many problems involve physical data, and these data contain observational error. For example, the speed of light in a vacuum is

$$c = (2.997925 + \epsilon) \cdot 10^{10} \text{ cm/sec}, \qquad |\epsilon| \leq 0.000003 \qquad (3.3)$$

Because the physical data contain an error, calculations based on the data will contain the effect of this observational error. Numerical analysis cannot remove the error in the data, but it can look at its propagated effect in a calculation. Numerical analysis also can suggest the best form for a calculation that will minimize the propagated effect of the errors in the data. Propagation of errors in arithmetic calculations is considered in section 3.2.

(E4) Machine errors occur when using computers or calculators, as for example with the rounding errors introduced in section 2.2. These errors are inevitable when using floating-point arithmetic; and they form the main source of error with some problems, for example, the solution of systems of linear equations. In section 3.3, we will analyze the effect of rounding errors for some summation procedures.

(E5) Mathematical approximation errors form the last class of errors. They will be the major form of error in which we will be interested in the following chapters. To illustrate this type of error, consider evaluating the integral

$$I = \int_0^1 e^{x^2} \, dx$$

There is no antiderivative for e^{x^2} in terms of elementary functions and, thus, the integral cannot be evaluated explicitly. Instead, we approximate the integral with a quantity that can be computed. For example, use the Taylor approximation

$$e^{x^2} \doteq 1 + x^2 + \frac{x^4}{2!} + \frac{x^6}{3!} + \frac{x^8}{4!}$$

Then

$$I \doteq \int_0^1 \left[1 + x^2 + \frac{x^4}{2!} + \frac{x^6}{3!} + \frac{x^8}{4!} \right] dx \qquad (3.4)$$

which can be evaluated easily. The error in (3.4) is called a mathematical approximation error; some authors call it a truncation error or discretization error. Such errors arise when we have a problem we cannot solve, and we approximate it with a new problem that we can solve. We try to estimate the size of such mathematical approximation errors and to make them sufficiently small.

We will now describe three important errors that occur commonly in practice. They are sometimes considered as sources of error, but we will think of them as deriving from the sources (E1) to (E5) described above. Also, the size of these errors can usually be minimized by using a properly chosen form for the computation being carried out.

Loss of Significance Errors

The idea of loss of significant digits is best understood by looking at some examples. Consider first the evaluation of

$$f(x) = x[\sqrt{x + 1} - \sqrt{x}] \qquad (3.5)$$

for an increasing sequence of values of x. The results of using a six digit decimal calculator are shown in Table 3.1. As x increases, there are fewer digits of accuracy in the computed value of $f(x)$.

To better understand what is happening, look at the individual steps of the calculation when $x = 100$. On the calculator,

$$\sqrt{100} = 10.0000, \qquad \sqrt{101} = 10.0499$$

Table 3.1 Values of (3.5) Using a Six Digit Calculator

x	Computed $f(x)$	True $f(x)$
1	.414210	.414214
10	1.54340	1.54347
100	4.99000	4.98756
1000	15.8000	15.8074
10000	50.0000	49.9988
100000	100.000	158.113

The first value is exact, and the second value is correctly rounded to six significant digits of accuracy. Next,

$$\sqrt{x + 1} - \sqrt{x} = \sqrt{101} - \sqrt{100} = 0.0499000 \qquad (3.6)$$

The true value should be 0.0498756. The calculation (3.6) has a *loss-of-significance error*. Three digits of accuracy in $\sqrt{x + 1} = \sqrt{101}$ were cancelled by subtraction of the corresponding digits in $\sqrt{x} = \sqrt{100}$. The loss of accuracy was a byproduct of the form of $f(x)$ and the finite precision arithmetic being used.

For this particular $f(x)$, there is a simple way to reformulate it so as to avoid the loss-of-significance error. Consider (3.5) as a fraction with a denominator of 1, and multiply numerator and denominator by $\sqrt{x + 1} + \sqrt{x}$, to get

$$f(x) = x\,\frac{\sqrt{x + 1} - \sqrt{x}}{1} \cdot \frac{\sqrt{x + 1} + \sqrt{x}}{\sqrt{x + 1} + \sqrt{x}} = \frac{x}{\sqrt{x + 1} + \sqrt{x}} \qquad (3.7)$$

The latter expression will not have any loss-of-significance errors in its evaluation. On our six digit decimal calculator, (3.7) gives

$$f(100) = 4.98756$$

the true answer to six digits.

As a second example, consider evaluating

$$f(x) = \frac{1 - \cos(x)}{x^2} \qquad (3.8)$$

for a sequence of values of x approaching 0. The results of using a popular ten digit decimal hand calculator are shown in Table 3.2. To understand the loss of accuracy, look at the individual steps of the cal-

Table 3.2 Values of (3.8) on a Ten Digit Calculator

x	Computed $f(x)$	True $f(x)$
.1	.4995834800	.4995834722
.01	.4999950000	.4999958333
.001	.5001000000	.4999999583
.0001	.5100000000	.4999999996
.00001	0.0	.5000000000

culation when $x = 0.01$. First, on the calculator

$$\cos(0.01) = 0.9999500005$$

This has nine significant digits of accuracy, and it is off in the tenth digit by one unit. Next, compute

$$1 - \cos(0.01) = 0.0000499995$$

This has only five significant digits, with four digits being lost in the subtraction. Division by $x^2 = 0.0001$ gives the entry in the table.

To avoid the loss of significant digits, we use another formulation for $f(x)$, avoiding the subtraction of nearly equal quantities. Using the Taylor approximation in (1.13), we get

$$\cos(x) = 1 - \frac{x^2}{2!} + \frac{x^4}{4!} - \frac{x^6}{6!} + R_6(x)$$

$$R_6(x) = \frac{x^8}{8!} \cos(\xi)$$

with ξ an unknown number between 0 and x. Therefore,

$$f(x) = \frac{1}{x^2}\left\{1 - \left[1 - \frac{x^2}{2!} + \frac{x^4}{4!} - \frac{x^6}{6!} + R_6(x)\right]\right\}$$

$$= \frac{1}{2!} - \frac{x^2}{4!} + \frac{x^4}{6!} - \frac{x^6}{8!}\cos(\xi)$$

Thus $f(0) = \frac{1}{2}$. And for $|x| \leq 0.1$,

$$\left|\frac{x^6}{8!}\cos(\xi)\right| \leq \frac{10^{-6}}{8!} = 2.5 \times 10^{-11} \qquad (3.9)$$

Thus

$$f(x) \doteq \frac{1}{2!} - \frac{x^2}{4!} + \frac{x^4}{6!}, \qquad |x| \leq 0.1$$

with an accuracy given by (3.9). This gives a much better way of evaluating $f(x)$ for small values of x.

When two nearly equal quantities are subtracted, leading significant digits will be lost. Sometimes this is easily recognized, as in the above two examples (3.5) and (3.8), and then ways can usually be found to avoid the loss of significance. More often, the loss of significance will be subtle and difficult to detect. One common situation is in calculating sums containing a number of terms, as when using a Taylor polynomial to approximate a function $f(x)$. If the value of the sum is relatively small when compared to some of the terms being summed, then there are probably some significant digits of accuracy being lost in the summation process.

As an example of this last phenomenon, consider using the Taylor series approximation (1.11) for e^x to evaluate e^{-5}:

$$e^{-5} = 1 + \frac{(-5)}{1!} + \frac{(-5)^2}{2!} + \frac{(-5)^3}{3!} + \frac{(-5)^4}{4!} + \ldots \qquad (3.10)$$

Imagine using a computer with four digit decimal floating-point arithmetic, so that the terms of this series must all be rounded to four significant digits. In Table 3.3, we give these terms, along with the associated exact sum of these terms through the given degree. The true value of e^{-5} is 0.006738, to four significant digits, and this is quite different from the final sum in the table. Also, if (3.10) is calculated exactly for terms of degree ≤ 25, then the correct value of e^{-5} is obtained to four digits.

In this example, the terms become relatively large, but they are then added to form a much smaller number e^{-5}. This means there are loss-of-significance errors in the calculation of the sum. To avoid the problem in this case is quite easy. Either use

$$e^{-5} = 1/e^5$$

Table 3.3 Calculation of (3.10) Using Four Digit Decimal Arithmetic

Degree	Term	Sum	Degree	Term	Sum
0	1.000	1.000	13	−0.1960	−0.04230
1	−5.000	−4.000	14	0.7001E-1	0.02771
2	12.50	8.500	15	−0.2334E-1	0.004370
3	−20.83	−12.33	16	0.7293E-2	0.01166
4	26.04	13.71	17	−0.2145E-2	0.009518
5	−26.04	−12.33	18	0.5958E-3	0.01011
6	21.70	9.370	19	−0.1568E-3	0.009957
7	−15.50	−6.130	20	0.3920E-4	0.009996
8	9.688	3.558	21	−0.9333E-5	0.009987
9	−5.382	−1.824	22	0.2121E-5	0.009989
10	2.691	0.8670	23	−0.4611E-6	0.009989
11	−1.223	−0.3560	24	0.9607E-7	0.009989
12	0.5097	0.1537	25	−0.1921E-7	0.009989

and form e^5 with a series not involving cancellation of positive and negative terms; or simply form $e^{-1} = 1/e$ and multiply it times itself to form e^{-5}. With other series, there may not be such a simple solution.

Noise in Function Evaluation

Consider evaluating a function $f(x)$ for all points x in some interval $a \leq x \leq b$. If the function is continuous, then the graph of this function is a continuous curve. Next, consider the evaluation of $f(x)$ on a computer using floating-point arithmetic with rounding or chopping. Arithmetic operations (e.g., additions and multiplications) cause errors in the evaluation of $f(x)$, generally quite small ones. If we look very carefully at the graph of the computed values of $f(x)$, it will no longer be a continuous curve, but rather a "fuzzy" curve reflecting the errors in the evaluation process.

Example

To illustrate these comments, we evaluate

$$f(x) = x^3 - 3x^2 + 3x - 1 \qquad (3.11)$$

on a microcomputer that contains about nine decimal digits in its mantissa and that uses binary arithmetic with rounding. Figure 3.1 contains the graph of the computed values of $f(x)$ for $0 \leq x \leq 2$, and it appears to be a smooth continuous curve. Next we look at a small segment of this curve, for $0.998 \leq x \leq 1.002$. The plot of the computed values of

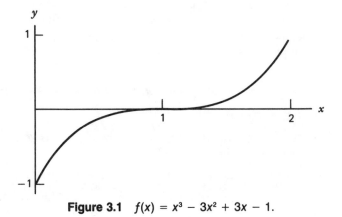

Figure 3.1 $f(x) = x^3 - 3x^2 + 3x - 1$.

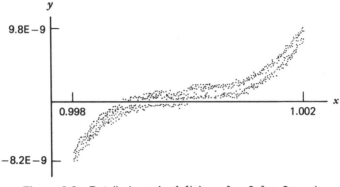

Figure 3.2 Detailed graph of $f(x) = x^3 - 3x^2 + 3x - 1.$

$f(x)$ are given in Figure 3.2, for 800 evenly spaced values of x in [0.998, 1.002]. Note that the graph of $f(x)$ is not a continuous curve, but a fuzzy band of seemingly random values. This is true of all parts of the computed curve of $f(x)$, but it becomes evident only when you look at the curve very closely.

Functions $f(x)$ on a computer should be thought of as a fuzzy band of random values, with the thickness of the band quite small in most cases. The implications of this are minimal in most cases, but there are situations where it will be important. For example, a rootfinding program might consider (3.11) to have a very large number of roots in [0.998, 1.002] on the basis of the many sign changes, as shown in Figure 3.2.

To help in examining the noise in function evaluations on your computer, we give the following program to evaluate the function in (3.11). The interval of evaluation is $1 - \delta \le x \le 1 + \delta$; and the appropriate size for δ depends on the length of the floating-point mantissa and on whether chopping or rounding is used in arithmetic operations.

```
C       TITLE: EXAMPLE OF NOISE IN EVALUATING A FUNCTION.
C
C       IN THE PROGRAM, THE FUNCTION
C          F(X) = -1 + 3*X - 3*X**2 + X**3
C       IS EVALUATED IN TWO FORMS: NESTED AND DIRECT.
C       THE INTERVAL OF EVALUATION IS
C          [A,B]  = [1-DELTA,1+DELTA],
C       WITH DELTA SPECIFIED IN THE DATA STATEMENT.
C       FOR OTHER COMPUTERS, THIS VALUE MAY HAVE TO BE
C       CHOSEN SMALLER, IN ORDER TO OBTAIN THE DESIRED
```

```
C       DEMONSTRATION OF NOISE IN EVALUATING F(X).
C
        DATA N/100/, DELTA/.01/
C
C       INITIALIZATION.
        A=1.0-DELTA
        B=1.0+DELTA
        H=2.0*DELTA/N
        PRINT *,'FOLLOWING ARE THE VALUES OF X, F(X):NESTED'
        PRINT *,'EVALUATION, AND F(X):DIRECT EVALUATION.'
C
C       BEGIN LOOP FOR EVALUATING FUNCTION ON [A,B].
        DO 10 J=0,N
          X=A+J*H
          FNESTD=-1.0+X*(3.0+X*(-3.0+X))
          FDIRCT=-1.0+3.0*X-3.0*X*X+X*X*X
          PRINT *,X,FNESTD,FDIRCT
10        CONTINUE
        STOP
        END
```

Underflow and Overflow Errors

From the definition of floating-point numbers given in section 2.2, there are upper and lower limits for the magnitudes of the numbers that can be expressed in floating-point form. Attempts to create numbers that are too small lead to what are called *underflow* errors. The default option on most computers is to set the number to zero and proceed with the computation.

For example, consider evaluating

$$f(x) = x^{10} \tag{3.12}$$

for x near 0. When working on an IBM 360/370 computer, the smallest nonzero positive number expressible in floating point is

$$m = 16^{-65} \doteq 5.4 \times 10^{-79} \tag{3.13}$$

Thus $f(x)$ will be set to zero if

$$
\begin{aligned}
&x^{10} < m \\
&|x| < \sqrt[10]{m} \doteq 1.49 \times 10^{-8}
\end{aligned} \tag{3.14}
$$

Similar results will be valid for other computers, with the actual bound dependent on the exponent range of the floating-point format.

Attempts to use numbers that are too large for floating-point format will lead to *overflow* errors, and these are generally fatal errors on most computers. Usually an overflow error is an indication of a more significant problem or error in the program. In a few cases, it is possible to eliminate an overflow error by just reformulating the expression being evaluated. For example, consider evaluating

$$z = \sqrt{x^2 + y^2}$$

If x or y is very large, then $x^2 + y^2$ might create an overflow error, even though z would probably be within the floating-point range of the machine. To avoid this, let

$$z = \begin{cases} |x| \sqrt{1 + (y/x)^2}, & 0 \le |y| \le |x| \\ |y| \sqrt{1 + (x/y)^2}, & 0 \le |x| \le |y| \end{cases}$$

In both cases, the argument of the square root is of the form $1 + w^2$ with $|w| \le 1$; this will not cause any overflow error, except when z is too large.

PROBLEMS

1. Calculate the error, relative error, and number of significant digits in the following approximations $x_A \approx x_T$.
   (a) $x_T = 28.254$, $x_A = 28.271$
   (b) $x_T = 0.028254$, $x_A = 0.028271$
   (c) $x_T = e$, $x_A = 19/7$
   (d) $x_T = \sqrt{2}$, $x_A = 1.414$
   (e) $x_T = \log (2)$, $x_A = 0.7$

2. (a) In the population growth model of (3.2), $N(t) = N_0 e^{kt}$, give a physical meaning to the constant N_0.
   (b) Show that $N(t)$ satisfies

$$\frac{N(t + 1)}{N(t)} = \text{constant}$$

   Thus $N(t + 1) = (\text{constant}) \cdot N(t)$, and the population increases by a constant ratio with every increase in t of 1 unit. Is this reasonable physically?

3. A slightly more sophisticated model for population growth is given by

$$N(t) = \frac{N_c}{1 + e^{-bt}}, \qquad t \ge 0$$

for some positive constants N_c and b. Give a major difference between this model and the one in problem 2. What is the physical meaning of the constant N_c?

Hint: Look at the behavior of each model as t becomes larger.

4. Bound the error in (3.4), using the remainder formula for the Taylor polynomial being used.

5. In some situations, loss-of-significance errors can be avoided by rearranging the function being evaluated, as was done with $f(x)$ in (3.7). Do something similar for the following cases, in some cases using trigonometric identities.

   (a) $\dfrac{1 - \cos (x)}{x^2}$ (b) $\log (x + 1) - \log (x)$

   (c) $\sin (a + x) - \sin (a)$ (d) $\sqrt[3]{1 + x} - 1$

6. Use Taylor polynomial approximations to avoid the loss-of-significance errors in the following formulas when x is near 0.

   (a) $\dfrac{e^x - 1}{x}$ (b) $\dfrac{e^x - e^{-x}}{2x}$ (c) $\dfrac{\log (1 - x) + xe^{x/2}}{x^3}$

7. (a) Solve the equation $x^2 - 26x + 1 = 0$. Use five digit decimal arithmetic to find numerical values for the roots of the equation; for example, you will need $\sqrt{168} \doteq 12.961$ or $\sqrt{672} \doteq 25.923$. Identify any loss-of-significance error that you encounter.

   (b) Find both roots accurately using only five digit decimal arithmetic.

   Hint: Use $13 - \sqrt{168} = \dfrac{1}{13 + \sqrt{168}}$.

8. Repeat problem 7 for the equation $x^2 - 40x + 1 = 0$.

9. Discuss the possible loss-of-significance error that may be encountered in solving the quadratic equation $ax^2 + bx + c = 0$. How might that loss-of-significance error be avoided?

10. Consider evaluating $\cos (x)$ for large x by using the Taylor approximation

$$\cos (x) \approx 1 - \frac{x^2}{2!} + \ldots + (-1)^n \frac{x^{2n}}{(2n)!}$$

To see the difficulty involved in using this approximation, use it to evaluate $\cos (2\pi) = 1$:

$$\cos (2\pi) \approx 1 - \frac{(2\pi)^2}{2!} + \frac{(2\pi)^4}{4!} - \ldots + (-1)^n \frac{(2\pi)^{2n}}{(2n)!}$$

Assume we are using decimal arithmetic, say a four digit decimal floating-point arithmetic with rounding. If $2n = 20$, then the error in this approximation is 0.00032 and, thus, the polynomial should be sufficiently accurate relative to the floating-point arithmetic being used. Now evaluate each term and round it to four significant digits, to find its exact floating-point representation. For example, the first three such terms are 1.000, $-19.74, 64.94$. Having found these 11 terms, add them up exactly. How closely do they approximate $\cos(2\pi) = 1.000$? Explain the source of the inaccuracy.

⑪ Evaluate the function $f(x) = x^3 - 3x^2 + 3x - 1$ of (3.11) on your computer. Evaluate it at a large number of points on some small interval about $x = 1$, attempting to obtain a behavior similar to that shown in Figure 3.2. Also, evaluate $f(x)$ in the form shown above and in the nested form

$$f(x) = -1 + x[3 + x(-3 + x)]$$

Note: On machines with a mantissa longer than 32 binary bits, the interval (0.998, 1.002) of Figure 3.2 will have to be made shorter.

12. For the polynomial

$$f(x) = x^5 - 4.2x^4 + 5.6x^3 - 6.1x^2 + 8.7x - 6.3$$

write a computer program to evaluate it in the given form and in the nested form. Evaluate both forms of $f(x)$ at a fairly large number of evenly spaced points on some subinterval, say 90 points on $1 \leq x \leq 1.1$. Both sets of values will contain noise, but generally it will be different because of the different formulations being used. Subtract the nested form from the value obtained with the original form, and then print this value. The printed values should be a composite of the noise obtained with the two different modes of evaluation.

13. To generate an overflow error on your computer, write a program to repeatedly square a number $x > 1$ and print the result. Eventually you will exceed your machine's exponent limit for floating-point numbers.

3.2 PROPAGATION OF ERROR

If a calculation is done with numbers that contain an error, then the resultant answer will be affected by these errors. We will look first at the effect of using such numbers with the ordinary arithmetic operations, and later we will consider the effect on more general function evaluations. Let x_A and y_A denote the numbers used in the calculation, and let x_T and y_T be the corresponding true values. We wish to bound

$$E = (x_T \ \omega \ y_T) - (x_A \ \omega \ y_A) \tag{3.15}$$

where ω denotes one of the operations " + ", " − ", " · ", or " ÷ ". The error E is called the *propagated error*.

The first technique used to bound E is known as *interval arithmetic*. Suppose we know bounds for $x_T - x_A$ and $y_T - y_A$. Then using these bounds and $x_A \ \omega \ y_A$, we look for an interval guaranteed to contain $x_T \ \omega \ y_T$.

Example

Let $x_A = 3.14$ and $y_A = 2.651$ be correctly rounded from x_T and y_T, to the number of digits shown. Then

$$|x_A - x_T| \leq 0.005, \qquad |y_A - y_T| \leq 0.0005$$

or, equivalently,

$$3.135 \leq x_T \leq 3.145, \qquad 2.6505 \leq y_T \leq 2.6515 \tag{3.16}$$

For the operation of addition,

$$x_A + y_A = 5.791 \tag{3.17}$$

For the true value, use (3.16) to obtain the bounding interval

$$3.135 + 2.6505 \leq x_T + y_T \leq 3.145 + 2.6515$$
$$5.7855 \leq x_T + y_T \leq 5.7965 \tag{3.18}$$

To obtain a bound for the propagated error, subtract (3.17) from (3.18) to get

$$-0.0055 \leq (x_T + y_T) - (x_A + y_A) \leq 0.0055$$

With division,

$$\frac{x_A}{y_A} = \frac{3.14}{2.651} \doteq 1.184459 \tag{3.19}$$

Also,

$$\frac{3.135}{2.6515} \leqq \frac{x_T}{y_T} \leqq \frac{3.145}{2.6505}$$

Dividing the fractions and rounding to seven digits,

$$1.182350 \leqq \frac{x_T}{y_T} \leqq 1.186569 \qquad (3.20)$$

For the error,

$$-0.002109 \leqq \frac{x_T}{y_T} - 1.184459 \leqq 0.002110$$

This technique of obtaining an interval that is guaranteed to contain the true answer is called interval arithmetic. It is a useful technique, and it has been implemented on computers, both using software and hardware. But for calculations involving arithmetic operations, interval arithmetic must be implemented with a great deal of care or else it will lead to predicted error bounds that are far in excess of the true error. For this reason, today it is not a widely used technique for the control or bounding of errors in practical computations.

When using floating-point arithmetic on a computer, the calculation of $x_A \omega y_A$ will involve an additional rounding or chopping error, just as in (3.19). The computed value of $x_A \omega y_A$ will involve the propagated error plus a rounding or chopping error. To be more precise, let $\omega*$ denote the complete operation as carried out on the computer, including any rounding or chopping. Then the total error is given by

$$(x_T \omega y_T) - (x_A \omega* y_A)$$
$$= [(x_T \omega y_T) - (x_A \omega y_A)] + [(x_A \omega y_A) - (x_A \omega* y_A)] \qquad (3.21)$$

The first term on the right is the propagated error; the second term is the error in computing $(x_A \omega y_A)$. The propagated error in using multiplication and division is examined in problems 6 and 7.

Because the error in $x_A \omega* y_A$ is usually a rounding or chopping error, we will assume

$$x_A \omega* y_A = fl(x_A \omega y_A) \qquad (3.22)$$

This says that the arithmetic calculation $x_A \omega y_A$ is to be carried out exactly and then rounded or chopped to standard floating-point length. For the error, use (2.21) to write

$$fl(x_A \; \omega \; y_A) = (1 + \epsilon)(x_A \; \omega \; y_A)$$

with ϵ bounded as in (2.15) to (2.16) or (2.22). Combining this with (3.22), we get

$$(x_A \; \omega \; y_A) - (x_A \; \omega^* \; y_A) = -\epsilon(x_A \; \omega \; y_A)$$
$$\frac{(x_A \; \omega \; y_A) - (x_A \; \omega^* \; y_A)}{x_A \; \omega \; y_A} = -\epsilon \qquad (3.23)$$

Thus the process of rounding or chopping introduces a relatively small new error into $x_A \; \omega^* \; y_A$ as compared to $x_A \; \omega \; y_A$.

Propagated Error in Function Evaluation

Consider evaluating $f(x)$ at the approximate value x_A rather than at x_T. Then how well does $f(x_A)$ approximate $f(x_T)$? There will also be an additional error introduced in the actual evaluation of $f(x_A)$, resulting in the noise phenomena described in section 3.1. But we will be concerned here with just comparing the exact value of $f(x_A)$ with the exact value of $f(x_T)$.

Using the mean-value theorem (A.3) of Appendix A,

$$f(x_T) - f(x_A) = f'(c)(x_T - x_A) \qquad (3.24)$$

with c an unknown point between x_A and x_T. Since x_A and x_T are generally very close together, we have

$$f(x_T) - f(x_A) \doteq f'(x_T)(x_T - x_A) \doteq f'(x_A)(x_T - x_A) \qquad (3.25)$$

In most cases, it is better to consider the relative error,

$$\text{Rel}\,(f(x_A)) \doteq \frac{f'(x_T)}{f(x_T)}\,(x_T - x_A) = \frac{f'(x_T)}{f(x_T)}\,x_T\,\text{Rel}\,(x_A) \qquad (3.26)$$

Example

In chemistry, one studies the ideal gas law,

$$PV = nRT$$

in which R is a constant for all gases. In the MKS measurement system,

$$R = 8.3143 + \epsilon, \quad |\epsilon| \leq 0.0012$$

To see the effect of this uncertainty in R in a sample computation, consider evaluating T, assuming $P = V = n = 1$. Then

$$T = 1/R$$

Define

$$f(x) = 1/x$$

and evaluate it at $x = R$. Letting

$$x_T = R, \qquad x_A = 8.3143$$

we have

$$|x_T - x_A| \leq 0.0012$$

For the error

$$E = \frac{1}{R} - \frac{1}{8.3143}$$

we have from (3.25) that

$$|E| = |f(x_T) - f(x_A)| \doteq |f'(x_A)|\,|x_T - x_A|$$

$$\leq \left(\frac{1}{x_A^2}\right)(0.0012)$$

$$\doteq 1.74 \cdot 10^{-5}$$

For the relative error,

$$\left|\frac{E}{f(x_T)}\right| = \left|\frac{f(x_T) - f(x_A)}{f(x_T)}\right| \leq \frac{1.74 \cdot 10^{-5}}{0.1202}$$

$$\doteq 0.000144$$

Thus the uncertainty in R has resulted in a relatively small error in the computed value

$$\frac{1}{R} \doteq \frac{1}{8.3143}$$

Example

We apply (3.24) to the evaluation of

$$f(x) = b^x \tag{3.27}$$

where b is a positive constant. Using

$$f'(x) = (\log b)b^x$$

formulas (3.25), (3.26) yield

$$b^{x_T} - b^{x_A} \doteq (\log b)b^{x_T} (x_T - x_A)$$
$$\text{Rel } (b^{x_A}) \doteq (\log b)x_T \text{ Rel}(x_A) \tag{3.28}$$

This example is also of interest for another reason. If the quantity

$$K = (\log b)x_T \tag{3.29}$$

is large in size, then the relative error in b^{x_A} will be much larger than the relative error in x_A. For example, if Rel $(x_A) = 10^{-7}$ and $K = 10^4$, then Rel $(b^{x_A}) \doteq 10^{-3}$, a significant decrease in the accuracy of b^{x_A} as compared to x_A, independent of how b^{x_A} is actually computed. The quantity in (3.29) is called a *condition number*. It relates the relative accuracy of the input to a problem (x_A in this case) to the relative accuracy of the output (b^{x_A}). We will return to this concept at other points in the text.

PROBLEMS

1. Let all of the numbers given below be correctly rounded to the number of digits shown. For each calculation, determine the smallest interval in which the result, using true instead of rounded values, must lie.
   (a)   1.1062 + 0.947   (b)   23.46 − 12.753
   (c)   (2.747)(6.83)   (d)   8.473/0.064

2. Referring to Chapter 2, Section 2.2, find the bounds for the ϵ in (3.23) for both **IBM 360/370** and **Control Data 6000/7000** series computers.

3. In the following function evaluations $f(x_A)$, assume the numbers x_A are correctly rounded to the number of digits shown. Bound the error $f(x_T) - f(x_A)$ and the relative error in $f(x_A)$.
    (a) $\cos (1.473)$ (b) $\tan^{-1} (2.62)$ (c) $\log (1.4712)$
    (d) $e^{2.653}$ (e) $\sqrt{0.0425}$

4. For the function $f(x) = \sqrt{x}$, $x > 0$, estimate $f(x_T) - f(x_A)$ and $\text{Rel}(f(x_A))$.

5. Apply (3.25) and (3.26) to $f(\pi) - f(22/7)$, for arbitrary differentiable $f(x)$.

6. This problem and the next one will explore the propagated error portion of (3.21). For the multiplication $x_A y_A$, the relative error as compared to $x_T y_T$ is

$$\text{Rel} (x_A y_A) = \frac{x_T y_T - x_A y_A}{x_T y_T}$$

Let $x_T = x_A + \epsilon$, $y_T = y_A + \eta$. Show

$$\text{Rel} (x_A y_A) = \frac{\epsilon}{x_T} + \frac{\eta}{y_T} - \left(\frac{\epsilon}{x_T}\right)\left(\frac{\eta}{y_T}\right)$$

$$= \text{Rel} (x_A) + \text{Rel} (y_A) - \text{Rel} (x_A)\text{Rel} (y_A)$$

And when both $\text{Rel} (x_A)$ and $\text{Rel} (y_A)$ are small compared to 1,

$$\text{Rel} (x_A y_A) \doteq \text{Rel} (x_A) + \text{Rel} (y_A)$$

This shows that relative errors propagate slowly with multiplication.

7. Using the method of problem 6, show

$$\text{Rel} (x_A/y_A) = \frac{\text{Rel} (x_A) - \text{Rel} (y_A)}{1 - \text{Rel} (y_A)}$$

$$\doteq \text{Rel} (x_A) - \text{Rel} (y_A)$$

with the latter holding for cases where $\text{Rel} (y_A)$ is small compared to 1.

8. To illustrate (3.28) and (3.29), compare π^{100} to $\pi^{100.1}$. Calculate these directly, as accurately as you can. (Most calculators carry enough decimal places to obtain sufficiently accurate answers for these ex-

1.20618 .69764032

ponentials.) Then calculate Rel $(\pi^{100.1})$ directly and using (3.28). Also give the condition number K of (3.29)

9. Let $f(x) = (x - 1)(x - 2) \ldots (x - n)$. Note that $f(1) = 0$. Estimate $f(1 + 10^{-4})$ by using (3.25) with $x_T = 1$, for $n = 2, 3, 4, 5, 6, 7, 8$. *Hint:* Do not calculate $f'(x)$ by first multiplying out $f(x)$. Rather, use the product rule for derivatives to evaluate $f'(x)$, and then obtain $f'(1)$.

3.3 SUMMATION

There are many situations in which sums of a fairly large number of terms must be calculated. We will study the errors introduced when doing summation on a computer and look at ways to minimize the error in the final computed sum.

Let the sum be denoted by

$$S = a_1 + a_2 + \ldots + a_n = \sum_{j=1}^{n} a_j \tag{3.30}$$

where each a_j is a floating-point number. Adding these values in the machine amounts to calculating a sequence of $n - 1$ additions, each of which will probably involve a rounding or chopping error. More precisely, define

$$S_2 = fl(a_1 + a_2)$$

the floating-point version of $a_1 + a_2$ [recall the use of $fl(x)$, introduced preceding (2.14)]. Next, define

$$S_3 = fl(a_3 + S_2)$$
$$S_4 = fl(a_4 + S_3)$$
$$\cdot$$
$$\cdot \tag{3.31}$$
$$\cdot$$
$$S_n = fl(a_n + S_{n-1})$$

S_n is the computed version of S.

From (2.14),

$$S_2 = (a_1 + a_2)(1 + \epsilon_2)$$
$$S_3 = (a_3 + S_2)(1 + \epsilon_3)$$

$$.$$
$$.$$
$$.$$

$$S_n = (a_n + S_{n-1})(1 + \epsilon_n)$$

(3.32)

Assuming the computer is binary, each ϵ_j is a small number satisfying (2.15) if the computer uses chopping and (2.16) if it rounds. The terms in (3.32) can be combined, manipulated, and estimated to give

$$
\begin{aligned}
S - S_n \doteq\ &-a_1(\epsilon_2 + \ldots + \epsilon_n) \\
&-a_2(\epsilon_2 + \ldots + \epsilon_n) \\
&-a_3(\epsilon_3 + \ldots + \epsilon_n) \\
&-a_4(\epsilon_4 + \ldots + \epsilon_n) \\
&- \ldots \\
&-a_n\epsilon_n
\end{aligned}
$$

(3.33)

Looking carefully at the formula and trying to minimize the total error $S - S_n$, the following appears to be a reasonable strategy: arrange the terms $a_1, a_2, \ldots, a_n$ before summing so that they are increasing in size,

$$|a_1| \leq |a_2| \leq |a_3| \leq \ldots \leq |a_n|$$

(3.34)

Then the terms on the right side of (3.33) with the largest number of ϵ_j's are multiplied by the smaller values among the a_j's. This should make $S - S_n$ smaller, without much additional cost in most cases.

Example

Define the terms a_j of the sum S as follows: convert the fraction $1/j$ to a decimal fraction, round it to four significant digits, and let this be a_j. To make more clear the errors in the calculation of S, we use a decimal machine that has four digits in the mantissa of a floating-point number. Adding S from the largest term to the smallest term is denoted by "LS" in the following tables; and adding from smallest to largest is denoted by "SL." The column "True" gives the true sum S, rounded to four digits. Table 3.4 gives the summation results for a decimal machine that uses chopping in all of its floating-point operations. Table 3.5 gives the analogous results for a machine using rounding.

Table 3.4   Calculating S on a Machine Using Chopping

n	True	SL	Error	LS	Error
10	2.929	2.928	.001	2.927	.002
25	3.816	3.813	.003	3.806	.010
50	4.499	4.491	.008	4.479	.020
100	5.187	5.170	.017	5.142	.045
200	5.878	5.841	.037	5.786	.092
500	6.793	6.692	.101	6.569	.224
1000	7.486	7.284	.202	7.069	.417

Table 3.5   Calculating S on a Machine Using Rounding

n	True	SL	Error	LS	Error
10	2.929	2.929	0	2.929	0
25	3.816	3.816	0	3.817	− .001
50	4.499	4.500	− .001	4.498	.001
100	5.187	5.187	0	5.187	0
200	5.878	5.878	0	5.876	.002
500	6.793	6.794	− .001	6.783	.010
1000	7.486	7.486	0	7.449	.037

In both tables, it is clear that the strategy of summing S from the smallest term to the largest is superior to the opposite procedure of summing from the largest term to the smallest. However, with the machine that rounds, it takes a fairly large number of terms before the order of summation makes any essential difference.

Rounding versus Chopping

A more important difference in the errors shown in the tables is that which exists between rounding and chopping. Rounding results in a far smaller error in the calculated sum than does chopping. To understand why this happens, return to the formula (3.33). As a typical case from it, consider the first term on the right side.

$$T \equiv -a_1(\epsilon_2 + \ldots + \epsilon_n) \tag{3.35}$$

Assume that we are using rounding with the four digit decimal machine of the above example. Using (2.22) with $\beta = 10$ and $n = 4$ for our decimal machine, we have that all ϵ_j satisfy

$$-0.0005 \leqq \epsilon_j \leqq 0.0005 \tag{3.36}$$

Rounding errors can usually be treated as random in nature, subject to this bounding interval. Thus the positive and negative values of the ϵ_j's in (3.35) will tend to cancel, and the sum T will be nearly zero. Using advanced methods from probability theory, it can be shown that (3.35) is very likely to satisfy

$$|T| \leqq (1.49)(0.0005) \cdot \sqrt{n}|a_1|$$

Thus (3.35) tends to be small until n becomes quite large, and the same is true of the total error on the right of (3.33).

For our decimal machine with chopping, (2.22) becomes

$$-0.001 \leqq \epsilon_j \leqq 0 \tag{3.37}$$

and the errors are all of one sign. Again, the chopping errors will vary randomly in this interval. But now the average value of the ϵ_j's will be -0.0005, the middle of the interval, and the likely value of (3.35) will be

$$-a_1(n - 1)(-0.0005) \tag{3.38}$$

This grows with n, where the corresponding result for the case of rounding was zero. Thus the error (3.35) and (3.33) will grow much more rapidly when chopping is used rather than rounding.

Example

As stated earlier in section 2.2, PRIME computers have both rounding and chopping available. To illustrate the difference, consider evaluating

$$S = \sum_{j=1}^{n} \frac{1}{j} \tag{3.39}$$

in single-precision arithmetic. For this calculation, errors occur in both the calculation of the floating-point form of $1/j$ and in the summation process. Table 3.6 contains the errors for the two modes of calculation. The column "RND" indicates rounding was used, and "CHP" denotes the case with chopping. The value in column "True" was calculated using double-precision arithmetic to evaluate (3.39). Also, all sums were performed from the smallest term to the largest.

Table 3.6 Calculation of (3.39): Rounding versus Chopping

n	True	RND Error	CHP Error
10	2.92896825	−1.76E-7	3.01E-7
50	4.49920534	7.00E-7	3.56E-6
100	5.18737752	−4.12E-7	6.26E-6
500	6.79282343	−1.32E-6	3.59E-5
1000	7.48547086	8.88E-8	7.35E-5

A DO-loop Error

An important example of accumulated errors in summation is the computation of independent variables in a DO-loop computation. Suppose we wish to calculate

$$x = a + jh \qquad (3.40)$$

for $j = 0, 1, 2, \ldots, n$, for given $h > 0$. This is then used in a further computation, perhaps to evaluate some $f(x)$, as in the earlier program of section 3.1. The question we wish to consider here is whether x should be computed as in (3.40) or by using the statement

$$x = x + h \qquad (3.41)$$

in the loop, having initially set $x = a$ before beginning the loop. These are mathematically equivalent ways to compute x, but they are usually not computationally equivalent.

The difficulty with computing x arises generally when h does not have a finite binary expansion that can be stored in the given floating-point mantissa, for example, $h = 0.1, 0.2$, etc. The computation (3.40) will involve two arithmetic operations and, thus, only two chopping or rounding errors, for each value of x. In contrast, the repeated use of (3.41) will involve a succession of additions, in fact, j of them for the x of (3.40). Thus as x increases in size, the use of (3.41) involves a larger number of rounding or chopping errors, leading to a different quantity than in

Table 3.7 Evaluation of $x = j*h$, $h = 0.1$

j	x	Error Using (3.40)	Error Using (3.41)
10	1	5.96E-8	5.36E-7
20	2	1.19E-7	1.55E-6
30	3	2.98E-7	2.68E-6
40	4	2.38E-7	3.58E-6
50	5	6.56E-7	9.24E-6
60	6	5.96E-7	1.49E-5
70	7	5.36E-7	2.06E-5
80	8	4.77E-7	2.62E-5
90	9	1.37E-6	4.14E-5
100	10	1.31E-6	5.66E-5

(3.40). Thus (3.40) is the preferred way to evaluate x. To provide a specific example, we give a program to compute the value of x in the two possible ways. And to check the accuracy, we also compute the true desired value of x using a double-precision computation. This program was run on a PRIME computer, with chopping, and the results are shown in Table 3.7. There is a significant improvement with (3.40) over (3.41).

```
C       TITLE: DEMONSTRATION OF COMPUTING X=J*H
C
C       TWO METHODS ARE USED TO COMPUTE X=J*H,
C       USING (3.40) AND (3.41) IN THE TEXT.
C
        DOUBLE PRECISION DX,DH
        DATA N/100/, H/0.1/
C
C       INITIALIZE.
        XSUM=0.0
        DH=H

        PRINT *, 'FOLLOWING ARE THE VALUES OF J, X=J*H, AND'
        PRINT *, 'THE ERRORS IN THE COMPUTATION OF X USING'
        PRINT *, '(3.40) AND (3.41), RESPECTIVELY.'
        PRINT *, ' '
C
        DO 10 J=1,N
          XSUM=XSUM+H
          XPROD=J*H
          DX=J*DH
          ERRSUM=DX-XSUM
          ERRPRD=DX-XPROD
          PRINT *, J,DX,ERRPRD,ERRSUM
10        CONTINUE
        STOP
        END
```

Calculation of Inner Products

A sum of the form

$$S = a_1b_1 + a_2b_2 + \ldots + a_nb_n = \sum_{j=1}^{n} a_jb_j \qquad (3.42)$$

is called an *inner product*. Typically the terms a_j and b_j are elements of arrays A and B, and S is the inner product of A and B. Such sums occur quite often in solving certain kinds of problems, particularly those involving systems of simultaneous linear equations.

If we calculate S in single precision, then there will be a single-precision rounding (or chopping) error for each multiplication and each addition. Thus there will be $2n - 1$ single-precision rounding errors involved in calculating S. The consequences of these errors can be analyzed in the manner of (3.33), and we could derive an optimal strategy for the calculation of (3.42). Instead we will look at a simpler alternative, using the double-precision arithmetic of the computer.

Convert each a_i and b_i to double precision by extending their mantissas with zeros. Multiply them in double precision, sum them in double precision; and when done, round (or chop) the answer to single precision to obtain the calculated value of S. For machines with built-in double-precision arithmetic, this procedure is a simple and rapid way to obtain more accurate inner products, and there need be no increase in storage space for the arrays A and B. The accuracy is improved since there will be only one single-precision rounding error, regardless of the size of n.

Following is a function subprogram SUMPRD, written in Fortran 77.

```
        FUNCTION SUMPRD(A,B,N)
C
C       THIS CALCULATES THE INNER PRODUCT
C
C                    I=N
C          SUMPRD=   SUM    A(I)*B(I)
C                    I=1
C
C       THE PRODUCTS AND SUMS ARE DONE IN DOUBLE
C       PRECISION, AND THE FINAL RESULT IS CONVERTED
C       BACK TO SINGLE PRECISION.
C
        DOUBLE PRECISION DSUM
        DIMENSION A(*),B(*)
C
        DSUM=0.0D0
        DO 10 I=1,N
10          DSUM=DSUM+DBLE(A(I))*DBLE(B(I))
        SUMPRD=DSUM
        RETURN
        END
```

In it, the elements of A and B are converted to double precision using the built-in function DBLE, which converts numbers to their equivalent forms in double precision. Also, the DIMENSION statement allows the dimension to be unspecified for arrays that are arguments of the subprogram. But this can only be used with one dimensional arrays and with the last dimension of multidimensional arrays.

PROBLEMS

1. Write a computer program to evaluate

$$S = \sum_{j=1}^{n} \frac{1}{j}$$

for arbitrary n. Do the calculation by both the methods LS and SL. For LS, compute and sum the terms from the largest term first to the smallest term last. For SL, do the calculation in the reverse order. Calculate a true value for S using double-precision arithmetic, and compare it with the values obtained by LS and SL.

2. Repeat problem 1 for the series

(a) $\quad S = \sum_{j=1}^{n} \frac{1}{j(j+1)} = \frac{n}{n+1}$

(b) $\quad \sum_{j=1}^{n} \frac{1}{j(j+2)} = \frac{3}{4} - \frac{2n+3}{2(n+1)(n+2)}$

In these cases, the true value is known exactly.

3. Derive the formula (3.33) for the cases $n = 2$, 3, and 4 from the formulas (3.32).

   Hint: If the ϵ_i's are small, as indicated in (2.15) to (2.16), or (2.22), then $\epsilon_i \epsilon_j$ is very small compared to ϵ_i and can therefore be neglected when added to ϵ_i.

4. Implement the Fortran subprogram SUMPRD given in this section. Apply it to the sums of problem 2, regarding the sum as an inner product of two arrays. For (a), write

$$A = \left[1, \frac{1}{2}, \frac{1}{3}, \ldots, \frac{1}{n} \right]$$

$$B = \left[\frac{1}{2}, \frac{1}{3}, \ldots, \frac{1}{n+1} \right]$$

Also, compute the same inner product using only single-precision arithmetic and compare it to your answer obtained using SUMPRD.

FOUR

ROOTFINDING

Calculating the roots of an equation

$$f(x) = 0 \qquad (4.1)$$

is a common problem in applied mathematics. We will look at some simple numerical methods for solving this equation and also consider some possible difficulties. We begin by giving a simple example arising in financial planning.

Example

The equation

$$P_{in}[(1 + r)^{N_{in}} - 1] - P_{out}[1 - (1 + r)^{-N_{out}}] = 0 \qquad (4.2)$$

arises in the calculation of annuities. An amount P_{in} is put into an account each time period (for example, one year) for N_{in} periods. Then an amount of P_{out} is drawn out of the account each time period for N_{out} periods. During these $N_{in} + N_{out}$ time periods, the account earns compound interest at a rate of r per time period (for example, $r = 0.09$ means 9%). In order that the account contains exactly zero at the end of $N_{in} + N_{out}$

time periods, equation (4.2) must be satisfied. Given any four of the quantities P_{in}, P_{out}, N_{in}, N_{out}, and r, find the fifth one. This is straight-forward in all cases but one: Finding r is a rootfinding problem for which we must use the methods of this chapter.

The function $f(x)$ will usually have at least one continuous derivative, and often we will have some estimate of the root that is being sought. Using this information, most numerical methods for (4.1) compute a sequence of increasingly accurate estimates of the root. These methods are called *iteration methods*. We will study three different iteration methods in the first three sections of the chapter; and section 4.4 considers difficulties that occur in solving (4.1) for some special types of functions $f(x)$.

4.1 THE BISECTION METHOD

Suppose $f(x)$ is continuous on an interval $[a, b]$ and that

$$f(a)f(b) < 0 \tag{4.3}$$

Then $f(x)$ changes sign on $[a, b]$, and $f(x) = 0$ has at least one root on the interval. The simplest numerical procedure for finding a root is to repeatedly halve the interval $[a, b]$, keeping the half on which $f(x)$ changes sign. This procedure is called the *bisection method*. It is guaranteed to converge to a root, denoted here by α.

To be more precise in our definition, suppose we are given an interval $[a, b]$ satisfying (4.3) and an error tolerance $\epsilon > 0$. Then the bisection method consists of the following steps:

B1. Define $c := (a + b)/2$.

B2. If $b - c \le \epsilon$, then accept c as the root and stop.

B3. If sign $[f(b)]$ sign $[f(c)] \le 0$, then set $a := c$.
Otherwise, set $b := c$. Return to step B1.

The interval $[a, b]$ is halved with each loop thru steps B1 to B3. The test B2 will be satisfied eventually, and with it the condition $|\alpha - c| \le \epsilon$ will be satisfied. Justification is given in (4.7) below.

Example

Find the largest root of

$$f(x) \equiv x^6 - x - 1 = 0 \tag{4.4}$$

Table 4.1 Bisection Method for (4.4)

n	a	b	c	$b - c$	$f(c)$
1	1.0000	2.0000	1.5000	0.5000	8.8906
2	1.0000	1.5000	1.2500	0.2500	1.5647
3	1.0000	1.2500	1.1250	0.1250	−0.0977
4	1.1250	1.2500	1.1875	0.0625	0.6167
5	1.1250	1.1875	1.1562	0.0312	0.2333
6	1.1250	1.1562	1.1406	0.0156	0.0616
7	1.1250	1.1406	1.1328	0.0078	−0.0196
8	1.1328	1.1406	1.1367	0.0039	0.0206
9	1.1328	1.1367	1.1348	0.0020	0.0004
10	1.1328	1.1348	1.1338	0.00098	−0.0096

accurate to within $\epsilon = 0.001$. With a graph, it is easy to check that $1 < \alpha < 2$. We choose $a = 1$, $b = 2$; then $f(a) = -1$, $f(b) = 61$, and (4.3) is satisfied. The results of the algorithm B1 to B3 are shown in Table 4.1. The entry n indicates that the associated row corresponds to iteration number n of steps B1 to B3.

Error Bounds

Let a_n, b_n, and c_n denote the nth computed values of a, b, and c, respectively. Then easily we get

$$b_{n+1} - a_{n+1} = \frac{1}{2}(b_n - a_n), \qquad n \geq 1 \qquad (4.5)$$

and it is straightforward to deduce that

$$b_n - a_n = \frac{1}{2^{n-1}}(b - a), \qquad n \geq 1 \qquad (4.6)$$

where $b - a$ denotes the length of the original interval we started with. Since the root α is in either the interval $[a_n, c_n]$ or $[c_n, b_n]$, we know

$$|\alpha - c_n| \leq c_n - a_n = b_n - c_n = \frac{1}{2}(b_n - a_n) \qquad (4.7)$$

This is the error bound for c_n that is used in step B2 of the earlier algorithm. Combining it with (4.6), we obtain the further bound

$$|\alpha - c_n| \leq \frac{1}{2^n} (b - a) \tag{4.8}$$

This shows that the iterates c_n converge to α as $n \to \infty$.

To see how many iterations will be necessary, suppose we want to have

$$|\alpha - c_n| \leq \epsilon$$

This will be satisfied if

$$\frac{1}{2^n} (b - a) \leq \epsilon$$

Taking logarithms of both sides, we can solve this to give

$$n \geq \frac{\ln\left(\dfrac{b - a}{\epsilon}\right)}{\ln 2} \tag{4.9}$$

For our example (4.4), this results in

$$n \geq \frac{\ln\left(\dfrac{1}{0.001}\right)}{\ln 2} = 9.97$$

Thus we must have $n = 10$ iterates, exactly the number computed.

There are several advantages to the bisection method. The principal one is that the method is guaranteed to converge. In addition, the error bound, given in (4.7), is guaranteed to decrease by half with each iteration. Many other numerical methods have variable rates of decrease for the error, and these may be worse than the bisection method for some equations. The principal disadvantage of the bisection method is that it generally converges slower than most other methods. For functions $f(x)$ that have a continuous derivative, other methods are usually faster. These methods may not always converge; but when they do converge, they are almost always much faster than bisection.

The Program BISECT

A subroutine BISECT, implementing the bisection method, is given below. The comment statements at the beginning of the programs should
be self-explanatory as to the purpose of the program and the use of the
subroutine. The form of the program is quite standard. The parameter
F is the name of a user-defined function subprogram. The actual name
of the user-supplied function must be declared in an EXTERNAL statement in the program calling BISECT. Named constants in BISECT are
given in the PARAMETER statement. If the precision of the program
is to be changed, then the constants can be easily changed to double
precision since they are all given in this one statement.

Lacking in the subroutine is any attempt to check whether the error
tolerance EPS is realistic in terms of the length of the mantissa in the
computer arithmetic being used. This was left out in order to keep the
program simple; but a realistic rootfinding program in a computer library would need to include such a check.

```
C       THIS IS A DEMONSTRATION PROGRAM FOR THE
C       ROOTFINDING SUBROUTINE 'BISECT'.
C
        EXTERNAL FCN
C
C       INPUT PROBLEM PARAMETERS.
10      PRINT *, ' WHAT ARE A,B,EPSILON?'
        PRINT *, ' TO STOP, LET EPSILON=0.'
        READ *,A,B,EPSLON
        IF(EPSLON .EQ. 0.0) STOP
C
C       CALCULATE ROOT
        CALL BISECT(FCN,A,B,EPSLON,ROOT,IER)
C
C       PRINT ANSWERS.
        PRINT 1000, A,B,EPSLON
        PRINT 1001, ROOT,IER
        GO TO 10
1000    FORMAT(/,' A=',E11.4,5X,'B='E11.4,5X,'EPSILON=',E9.3)
1001    FORMAT(' ROOT=',E14.7,5X,' IER=',I1)
        END

        FUNCTION FCN(X)
C
        FCN=X-TAN(X)
        RETURN
        END
```

```
        SUBROUTINE BISECT(F,A,B,EPS,ROOT,IER)
C
C       THE PROGRAM USES THE BISECTION METHOD TO SOLVE
C       THE EQUATION
C                      F(X) = 0.
C       THE SOLUTION IS TO BE BETWEEN 'A' AND 'B', AND
C       IT IS ASSUMED THAT
C                   F(A)*F(B) .LE. O.
C       THE SOLUTION IS RETURNED IN 'ROOT', AND IT IS
C       TO BE IN ERROR BY AT MOST 'EPS'.
C
C       'IER' IS AN ERROR INDICATOR.
C       IF IER=0 ON COMPLETION OF THE ROUTINE, THEN THE
C       SOLUTION HAS BEEN COMPUTED SATISFACTORILY.
C       IF IER=1, THEN F(A)*F(B) WAS GREATER THAN 0,
C       CONTRARY TO ASSUMPTION.
C
        PARAMETER(ZERO=0.0,ONE=1.0,TWO=2.0)
C
C       INITIALIZE
        FA=F(A)
        FB=F(B)
        SFA=SIGN(ONE,FA)
        SFB=SIGN(ONE,FB)
        IF(SFA*SFB .GT. ZERO) THEN
C           THE CHOICE OF A AND B IS IN ERROR.
            IER=1
            RETURN
          END IF
C
C       CREATE A NEW VALUE OF C, THE MIDPOINT OF [A,B].
10      C=(A+B)/TWO
        IF( ABS(B-C) .LE. EPS) THEN
C           C IS AN ACCEPTABLE SOLUTION OF F(X)=0.
20          ROOT=C
            IER=0
            RETURN
          END IF
C
C       THE VALUE OF C WAS NOT SUFFICIENTLY ACCURATE.
C       BEGIN A NEW ITERATION.
        FC=F(C)
        IF(FC .EQ. ZERO) GO TO 20
        SFC=SIGN(ONE,FC)
        IF(SFB*SFC .GT. ZERO) THEN
C           THE SOLUTION IS IN [A,C].
            B=C
            SFB=SFC
          ELSE
C           THE SOLUTION IS IN [C,B].
            A=C
            SFA=SFC
          END IF
        GO TO 10
        END
```

PROBLEMS

1. Use the bisection method with a hand calculator or computer to find the indicated roots of the following equations. Use an error tolerance of $\epsilon = 0.001$.

(a) The real root of $x^3 - x^2 - x - 1 = 0$.

(b) The root of $x = 1 + 0.3 \cos (x)$.

(c) The smallest positive root of $\cos (x) = \frac{1}{2} + \sin (x)$.

2. To help determine the roots of $x = \tan (x)$, graph $y = x$ and $y = \tan (x)$, and look at the intersection points of the two graphs.

(a) Find the smallest nonzero positive root of $x = \tan (x)$, with an accuracy of $\epsilon = 0.0001$.

Note: The root is greater than $\pi/2$.

(b) Solve $x = \tan (x)$ for the root which is closest to $x = 100$.

3. Consider equation (4.2) with $P_{in} = 100$, $P_{out} = 1500$, $N_{in} = 30$, and $N_{out} = 20$. Find r within an accuracy of $\epsilon = 0.0001$.

4. Show that $x = a + b \cos (x)$ has at least one root.

Hint: Find an interval on which $f(x) = x - a - b \cos (x)$ changes sign.

5. Implement the subroutine BISECT or another one of your own design. Use it to solve the equations in problems 1 and 2 with an accuracy of $\epsilon = 10^{-5}$.

6. Using the program of problem 5, solve the equation

$$f(x) \equiv x^3 - 3x^2 + 3x - 1 = 0$$

with an accuracy of $\epsilon = 10^{-6}$. Experiment with different ways of evaluating $f(x)$; for example, use (i) the given form, (ii) reverse its order, and (iii) also use the nested form of (1.21) in Chapter 1. Try various initial intervals $[a, b]$, for example, $[0, 1.5]$, $[0.5, 2.0]$, and $[0.5, 1.1]$. Explain the results. Note that $\alpha = 1$ is the only root of $f(x) = 0$.

7. Let the initial interval used in the bisection method have length $b - a = 3$. Find the number of midpoints c_n that must be calculated with the bisection method to obtain an approximate root within an error tolerance of 10^{-9}.

8. Imagine finding a root α satisfying $0.5 < \alpha < 1$. If you are using a binary computer with m binary digits in its mantissa, what is the smallest error tolerance that makes sense in finding an approximation to α? If the original interval $[a, b] = [0.5, 1]$, how many interval halvings are needed to find an approximation c_n to α with the maximum accuracy possible for this computer?

4.2 NEWTON'S METHOD

Consider the sample graph of $y = f(x)$ shown in Figure 4.1. The root α occurs where the graph crosses the x-axis. We will usually have an estimate of α, and it will be denoted here by x_0. To improve on this estimate, consider the straight line that is tangent to the graph at the point $(x_0, f(x_0))$. If x_0 is near α, this tangent line should be nearly coincident with the graph of $y = f(x)$ for points x about α. Then the root of the tangent line should nearly equal α. This root is denoted here by x_1.

To find a formula for x_1, consider the slope of the tangent line. Using the derivative $f'(x)$, we know from calculus that the slope of the tangent line at $(x_0, f(x_0))$ is $f'(x_0)$. We can also calculate the slope using the fact that the tangent line contains the two points $(x_0, f(x_0))$ and $(x_1, 0)$. This leads to the slope being equal to

$$\frac{f(x_0) - 0}{x_0 - x_1}$$

the difference in the y-coordinates divided by the difference in the x-

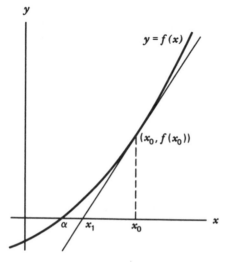

Figure 4.1 Schematic for Newton's method.

coordinates. Equating the two different formulas for the slope, we obtain

$$f'(x_0) = \frac{f(x_0)}{x_0 - x_1}$$

This can be solved to give

$$x_1 = x_0 - \frac{f(x_0)}{f'(x_0)} \tag{4.10}$$

Since x_1 should be an improvement over x_0 as an estimate of α, this entire procedure can be repeated with x_1 as the initial guess. This leads to the new estimate

$$x_2 = x_1 - \frac{f(x_1)}{f'(x_1)}$$

Repeating this process, we obtain a sequence of numbers $x_1, x_2, x_3, \ldots$ that we hope will approach the root α. These numbers are called iterates, and they are defined recursively by the following general *iteration formula:*

$$x_{n+1} = x_n - \frac{f(x_n)}{f'(x_n)}, \qquad n = 0, 1, 2, \ldots \tag{4.11}$$

This is Newton's method for solving $f(x) = 0$.

Example

Solve equation (4.4), which was used earlier as an example for the bisection method. With it,

$$f(x) = x^6 - x - 1, \qquad f'(x) = 6x^5 - 1$$

and the iteration is given by

$$x_{n+1} = x_n - \frac{x_n^6 - x_n - 1}{6x_n^5 - 1}, \qquad n \geq 0 \tag{4.12}$$

We use an initial guess of $x_0 = 1.5$. The results are shown in Table 4.2. The column "$x_n - x_{n-1}$" is an estimate of the error $\alpha - x_{n-1}$; justification for this is given later in the section.

Table 4.2   Newton's Method for $x^6 - x - 1 = 0$

n	x_n	$f(x_n)$	$x_n - x_{n-1}$
0	1.5	8.89E + 1	
1	1.30049088	2.54E + 1	-2.00E-1
2	1.18148042	5.38E -1	-1.19E-1
3	1.13945559	4.92E -2	-4.20E-2
4	1.13477763	5.50E -4	-4.68E-3
5	1.13472415	6.80E -8	-5.35E-5
6	1.13472414	-4.00E-9	-1.00E-8

The true root is $\alpha = 1.134724138$, and x_6 equals α to nine significant digits. Compare this with the results shown in Table 4.1 for the bisection method. Also, Newton's method may converge slowly at first; but as the iterates come closer to the root, the speed of convergence increases, as shown in the table.

Example

We will look at a procedure that was used to carry out division on some of the early computers. These computers had hardware arithmetic for addition, subtraction, and multiplication. But division had to be implemented by software; this is also true with many modern day microcomputers.

Suppose we want to form a/b. It can be done by multiplying a and $1/b$, with the latter produced approximately using Newton's method. More specifically, we want to solve

$$f(x) \equiv b - \frac{1}{x} = 0 \tag{4.13}$$

where we assume $b > 0$. The root is $\alpha = 1/b$. The derivative is

$$f'(x) = \frac{1}{x^2}$$

and Newton's method is given by

$$x_{n+1} = x_n - \frac{b - \dfrac{1}{x_n}}{\dfrac{1}{x_n^2}}$$

Simplifying, we get

$$x_{n+1} = x_n(2 - bx_n), \qquad n \geq 0 \tag{4.14}$$

This involves only multiplication and subtraction and, thus, there is no difficulty in implementing it on the computers discussed above. The initial guess, of course, should be chosen with $x_0 > 0$.

For the error, it can be shown that

$$\text{Rel}\ (x_{n+1}) = [\text{Rel}\ (x_n)]^2, \qquad n \geq 0 \tag{4.15}$$

where

$$\text{Rel}\ (x_n) = \frac{\alpha - x_n}{\alpha}$$

the relative error in x_n as an approximation to $\alpha = 1/b$. From (4.15), we must have

$$|\text{Rel}\ (x_0)| < 1 \tag{4.16}$$

Otherwise, the error in x_n will not decrease to zero as n increases. This condition means

$$-1 < \frac{(1/b) - x_0}{1/b} < 1$$

and this reduces to the equivalent condition

$$0 < x_0 < \frac{2}{b} \tag{4.17}$$

The iteration (4.14) converges to $\alpha = 1/b$ if and only if the initial guess x_0 satisfies (4.17). Figure 4.2 shows a sample case. From Figure 4.2, it is fairly easy to justify (4.17), since if it is violated, the calculated value of x_1 would be negative.

The result (4.15) shows that the convergence is very rapid, once we have a somewhat accurate initial guess. For example, suppose

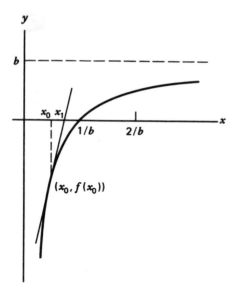

Figure 4.2 Iterative solution of
$b - (1/x) = 0$.

$|\text{Rel}(x_0)| = 0.1$, which corresponds to a 10% error in x_0. Then from (4.15),

$$\text{Rel}(x_1) = 10^{-2}, \qquad \text{Rel}(x_2) = 10^{-4},$$
$$\text{Rel}(x_3) = 10^{-8}, \qquad \text{Rel}(x_4) = 10^{-16} \tag{4.18}$$

Thus x_3 or x_4 should be sufficiently accurate for most purposes.

Error Analysis

Assuming $f(x)$ has at least two continuous derivatives for all x in some interval about the root α, use Taylor's theorem to write

$$f(\alpha) = f(x_n) + (\alpha - x_n)f'(x_n) + \frac{1}{2}(\alpha - x_n)^2 f''(c_n)$$

with c_n an unknown point between α and x_n. Note that $f(\alpha) = 0$ by assumption, and then divide by $f'(x_n)$ to obtain

$$0 = \frac{f(x_n)}{f'(x_n)} + \alpha - x_n + (\alpha - x_n)^2 \frac{f''(c_n)}{2f'(x_n)}$$

From (4.11), the first term on the right side is $x_n - x_{n+1}$, and we have

$$0 = x_n - x_{n+1} + \alpha - x_n + (\alpha - x_n)^2 \frac{f''(c_n)}{2f'(x_n)}$$

Solving for $\alpha - x_{n+1}$, we have

$$\alpha - x_{n+1} = \left[\frac{-f''(c_n)}{2f'(x_n)} \right] (\alpha - x_n)^2 \qquad (4.19)$$

This formula says that the error in x_{n+1} is proportional to the square of the error in x_n. When the initial error is sufficiently small, this shows that the error in the succeeding iterates will decrease very rapidly, just as in (4.18). Formula (4.19) can also be used to give a formal mathematical proof of the convergence of Newton's method, but we omit it.

Example

For the earlier iteration (4.12), $f''(x) = 30x^4$. If we are near the root α, then

$$\frac{-f''(c_n)}{2f'(x_n)} \doteq \frac{-f''(\alpha)}{2f'(\alpha)} = \frac{-30\alpha^4}{2(6\alpha^5 - 1)} \doteq -2.42$$

Thus for the error in (4.12),

$$\alpha - x_{n+1} \doteq -2.42(\alpha - x_n)^2 \qquad (4.20)$$

This explains the rapid convergence of the final iterates in Table 4.2. For example, consider the case of $n = 3$, with $\alpha - x_3 = -4.73\mathrm{E}-3$. Then (4.20) predicts

$$\alpha - x_4 \doteq -2.42(4.73\mathrm{E}-3)^2 = -5.42\mathrm{E}-5$$

which compares well to the actual error of $\alpha - x_4 = -5.35\mathrm{E}-5$.

Assuming that the iterate x_n is near to the root α, the multiplier on the right of (4.19) can be written as

$$\frac{-f''(c_n)}{2f'(x_n)} \doteq \frac{-f''(\alpha)}{2f'(\alpha)} \equiv M \qquad (4.21)$$

Thus

$$\alpha - x_{n+1} \doteq M(\alpha - x_n)^2, \qquad n \geq 0 \qquad (4.22)$$

Multiply both sides by M to get

$$M(\alpha - x_{n+1}) \doteq [M(\alpha - x_n)]^2 \qquad (4.23)$$

Assuming that all the iterates are near α, then inductively we can show

$$M(\alpha - x_n) \doteq [M(\alpha - x_0)]^{2^n}, \qquad n \geq 0$$

Since we want $\alpha - x_n$ to converge to zero, this says that we must have

$$|M(\alpha - x_0)| < 1$$
$$|\alpha - x_0| < \frac{1}{M} = \left| \frac{2f'(\alpha)}{f''(\alpha)} \right| \qquad (4.24)$$

If the quantity M is very large, then x_0 will have to be chosen very close to α in order to obtain convergence. In such situations, the bisection method is probably an easier method to use. Examples of this situation are given in problems 5 and 6 at the end of the section.

The choice of x_0 can be very important in determining whether or not Newton's method will converge. Unfortunately, there is no one strategy that is always effective in choosing x_0. In some cases, probably most, a choice of x_0 arises from the physical situation that led to the rootfinding problem. In other cases, graphing $y = f(x)$ will probably be needed, possibly combined with the bisection method for a few iterates.

Error Estimation

We are computing a sequence of iterates x_n, and we would like to estimate their accuracy in order to know when to stop the iteration. To estimate $\alpha - x_n$, note that since $f(x) = 0$, we have

$$f(x_n) = f(x_n) - f(\alpha)$$
$$= f'(\xi_n)(x_n - \alpha)$$

for some ξ_n between x_n and α, by the mean-value theorem. Solving for the error, we obtain

$$\alpha - x_n = \frac{-f(x_n)}{f'(\xi_n)} \doteq \frac{-f(x_n)}{f'(x_n)}$$

provided x_n is so close to α that $f'(x_n) \doteq f'(\xi_n)$. From (4.11), this becomes

$$\alpha - x_n \doteq x_{n+1} - x_n \qquad (4.25)$$

This is the standard error estimation formula for Newton's method, and it is usually fairly accurate. However, it is not valid if $f'(\alpha) = 0$, a case that is discussed in section 4.4 of this chapter.

Example

Consider the error in the entry x_3 of Table 4.2.

$$\begin{aligned} \alpha - x_3 &= -4.73\mathrm{E} - 3 \\ x_4 - x_3 &= -4.68\mathrm{E} - 3 \end{aligned} \qquad (4.26)$$

This confirms the accuracy of (4.25) for that case.

The Program NEWTON

Subroutine NEWTON is an implementation of Newton's method, with error estimation and a safeguard against an infinite loop. The input parameter ITMAX is an upper limit on the number of iterates to be computed; this prevents the occurrence of an infinite loop. The extensive use of comments is typical of Fortran programs that are widely distributed. Of course, these comments can be omitted if you type this program for your own use.

```
      SUBROUTINE NEWTON(F,DF,XINIT,EPS,ITMAX,ROOT,IER)
C
C        THIS ROUTINE CALCULATES A ROOT OF
C                  F(X) = 0
C        USING NEWTON'S METHOD.
C
C        INPUT PARAMETERS:
C        (1) F,DF ARE THE NAMES OF FUNCTION SUBPROGRAMS FOR
C             COMPUTING F(X) AND DF(X)=F'(X). THE ACTUAL NAMES
C             USED IN CALLING SUBROUTINE NEWTON MUST BE DECLARED
C             IN AN EXTERNAL STATEMENT IN THE CALLING PROGRAM.
C        (2) XINIT IS AN INITIAL GUESS OF THE SOLUTION.
C        (3) EPS IS THE DESIRED ERROR TOLERANCE. THE TEST FOR
C             CONVERGENCE IS
C                  ABS(X1-XO) .LE. EPS,
C             WHERE X1 IS THE CURRENT NEWTON ITERATE AND XO
C             IS THE PRECEDING ITERATE.
```

```
C       (4) ITMAX IS AN UPPER LIMIT ON THE NUMBER OF ITERATES
C           TO BE COMPUTED. WHEN THE NUMBER OF ITERATES
C           REACHES THIS NUMBER, THE PROGRAM IS TERMINATED.
C           WARNING: ITMAX IS ALTERED ON OUTPUT.
C
C       OUTPUT PARAMETERS:
C       (1) ROOT CONTAINS THE COMPUTED VALUE OF THE SOLUTION.
C           REGARDLESS OF HOW THE ROUTINE IS TERMINATED,
C           SUCCESSFULLY OR OTHERWISE, ROOT WILL CONTAIN
C           THE MOST RECENTLY COMPUTED NEWTON ITERATE.
C       (2) ITMAX IS SET TO THE NUMBER OF ITERATES
C           COMPUTED IN THE ROUTINE.
C       (3) IER IS AN ERROR INDICATOR.
C           =0 MEANS A SUCCESSFUL COMPLETION OF NEWTON.
C           =1 MEANS ITMAX ITERATES WERE COMPUTED AND THE
C              ROUTINE WAS ABORTED.
C           =2 MEANS THAT THE DERIVATIVE DF(X) BECAME ZERO
C              AT SOME NEWTON ITERATE, AND THE ROUTINE
C              WAS ABORTED.
C
        PARAMETER(ZERO=0.0)
C
C       INITIALIZE.
        XO=XINIT
        ITNUM=1
C
C       BEGIN MAIN LOOP.
10      DENOM=DF(XO)
        IF(DENOM .EQ. ZERO) THEN
C           DERIVATIVE EQUALS ZERO. TERMINATE ITERATION.
            IER=2
            ROOT=XO
            ITMAX=ITNUM-1
            RETURN
        END IF
C       COMPUTE NEWTON ITERATION.
        X1=XO-F(XO)/DENOM
        IF(ABS(X1-XO) .LE. EPS) THEN
C           ERROR TEST SATISFIED.
            IER=0
            ROOT=X1
            ITMAX=ITNUM
            RETURN
        END IF
        IF(ITNUM .LE. ITMAX) THEN
C           INITIALIZE FOR ANOTHER LOOP.
            ITNUM=ITNUM+1
            XO=X1
            GO TO 10
        ELSE
C           ITMAX ITERATES HAVE BEEN COMPUTED. TERMINATE.
            IER=1
            ROOT=X1
            RETURN
        END IF
        END
```

PROBLEMS

①. Carry out the Newton iteration (4.12) with the two initial guesses $x_0 = 1.0$ and $x_0 = 2.0$. Compare the results with Table 4.2.

②. Using Newton's method, find the roots of the equations in problem 1, section 4.1 in this chapter. Use an error tolerance of $\epsilon = 10^{-5}$.

③.(a) On most computers, the computation of $\sqrt{a}$ is based on Newton's method. Set up the Newton iteration for solving $x^2 - a = 0$, and show that it can be written in the form

$$x_{n+1} = \frac{1}{2}\left(x_n + \frac{a}{x_n}\right), \qquad n \geq 0$$

(b) Derive the error and relative error formulas

$$\sqrt{a} - x_{n+1} = -\frac{1}{2x_n}(\sqrt{a} - x_n)^2$$

$$\text{Rel}(x_{n+1}) = \frac{-\sqrt{a}}{2x_n}[\text{Rel}(x_n)]^2$$

(c) For x_0 near $\sqrt{a}$, the last formula becomes

$$\text{Rel}(x_{n+1}) \doteq -\frac{1}{2}[\text{Rel}(x_n)]^2, \qquad n \geq 0$$

Assuming $\text{Rel}(x_0) = 0.1$, use this formula to estimate the relative error in x_1, x_2, x_3, and x_4.

④. Give Newton's method for finding $\sqrt[m]{a}$, with $a > 0$.
   Hint: Solve $x^m - a = 0$.

5. (a) Repeat problem 2 of section 4.1, for finding roots of $x = \tan(x)$. Use an error tolerance of $\epsilon = 10^{-4}$.

   (b) The root near 100 will be difficult to find using Newton's method. To explain this, compute the quantity M of (4.21), and use it in the condition (4.24) for x_0.

⑥. The equation

$$f(x) \equiv x + e^{-Bx^2}\cos(x) = 0, \qquad B > 0$$

has a unique root, and it is in the interval $(-1, 0)$. Use Newton's method to find it as accurately as possible. Use values of $B = 1, 5, 10, 25, 50$. Among your choices of x_0, choose $x_0 = 0$, and explain the behavior observed in the iterates for the larger values of B. *Hint:* Draw a graph of $f(x)$ to better understand the behavior of the function.

7. Check the accuracy of the error approximation (4.25) that is used in Table 4.2, as was done in (4.26).

8. Use the iteration (4.14) to compute $1/3$. Use $x_0 = 0.2$.

9. Derive formula (4.15).
   Hint: Use

$$\text{Rel } (x_{n+1}) = \frac{\alpha - x_{n+1}}{\alpha} = 1 - bx_{n+1}$$

   Replace x_{n+1} using (4.14), and then compare the result to $[\text{Rel } (x_n)]^2$.

10. Solve the equation

$$x^3 - 3x^2 + 3x - 1 = 0$$

   on a computer and use Newton's method. Recalling problem 6 of section 4.1, experiment with the choice of initial guess x_0. Note any unusual behavior.

4.3 SECANT METHOD

The Newton method is based on approximating the graph of $y = f(x)$ with a tangent line and on then using the root of this straight line as an approximation to the root α of $f(x)$. From this perspective, other straight line approximations to $y = f(x)$ would also lead to methods for approx-

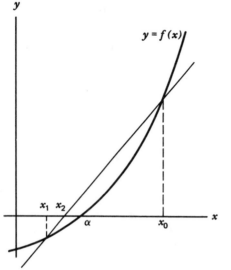

Figure 4.3 Schematic of secant method: $x_1 < \alpha < x_0$.

imating a root of $f(x)$. One such straight line approximation leads to the secant method.

Assume that two initial guesses to α are known and denote them by x_0 and x_1. They may occur on opposite sides of α, as in Figure 4.3, or on the same side of α, as in Figure 4.4. The two points $(x_0, f(x_0))$ and $(x_1, f(x_1))$, on the graph of $y = f(x)$, determine a straight line, called a secant line. This line is an approximation to the graph of $y = f(x)$, and its root x_2 is an approximation of α.

To derive a formula for x_2, we proceed in a manner similar to that used to derive Newton's method: match the slope determined by $\{(x_0, f(x_0)), (x_1, f(x_1))\}$ with the slope determined by $\{(x_1, f(x_1)), (x_2, 0)\}$. This gives

$$\frac{f(x_1) - f(x_0)}{x_1 - x_0} = \frac{0 - f(x_1)}{x_2 - x_1}$$

Solving for x_2, we get

$$x_2 = x_1 - f(x_1) \cdot \frac{x_1 - x_0}{f(x_1) - f(x_0)}$$

Having found x_2, we can drop x_0 and use x_1, x_2 as a new set of approximate values for α. This leads to an improved value x_3; and this process can

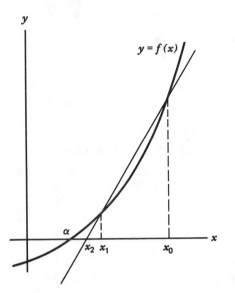

Figure 4.4 Schematic of secant method: $\alpha < x_1 < x_0$.

be continued indefinitely. Doing so, we obtain the general iteration formula

$$x_{n+1} = x_n - f(x_n) \cdot \frac{x_n - x_{n-1}}{f(x_n) - f(x_{n-1})}, \qquad n \ge 1 \qquad (4.27)$$

This is the *secant method*. It is called a two point method, since two approximate values are needed to obtain an improved value. The bisection method is also a two point method, but the secant method will almost always converge faster than bisection.

Example

We solve the equation

$$f(x) \equiv x^6 - x - 1 = 0$$

which was used previously as an example for both the bisection and Newton methods. The results are given in Table 4.3, including the quantity $x_n - x_{n+1}$ as an estimate of $\alpha - x_{n-1}$. The iterate x_8 is α rounded to nine significant digits. As with the Newton method (4.12) for this equation, the initial iterates do not converge rapidly. But as the iterates become closer to α, the speed of convergence increases.

Error Analysis

Using techniques from calculus, it is possible to show that the iterates x_n of (4.27) satisfy

$$\alpha - x_{n+1} = (\alpha - x_n)(\alpha - x_{n-1}) \left[\frac{-f''(\xi_n)}{2f'(\zeta_n)} \right] \qquad (4.28)$$

The unknown number ζ_n is between x_n and x_{n-1}, and the unknown number ξ_n is between the largest and the smallest of the numbers α, x_n, and x_{n-1}. The error formula closely resembles the Newton error formula (4.19). This should be expected since the secant method can be considered as an approximation of Newton's method, based on using

$$f'(x_n) \doteq \frac{f(x_n) - f(x_{n-1})}{x_n - x_{n-1}} \qquad (4.29)$$

Check that the use of this in the Newton formula (4.11) will yield (4.27).

Table 4.3 Secant Method for $x^6 - x - 1 = 0$

n	x_n	$f(x_n)$	$x_n - x_{n-1}$
0	2.0	61.0	
1	1.0	-1.0	-1.0
2	1.01612903	$-9.15E-1$	$1.61E-2$
3	1.19057777	$6.57E-1$	$1.74E-1$
4	1.11765583	$-1.68E-1$	$-7.29E-2$
5	1.13253155	$-2.24E-2$	$1.49E-2$
6	1.13481681	$9.54E-4$	$2.29E-3$
7	1.13472365	$-5.07E-6$	$-9.32E-5$
8	1.13472414	$-4.00E-9$	$4.90E-7$

The formula (4.28) can be used to obtain the further error result

$$\lim_{n\to\infty} \frac{|\alpha - x_{n+1}|}{|\alpha - x_n|^r} = \left|\frac{f''(\alpha)}{2f'(\alpha)}\right|^{r-1} \equiv c \tag{4.30}$$

where $r = (\sqrt{5} + 1)/2 \doteq 1.62$. Thus

$$|\alpha - x_{n+1}| \doteq c|\alpha - x_n|^{1.62} \tag{4.31}$$

as x_n approaches α. Compare this with the Newton estimate (4.22), in which the exponent is 2 rather than 1.62. Thus Newton's method converges more rapidly than the secant method. Also, the constant c in (4.31) plays the same role as M in (4.22), and they are related by

$$c = |M|^{r-1}$$

The restriction (4.24) on the initial guess for Newton's method can be replaced by a similar one for the secant iterates, but we omit it. Finally, the result (4.31) can be used to justify the error estimate

$$\alpha - x_{n-1} \doteq x_n - x_{n-1} \tag{4.32}$$

for iterates x_n that are sufficiently close to the root.

Example

For the iterate x_5 in Table 4.3,

$$\alpha - x_5 \doteq 2.19E-3$$
$$x_6 - x_5 \doteq 2.29E-3 \tag{4.33}$$

Comparison of Newton and Secant Methods

From the above discussion, Newton's method converges more rapidly than the secant method. Thus Newton's method should require fewer iterations to attain a given error tolerance. However, Newton's method requires two function evaluations per iteration, that of $f(x_n)$ and $f'(x_n)$. And the secant method requires only one evaluation, $f(x_n)$, if it is programmed carefully to retain the value of $f(x_{n-1})$ from the preceding iteration. Thus the secant method will require less time per iteration than the Newton method.

The decision as to which method should be used will depend on the above factors, including the difficulty or expense of evaluating $f'(x_n)$; and it will depend on intangible human factors such as convenience of use. Newton's method is very simple to program and to understand; but for many problems with a complicated $f'(x)$, the secant method will probably be faster in actual running time on a computer.

The derivations of both the Newton and secant methods illustrate a general principle of numerical analysis. When trying to solve a problem for which there is no direct or simple method of solution, approximate it by another problem that you can solve more easily. In both cases, we have replaced the solution of $f(x) = 0$ with the solution of a much simpler rootfinding problem for a straight line.

PROBLEMS

1. Using the secant method, find the roots of the equations in problem 1 of section 4.1. Use an error tolerance of $\epsilon = 10^{-5}$.

2. Solve problem 6 of section 4.2, using the secant method. As one choice of initial guesses, use $x_0 = -1$, $x_1 = 0$.

3. To experimentally confirm the error estimate (4.32), compute $\alpha - x_{n-1}$ for each entry in Table 4.3 and compare it to $x_n - x_{n-1}$. This was done for $x_{n-1} = x_5$ in (4.33).

4. Using the secant method, repeat problem 6 of section 4.1 and problem 10 of section 4.2.

5. Write formula (4.28) as

$$\alpha - x_{n+1} \doteq M(\alpha - x_n)(\alpha - x_{n-1}), \qquad n \geq 0$$

with M defined in (4.21). Then multiply both sides by M, obtaining

$$|M(\alpha - x_{n+1})| \doteq |M(\alpha - x_n)|\,|M(\alpha - x_{n-1})|$$

Let $B_n = |M(\alpha - x_n)|$, $n \geq 0$. To have x_n converge to α, we must have B_n converge to 0. The above formula yields

$$B_{n+1} = B_n B_{n-1}$$

For simplicity, assume $B_0 = B_1 = \delta$.

(a)   Compute $B_2, B_3, B_4, B_5, B_6, B_7$ in terms of δ.

(b)   If we write $B = \delta^{q_n}$, $n \geq 0$, give a formula for q_{n+1} in terms of q_{n-1} and q_n. What are q_0 and q_1?

(c)   Experimentally, confirm that

$$q_n \doteq \frac{1}{\sqrt{5}} r^{n+1} \doteq \frac{1}{\sqrt{5}} (1.618)^{n+1}$$

for larger values of n, say $n \geq 4$. The number r was used in (4.30).

(d)   Using (c), show that

$$\frac{B_{n+1}}{B_n^r}$$

is approximately constant. Find the constant. This result can be used to construct a proof of (4.30).

4.4 ILL-BEHAVED ROOTFINDING PROBLEMS

We will look at two classes of problems for which the methods of sections 4.1 through 4.3 do not perform well. Often there is little that a numerical analyst can do to improve these problems, but one should be aware of their existence and of the reason for their ill-behavior.

We begin with functions that have a *multiple root*. The root α of $f(x)$ is said to be of multiplicity m if

$$f(x) = (x - \alpha)^m g(x) \tag{4.34}$$

for some continuous function $g(x)$ with $g(\alpha) \neq 0$. Assuming that $f(x)$ is sufficiently differentiable, an equivalent definition is that

$$f(\alpha) = f'(\alpha) = \ldots = f^{(m-1)}(\alpha) = 0, \qquad f^{(m)}(\alpha) \neq 0 \tag{4.35}$$

A root of multiplicity $m = 1$ is called a simple root.

Example

a. $f(x) = (x - 1)^2(x + 2)$ has two roots. $\alpha = 1$ has multiplicity 2, and $\alpha = -2$ is a simple root.

b. $f(x) = x^3 - 3x^2 + 3x - 1$ has $\alpha = 1$ as a root of multiplicity 3. To see this, note that

$$f(1) = f'(1) = f''(1) = 0, \qquad f'''(1) = 6$$

The result follows from (4.35).

When the Newton and secant methods are applied to the calculation of a multiple root α, the convergence of $\alpha - x_n$ to zero is much slower than it would be for a simple root. In addition, there is a large *interval of uncertainty* as to where the root actually lies, because of the noise of evaluating $f(x)$.

The large interval of uncertainty for a multiple root is the most serious problem associated with numerically finding such a root. In Figure 3.2 of Chapter 3, we illustrated the noise in evaluating $f(x) = (x - 1)^3$, which has $\alpha = 1$ as a root of multiplicity 3. That graph also illustrates the large interval of uncertainty in finding α. To further illustrate the difference in the intervals of uncertainty between simple roots and multiple roots, see Figure 4.5. The dotted lines drawn about each side of the graph of $y = f(x)$ are meant to give the outer limits on the noise in evaluating $f(x)$. The intersection of the band of noise with the x-axis shows the interval in which the root α may be located, and it is much larger with the multiple root.

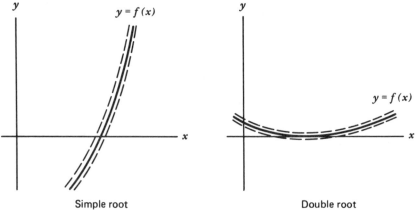

Figure 4.5 Interval of uncertainty in evaluation of a root.

Example

To illustrate the effect of a multiple root on a rootfinding method, we use Newton's method to calculate the root $\alpha = 1.1$ of

$$f(x) = (x - 1.1)^3(x - 2.1)$$
$$= 2.7951 + x(-8.954 + x(10.56 + x(-5.4 + x))) \quad (4.36)$$

The computer used is decimal with six digits in the mantissa, and it uses rounding. The function $f(x)$ is evaluated in the nested form of (4.36), and $f'(x)$ is evaluated similarly. The results are given in Table 4.4. The column "Ratio" gives the values of

$$\frac{\alpha - x_n}{\alpha - x_{n-1}} \quad (4.37)$$

and we can see that these values equal about $\frac{2}{3}$.

It is possible to show that when using Newton's method to calculate a root of multiplicity m, the ratios (4.37) will approach

$$\lambda = \frac{m - 1}{m}, \quad m \geq 1 \quad (4.38)$$

Thus as x_n approaches α,

$$\alpha - x_n \doteq \lambda(\alpha - x_{n-1}) \quad (4.39)$$

and the error decreases at about the constant rate. In our example, $\lambda = \frac{2}{3}$ since the root has multiplicity $m = 3$, which corresponds to the values in the last column of the table. The error formula (4.39) implies a much

Table 4.4 Newton's Method for (4.36)

n	x_n	$f(x_n)$	$\alpha - x_n$	Ratio
0	0.800000	0.03510	0.300000	
1	0.892857	0.01073	0.207143	0.690
2	0.958176	0.00325	0.141824	0.685
3	1.00344	0.00099	0.09656	0.681
4	1.03486	0.00029	0.06514	0.675
5	1.05581	0.00009	0.04419	0.678
6	1.07028	0.00003	0.02972	0.673
7	1.08092	0.0	0.01908	0.642

slower rate of convergence than is usual for Newton's method. With any root of multiplicity $m \geq 2$, the number $\lambda \geq \frac{1}{2}$; thus the bisection method is always at least as fast as Newton's method for multiple roots.

A further observation from the table is that the iterate x_7 is an exact root of $f(x)$ in the computer, even though it is very far from $\alpha = 1.1$. This is explained by the rounding errors that occur in the evaluation of $f(x)$. The resulting noise in evaluating $f(x)$ leads to a fairly large interval in which the root α might lie, just as illustrated earlier in Figure 4.5.

The only way to obtain accurate values for multiple roots is to analytically remove the multiplicity, obtaining a new function for which α is a simple root. Otherwise, there will be a large interval of uncertainty for the location of the root. To remove the multiplicity, first use Newton's method to determine the multiplicity m of α, using the results (4.38) and (4.39), together with the approximation

$$\lambda \doteq \frac{x_{n+1} - x_n}{x_n - x_{n-1}} \tag{4.40}$$

By finding λ, we can determine m from (4.38). Then analytically calculate

$$F(x) = f^{(m-1)}(x)$$

From (4.34), it can be shown that $F(x)$ will have α as a simple root. Solve $F(x) = 0$ to find α accurately.

Example

Differentiate (4.36) twice to obtain the new rootfinding problem

$$f''(x) \equiv 21.12 - 32.4x + 12x^2 = 0$$

This equation has $\alpha = 1.1$ as a simple root, and Newton's method will converge rapidly to a very accurate value.

Stability of Roots

With most functions $f(x)$, if a small error is made in calculating the function, then the root will change by a correspondingly small amount. However, there are a number of functions for which this is not true. With such functions, very small errors in evaluating $f(x)$ will lead to very large changes in the roots of the function. Finding the root of such a function is called an *ill-conditioned* or *unstable problem*. We give a well-known example of such a function, and then we return to an analysis of unstable rootfinding problems.

Example

Define

$$f(x) = (x - 1)(x - 2)(x - 3)(x - 4)(x - 5)(x - 6)(x - 7)$$
$$= x^7 - 28x^6 + 322x^5 - 1960x^4 + 6769x^3 - 13132x^2$$
$$+ 13068x - 5040 \quad (4.41)$$

Change the coefficient of x^6 from -28 to -28.002, and call the new function $\hat{f}(x)$. The change in the coefficient is relatively small,

$$|\text{Rel } (28.002)| = \frac{0.002}{28} = 7.14 \times 10^{-5} \quad (4.42)$$

The roots of $f(x)$ are clearly $\{1, 2, \ldots, 7\}$. The roots of $\hat{f}(x)$ are given in Table 4.5, correctly rounded to eight significant digits. Some of these roots are far from the corresponding roots of $f(x)$, even though the change from $f(x)$ to $\hat{f}(x)$ involves only a relatively small change. A similar change in some of the other coefficients of $f(x)$ will lead to the same kind of behavior in the roots.

To talk about the approximate evaluation of a function $f(x)$, introduce the perturbed function

$$F_\epsilon(x) = f(x) + \epsilon g(x) \quad (4.43)$$

The function $g(x)$ is assumed to be continuously differentiable, and ϵ is

Table 4.5 Roots of $f(x)$ and $\hat{f}(x)$

Root of $f(x)$	Root of $\hat{f}(x)$	Error
1	1.0000028	$-2.8E-6$
2	1.9989382	$1.1E-3$
3	3.0331253	$-3.3E-2$
4	3.8195692	0.18
5	5.4586758 + 0.54012578i	$-0.46 - 0.54i$
6	5.4586758 − 0.54012578i	$0.54 + 0.54i$
7	7.2330128	-0.23

to be a reasonably small number. For small values of ϵ, $F_\epsilon(x)$ and $f(x)$ will be nearly the same.

Example

For the preceding example (4.41), we would have

$$F_\epsilon(x) = f(x) + \epsilon g(x), \qquad g(x) = x^6, \qquad \epsilon = -0.002 \qquad (4.44)$$

The roots of $F_\epsilon(x)$ will depend on ϵ, and we denote such a root by $\alpha(\epsilon)$. The original root α of $f(x)$ is just $\alpha(0)$. To simplify our discussion, we will assume that $\alpha(0)$ is a simple root of $f(x)$ and, thus, $f'(\alpha(0)) \neq 0$. This discussion will be adequate for understanding the example (4.41). Using these assumptions, it can be shown by more advanced mathematical results that a Taylor polynomial approximation can be used to estimate $\alpha(\epsilon)$ if ϵ is sufficiently small. We will use

$$\alpha(\epsilon) \doteq \alpha(0) + \epsilon \alpha'(0) \qquad (4.45)$$

Thus we need to compute $\alpha'(0)$.

Since $\alpha(\epsilon)$ is a root of $F_\epsilon(x)$, we have

$$f(\alpha(\epsilon)) + \epsilon g(\alpha(\epsilon)) = 0 \qquad (4.46)$$

for all small values of ϵ. Take the derivative of both sides of this equation, using ϵ as the variable for the differentiation. This yields

$$f'(\alpha(\epsilon))\alpha'(\epsilon) + g(\alpha(\epsilon)) + \epsilon g'(\alpha(\epsilon))\alpha'(\epsilon) = 0 \qquad (4.47)$$

Substitute $\epsilon = 0$ to get

$$f'(\alpha(0))\alpha'(0) + g(\alpha(0)) = 0$$

and solve for $\alpha'(0)$,

$$\alpha'(0) = -\frac{g(\alpha(0))}{f'(\alpha(0))} \qquad (4.48)$$

Using this in (4.45), we get

$$\alpha(\epsilon) \doteq \alpha(0) - \epsilon \frac{g(\alpha(0))}{f'(\alpha(0))} \qquad (4.49)$$

for all sufficiently small values of ϵ. If the derivative $\alpha'(0)$ is very large in size, then the small change ϵ will be magnified greatly in its effect on the root.

Example

Consider the root $\alpha(0) = 4$ for the polynomial $f(x)$ of (4.41). Use the definition of $F_\epsilon(x)$ given in (4.44). Then

$$f'(4) = (4 - 1)(4 - 2)(4 - 3)(4 - 5)(4 - 6)(4 - 7) = -36$$
$$g(4) = 4^6 = 4096$$
$$\alpha'(0) = -\frac{4096}{-36} \doteq 114 \tag{4.50}$$
$$\alpha(\epsilon) \doteq 4 + 114\epsilon$$

This shows that the small change ϵ will be magnified greatly in its effect on the root $\alpha(0) = 4$ for $f(x)$. For the particular choice of (4.44),

$$\alpha(\epsilon) \doteq 4 + 114(-0.002) = 3.772,$$

which is approximately the actual root 3.820 given in Table 4.5.

Finding the roots of a polynomial such as (4.41) is an unstable problem. There is not much that can be done with such a problem except to go to higher precision arithmetic. The main difficulty lies in the original formulation of the mathematical equation to be solved, and often there is another way to approach the problem that will prevent the unstable behavior.

PROBLEMS

1. Use Newton's method to calculate the roots of

$$f(x) = x^5 + 0.9x^4 - 1.62x^3 - 1.458x^2 + 0.6561x + 0.59049$$

Print out the iterates and the function values. Produce the ratios of (4.37) by using the approximation (4.40),

$$\frac{\alpha - x_n}{\alpha - x_{n-1}} \doteq \frac{x_{n+1} - x_n}{x_n - x_{n-1}}$$

Repeat the problem for several choices of x_0. Make observations that seem important relative to the rootfinding problem.

Note: The above will first approach $\lambda = (m - 1)/m$, as in (4.38), but then they will depart from it due to noise in the evaluation of $f(x)$ as x_n approaches the root.

2. Do the calculation of $\alpha(\epsilon)$ for the roots $\alpha(0) = 3, 5$, and 7 of example (4.41), proceeding as in (4.50).

3. Do the perturbation calculation for another change in (4.41). Change the coefficient of x^4 from -1960 to -1960.14. What is the relative perturbation in the coefficient? Calculate $\alpha(\epsilon)$ for $\alpha(0) = 3$ and $\alpha(0) = 5$.

4. Consider the polynomial $f(x) = x^5 - 300x^2 - 126x + 5005$, which has a root $\alpha = 5$. Also consider the perturbed function

$$F_\epsilon(x) = f(x) + \epsilon x^5 = (1 + \epsilon)x^5 - 300x^2 - 126x + 5005$$

with ϵ a small number. Letting $\alpha(\epsilon)$ denote the perturbed root of $F_\epsilon(x) = 0$ corresponding to $\alpha(0) = 5$, estimate $\alpha(\epsilon) - 5$. Is finding $\alpha = 5$ for $f(x) = 0$ an unstable rootfinding problem?

5. Newton's method is used to find a root of $f(x) = 0$. The first few iterates are shown in the following table, giving a very slow speed of convergence. What can be said about the root α to explain this convergence? Knowing $f(x)$, how would you find an accurate value for α?

n	x_n	$x_n - x_{n-1}$
0	0.75	
1	0.752710	0.00271
2	0.754795	0.00208
3	0.756368	0.00157
4	0.757552	0.00118
5	0.758441	0.000889

FIVE

INTERPOLATION

Interpolation is the process of finding and evaluating a function whose graph goes through a set of given points. The points may arise as measurements in a physical problem, or they may be obtained from a known function. The interpolating function is usually chosen from a restricted class of functions, with polynomials being the most commonly used class. In sections 5.1 and 5.2 of this chapter, we look at two different formulations for polynomial interpolation, and in section 5.3 we analyze the error involved in polynomial interpolation. The final section of the chapter considers interpolation with another class of functions called spline functions, which have become quite popular in recent years.

Interpolation is used to solve problems from the more general area of approximation theory. To provide some perspective, we list a number of approximation problems and the possible role of interpolation in their solution. (1) Given a table of values of a function, for example $f(x) = \cos(x)$, we use interpolation to extend the table to arguments x not in the table. This was once an important problem, but it is no longer such because the use of computers and hand calculators has greatly reduced the need for tables. Nonetheless, it is an easy and familiar problem to most students, and it will be used to illustrate the ideas of interpolation introduced in the first three sections. (2) Given a set of data points, find a smooth nonoscillatory function $f(x)$ that fits the data either exactly or

91

approximately. The exact case leads to the study of interpolating spline functions; and the approximate fitting of the data is usually done with the method of least squares, which is taken up in section 8.6 of Chapter 8. (3) Given a known function $f(x)$, such as $f(x) = \log(x)$, we need to have a way of evaluating it on a computer. This is taken up in the following chapter, and interpolation is used as a means of obtaining a computable approximation. (4) To numerically integrate or differentiate a function, we often replace the function with a simpler approximating expression, that is then integrated or differentiated. These simpler expressions are almost always obtained by interpolation. Also, the most widely used numerical methods for solving differential equations are obtained from interpolating approximations.

5.1 POLYNOMIAL INTERPOLATION

Most people are introduced to interpolation in elementary algebra, where it is used to extend tables of logarithms or trigonometric functions to argument values not given in the table. The form of interpolation used is called *linear interpolation*, and it will be used here as an introduction to more general polynomial interpolation. Our approach will not be the same as that given in most beginning algebra texts, but later we will show the connection between the two approaches.

Given two points (x_0, y_0) and (x_1, y_1) with $x_0 \neq x_1$, draw a straight line through them, as in Figure 5.1. The straight line is the graph of the linear polynomial

$$P_1(x) = \frac{(x_1 - x)y_0 + (x - x_0)y_1}{x_1 - x_0} \tag{5.1}$$

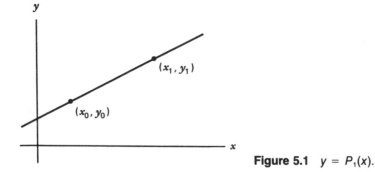

Figure 5.1 $y = P_1(x)$.

The reader should check that the graph of this function is the straight line determined by (x_0, y_0) and (x_1, y_1). We say that this function interpolates the value y_i at the point x_i, $i = 0, 1$; or

$$P_1(x_i) = y_i, \qquad i = 0, 1$$

Example

Let the data points be $(1, 1)$ and $(4, 2)$. The polynomial $P_1(x)$ is given by

$$P_1(x) = \frac{(4 - x)(1) + (x - 1)(2)}{3} \tag{5.2}$$

The graph of $y = P_1(x)$ is shown in Figure 5.2, along with that of $y = \sqrt{x}$ from which the data points were taken.

For table interpolation, the values of a function $f(x)$ are given at a large set of evenly spaced values of x. An example of a section of such a table is given in Table 5.1, for $f(x) = e^x$. To use linear interpolation to estimate the value of $f(x)$ when x is not in the table, locate two values x_0 and x_1 that enclose x in the table. Construct $P_1(x)$ to interpolate $f(x_i)$ at x_i, $i = 0, 1$. Then use $P_1(x)$ to estimate $f(x)$.

Example

Obtain an estimate of $e^{0.826}$ using linear interpolation in Table 5.1. Since $x = 0.826$, choose $x_0 = 0.82$ and $x_1 = 0.83$. The polynomial $P_1(x)$ is given

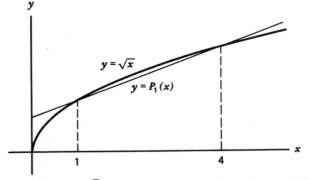

Figure 5.2 $y = \sqrt{x}$ and its linear interpolating polynomial (5.2).

Table 5.1 Values of e^x

x	e^x	x	e^x
.80	2.225541	.85	2.339647
.81	2.247908	.86	2.363161
.82	2.270500	.87	2.386911
.83	2.293319	.88	2.410900
.84	2.316367	.89	2.435130

by

$$P_1(x) = \frac{(0.83 - x)(2.270500) + (x - 0.82)(2.293319)}{0.01} \tag{5.3}$$

In particular,

$$P_1(0.826) = 2.2841914 \tag{5.4}$$

and for comparison, the true value is

$$e^{0.826} \doteq 2.2841638 \tag{5.5}$$

to eight significant digits.

The formulas (5.1) to (5.3) are not very convenient for actual calculations. Instead, rewrite (5.1) as

$$P_1(x) = y_0 + \mu(y_1 - y_0), \qquad \mu = \frac{x - x_0}{x_1 - x_0} \tag{5.6}$$

This is the form used in elementary algebra texts, and it is much easier to work with than (5.1). Check that (5.6) and (5.1) are the same function.

Example

Repeat the preceding example. With it,

$$\mu = \frac{0.826 - 0.82}{0.01} = 0.6, \qquad y_1 - y_0 = 0.022819$$

$$P_1(0.826) = 2.270500 + (0.6)(0.022819) = 2.2841914 \tag{5.7}$$

Quadratic Interpolation

Most data arise from graphs that are curved rather than straight. To better approximate such behavior, we look at polynomials of degree greater than one. Assume that three data points (x_0, y_0), (x_1, y_1), and (x_2, y_2) are given, with x_0, x_1, x_2 distinct points. We construct the quadratic polynomial that passes through these points as follows:

$$P_2(x) = y_0 L_0(x) + y_1 L_1(x) + y_2 L_2(x) \qquad (5.8)$$

with

$$L_0(x) = \frac{(x - x_1)(x - x_2)}{(x_0 - x_1)(x_0 - x_2)}, \qquad L_1(x) = \frac{(x - x_0)(x - x_2)}{(x_1 - x_0)(x_1 - x_2)}$$

$$L_2(x) = \frac{(x - x_0)(x - x_1)}{(x_2 - x_0)(x_2 - x_1)} \qquad (5.9)$$

Formula (5.8) is called *Lagrange's formula* for the quadratic interpolating polynomial; and polynomials L_0, L_1, and L_2 are called the Lagrange interpolation basis functions.

Each polynomial $L_i(x)$ has degree 2 and, thus, $P_2(x)$ has degree ≤ 2. In addition,

$$L_i(x_j) = 0, \qquad i \neq j$$
$$L_i(x_i) = 1$$

for $0 \leq i, j \leq 2$. These two statements are combined into the statement

$$L_i(x_j) = \delta_{ij}, \qquad 0 \leq i, j \leq 2 \qquad (5.10)$$

where δ_{ij} is called the *Kronecker delta function*,

$$\delta_{ij} = \begin{cases} 1, & i = j \\ 0, & i \neq j \end{cases}$$

With the use of these properties, we can easily show that $P_2(x)$ interpolates the data,

$$P_2(x_i) = y_i, \qquad i = 0, 1, 2$$

Example

Construct $P_2(x)$ for the data points $(0, -1)$, $(1, -1)$, and $(2, 7)$. Then

$$P_2(x) = \frac{(x - 1)(x - 2)}{2}(-1) + \frac{x(x - 2)}{-1}(-1) + \frac{x(x - 1)}{2}(7) \quad (7) \quad (5.11)$$

Its graph is shown in Figure 5.3.

With linear interpolation, it was obvious that there was only one straight line passing through two given data points. But with three data points, it is less obvious that there is only one quadratic interpolating polynomial whose graph passes through the points. To see that there is only one such polynomial, we assume that there is a second one and then show that it must equal $P_2(x)$. Let $Q_2(x)$ denote another polynomial of degree ≤ 2 whose graph passes through (x_i, y_i), $i = 0, 1, 2$. Define

$$r(x) = P_2(x) - Q_2(x) \quad (5.12)$$

Since P_2 and Q_2 both have degree ≤ 2, we also have degree $(r) \leq 2$. In addition, using the interpolating property of both P_2 and Q_2,

$$r(x_i) = P_2(x_i) - Q_2(x_i)$$
$$= y_i - y_i = 0$$

for $i = 0, 1, 2$. Thus $r(x)$ has three distinct roots x_0, x_1, and x_2, but its degree is ≤ 2. By the fundamental theorem of algebra [see I(e) of Appendix B], this is not possible unless $r(x) \equiv 0$, the zero polynomial. But

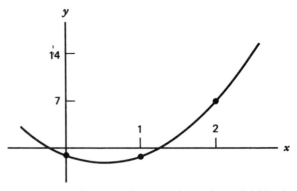

Figure 5.3 Quadratic interpolating polynomial (5.11).

from (5.12), that says $P_2(x) \equiv Q_2(x)$; thus there is only one polynomial of degree ≤ 2 that satisfies

$$P_2(x_i) = y_i, \quad i = 0, 1, 2 \tag{5.13}$$

It would be expected that a quadratic interpolation polynomial would yield better accuracy in table interpolation. As before, when given an x at which we wish to evaluate a table function $f(x)$, select in the table three evenly spaced points x_0, x_1, x_2 that enclose x. Choose them in order with

$$x_0 < x_1 < x_2, \quad x_1 - x_0 = x_2 - x_1, \quad x_0 < x < x_2$$

The Lagrange formula (5.8) is inconvenient for doing actual calculations; instead, we use the following generalization of (5.6):

$$P_2(x) = y_0 + \mu(y_1 - y_0) + \frac{1}{2}\mu(\mu - 1)[(y_2 - y_1) - (y_1 - y_0)] \tag{5.14}$$

where $\mu = (x - x_0)/(x_1 - x_0)$ as before. We leave it as problem 9 to show that (5.14) is just a rearrangement of (5.8).

Note that in computing $P_2(x)$ using (5.14), it can be written as

$$P_2(x) = P_1(x) + \frac{1}{2}\mu(\mu - 1)[(y_2 - y_1) - (y_1 - y_0)] \tag{5.15}$$

Thus $P_2(x)$ is obtained by adding a *correction term* onto $P_1(x)$. This is illustrated in (5.16), in which $P_1(0.826)$ is used from the earlier calculation (5.7).

Example

Calculate a quadratic interpolate to $e^{0.826}$ from Table 5.1. We choose $x_0 = 0.82$, $x_1 = 0.83$, $x_2 = 0.84$. Then

$$\mu = 0.6, \quad y_1 - y_0 = 0.022819, \quad y_2 - y_1 = 0.023048$$

$$P_2(x) = P_1(x) + \frac{1}{2}(0.6)(0.6 - 1)(0.023048 - 0.022819) \tag{5.16}$$

$$= 2.2841914 - 0.00002748 \doteq 2.2841639$$

to eight digits. Comparison to the true answer in (5.5) shows that $P_2(0.826)$ is a significant improvement over $P_1(0.826)$.

Higher Degree Interpolation

We now consider the general case. Assume we are given $n + 1$ data points $(x_0, y_0), \ldots, (x_n, y_n)$, with all of the x_i's distinct. The interpolating polynomial of degree $\leq n$ is given by

$$P_n(x) = y_0 L_0(x) + y_1 L_1(x) + \ldots + y_n L_n(x) \qquad (5.17)$$

with each $L_i(x)$ a polynomial of degree n given by

$$L_i(x) = \frac{(x - x_0) \ldots (x - x_{i-1})(x - x_{i+1}) \ldots (x - x_n)}{(x_i - x_0) \ldots (x_i - x_{i-1})(x_i - x_{i+1}) \ldots (x_i - x_n)}, \qquad (5.18)$$

for $0 \leq i \leq n$. The denominator equals the value of the numerator at $x = x_i$. From (5.18), it is easy to show that $L_i(x)$ satisfies

$$L_i(x_j) = \delta_{ij}, \qquad 0 \leq j \leq n \qquad (5.19)$$

for $i = 0, 1, \ldots, n$. Using (5.19), it follows by direct substitution of $x = x_j$ that

$$P_n(x_j) = y_j, \qquad j = 0, 1, \ldots, n \qquad (5.20)$$

Formula (5.17) is called Lagrange's formula for the degree n interpolating polynomial.

The fact that there is only one polynomial satisfying (5.20) among all polynomials of degree $\leq n$ is proven in a manner analogous to that used with quadratic interpolation, following (5.12). This is left as problem 12. Because of its importance, we state the results of this section as a formal mathematical theorem.

Theorem 5.1

Let $n \geq 0$, assume $x_0, x_1, \ldots, x_n$ are $n + 1$ distinct numbers, and let $y_0, \ldots, y_n$ be $n + 1$ given numbers, not necessarily distinct. Then among all polynomials of degree $\leq n$, there is exactly one polynomial $P_n(x)$ that satisfies

$$P_n(x_i) = y_i, \qquad i = 0, 1, \ldots, n$$

It is possible to give generalizations of formulas (5.6) and (5.14) for easier calculation of $P_n(x)$ in table interpolation. But very little table interpolation is done any longer, and linear or quadratic interpolation will usually be sufficient for the few cases that most students will encounter. There are two other more important uses for interpolation, and they are the main justification for our taking up the subject.

First, we use polynomial interpolation to produce approximations to a given function $f(x)$. Polynomials can be easily evaluated, and it is also trivial to integrate or differentiate them. Interpolating polynomials will be used in the next two chapters to produce approximate methods for evaluating, integrating, and differentiating arbitrary functions $f(x)$. The material of the next section also gives a more efficient way of evaluating interpolating polynomials, and it will be useful in the work of the following chapters.

The second important use of interpolation is in finding a *smooth curve* passing through a set of data points $\{(x_i, y_i) | i = 1, \ldots, N\}$ and, generally, we want a curve that does not have any oscillations or ripples in it. The solution to this problem leads us to *spline functions*, taken up in section 5.4; but the material on polynomial interpolation will still be useful in developing this new type of interpolation function.

We close with a final note of caution on the degree of the interpolation polynomial. For $n + 1$ data points, the interpolating polynomial $P_n(x)$ may have degree $< n$. For purposes of illustration, suppose the three data points (x_i, y_i), $i = 0, 1, 2$ lie on a straight line. Then $P_2(x) \equiv P_1(x)$, where $P_1(x)$ is the linear interpolating polynomial determined by the points (x_0, y_0) and (x_1, y_1). The reason for this lies in the uniqueness result proven following (5.12). The polynomial $P_1(x)$ interpolates to all three data points, by our assumption on them and, thus, it must equal $P_2(x)$, by the uniqueness of $P_2(x)$. A similar argument holds for higher order polynomial interpolation.

PROBLEMS

1. Given the data points $(0, 2)$, $(1, 1)$, find:
   (a) The straight line interpolating this data.
   (b) The function $f(x) = a + be^x$ interpolating this data.
       Hint: Find a and b so that $f(0) = 2$, $f(1) = 1$.
   (c) The function $f(x) = a/(b + x)$ interpolating this data.
   In each case, graph the interpolating function.

2. Find the function $P(x) = a + b \cos(\pi x) + c \sin(\pi x)$, which interpolates the data

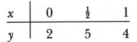

x	0	$\frac{1}{2}$	1
y	2	5	4

3. Using Table 5.1, find linear interpolates to
(a) $e^{0.865}$ (b) $e^{5/6}$
Compare the values obtained to the true values obtained from the function e^x on a hand calculator.

4. Show that formulas (5.6) and (5.1) are equivalent formulas for $P_1(x)$.

5. Write a computer program to do linear interpolation and to check its accuracy. Read x_0 and x_1 and then generate the data values using $y = e^x$ (or some other commonly available function). For a variety of values of x, both inside and outside $[x_0, x_1]$, compute $P_1(x)$, e^x, and their difference $E(x) = e^x - P_1(x)$. Plot the values of $E(x)$, to see how the error varies with x.

6. Show that the formula (5.11) simplifies to

$$P_2(x) = 4x^2 - 4x - 1$$

7. Produce the quadratic polynomial interpolating the data $\{(-2, -15), (-1, -8), (0, -3)\}$. Find the first zero of $P_2(x)$ to the right of $x = 0$. Does $P_2(x)$ have a maximum?

8. Using (5.8), find the polynomial $P_2(x)$ that interpolates the following data. In each case, simplify (5.8) as much as possible.
(a) $\{(0, 1), (1, 2), (2, 3)\}$ (b) $\{(0, 1), (1, 1), (2, 1)\}$
Comment on your results.

9. Show that (5.14) is a quadratic polynomial in x, and show that it satisfies the interpolating conditions (5.13).
Hint: The points $x = x_0, x_1, x_2$ correspond to $\mu = 0, 1, 2$, respectively.

10. Repeat problem 3 using quadratic interpolation. Use (5.14) for easier calculations.

11. Write out the complete formula (5.17) for $P_3(x)$, including all four of the polynomials $L_0(x)$, $L_1(x)$, $L_2(x)$, $L_3(x)$ for this case.

12. (a) Prove there is only one polynomial $P_3(x)$ among all polynomials of degree ≤ 3 that satisfy the interpolating conditions

$$P_3(x_i) = y_i, \quad i = 0, 1, 2, 3$$

where the x_i's are distinct.
Hint: Generalize the proof given in and following (5.12) for the uniqueness of $P_2(x)$.
(b) Give a proof of uniqueness for $P_n(x)$ in Theorem 5.1.

13. For $n = 3$, explain why

$$L_0(x) + L_1(x) + L_2(x) + L_3(x) = 1$$

for all x.

Note: It is unnecessary to actually multiply out and combine the functions $L_i(x)$ of (5.18).

(14) Find a polynomial $P(x)$ of degree ≤ 3 for which

$$P(x_1) = y_1 \qquad P(x_2) = y_2$$
$$P'(x_1) = y_1' \qquad P'(x_2) = y_2'$$

with $x_1 \neq x_2$ and y_1, y_2, y_1', y_2' given constants. The resulting polynomial is called the "cubic Hermite interpolating polynomial."
Hint: Write

$$P(x) = y_1 L_1(x) + y_1' L_2(x) + y_2 L_3(x) + y_2' L_4(x), \quad \text{with } L_1, L_2, L_3, L_4$$

cubic polynomials satisfying appropriate properties, in analogy with (5.19).

15. As a generalized interpolation problem, find the quadratic polynomial $q(x)$ for which

$$q(0) = -1, \qquad q(1) = -1, \qquad q'(1) = 4$$

16. Find the general solution to the interpolation problem

$$q(x_0) = y_0, \qquad q(x_1) = y_1, \qquad q'(x_1) = y_1'$$

with $x_0 \neq x_1$.
Hint: Write $q(x) = y_0 M_0(x) + y_1 M_1(x) + y_1' M_2(x)$
where degree $(M_i) \leq 2$, $i = 0, 1, 2$ and each $M_i(x)$ satisfies suitable interpolating conditions at the points x_0 and x_1. For example, $M_0(x)$ should satisfy

$$M_0(x_0) = 1, \qquad M_0(x_1) = M_0'(x_1) = 0$$

5.2 DIVIDED DIFFERENCES

We introduce a discrete version of the derivative of a function $f(x)$. Let x_0, x_1, be distinct numbers, and define

$$f[x_0, x_1] = \frac{f(x_1) - f(x_0)}{x_1 - x_0} \tag{5.21}$$

This is called the first order *divided difference* of $f(x)$. If $f(x)$ is differentiable on an interval containing x_0 and x_1, then the mean-value theorem implies

$$f[x_0, x_1] = f'(c) \tag{5.22}$$

for some c between x_0 and x_1. This is one justification for thinking of $f[x_0, x_1]$ as being an analog of the derivative of $f(x)$. Also, if x_0 and x_1 are close together, then

$$f[x_0, x_1] \doteq f'\left(\frac{x_0 + x_1}{2}\right) \tag{5.23}$$

which is usually a very accurate approximation. The analysis of this approximation is taken up in problem 2.

Example

Let $f(x) = \cos(x)$, $x_0 = 0.2$, $x_1 = 0.3$. Then

$$f[x_0, x_1] = \frac{\cos(0.3) - \cos(0.2)}{0.3 - 0.2} \doteq -0.2473009 \tag{5.24}$$

With reference to (5.22), this is

$$f'(c) \doteq -\sin(0.2498936)$$

and for (5.23),

$$f'\left(\frac{x_0 + x_1}{2}\right) = -\sin(0.25) \doteq -0.2474040,$$

a very good approximation to $f[x_0, x_1]$ in (5.24).

We define higher order divided differences recursively using lower order ones. Let x_0, x_1, and x_2 be distinct real numbers, and define

$$f[x_0, x_1, x_2] = \frac{f[x_1, x_2] - f[x_0, x_1]}{x_2 - x_0} \tag{5.25}$$

This is called the second order divided difference. For x_0, x_1, x_2, and x_3 distinct, define

$$f[x_0, x_1, x_2, x_3] = \frac{f[x_1, x_2, x_3] - f[x_0, x_1, x_2]}{x_3 - x_0} \tag{5.26}$$

the third order divided difference. In general, let $x_0, x_1, \ldots, x_n$ be $n + 1$

distinct numbers, and define

$$f[x_0, \ldots, x_n] = \frac{f[x_1, \ldots, x_n] - f[x_0, \ldots, x_{n-1}]}{x_n - x_0} \qquad (5.27)$$

This is the divided difference of order n, sometimes also called the *Newton divided difference*.

The relationship of the higher order divided differences to the derivatives of corresponding order is given by the following theorem. Its proof is given in the next section.

Theorem 5.2

Let $n \geq 1$ and assume $f(x)$ is n times continuously differentiable on some interval $\alpha \leq x \leq \beta$. Let $x_0, x_1, \ldots, x_n$ be $n + 1$ distinct numbers in $[\alpha, \beta]$. Then

$$f[x_0, x_1, \ldots, x_n] = \frac{1}{n!} f^{(n)}(c) \qquad (5.28)$$

for some unknown point c lying between the minimum and maximum of the numbers $x_0, \ldots, x_n$.

Example

Let $f(x) = \cos(x)$, $x_0 = 0.2$, $x_1 = 0.3$, $x_2 = 0.4$. Then $f[x_0, x_1]$ is given in (5.24), and

$$f[x_1, x_2] = \frac{\cos(0.4) - \cos(0.3)}{0.4 - 0.3} \doteq -0.3427550$$

From (5.25),

$$f[x_0, x_1, x_2] \doteq \frac{-0.3427550 - (-0.2473009)}{0.4 - 0.2} = -0.4772705 \qquad (5.29)$$

For the case $n = 2$, (5.28) becomes

$$f[x_0, x_1, x_2] = \frac{1}{2} f''(c) \qquad (5.30)$$

for some c between the minimum and maximum of x_0, x_1, and x_2. Taking

$f''(x) = -\cos(x)$ with $x = x_1$, we have

$$\frac{1}{2} f''(0.3) = -\frac{1}{2} \cos(0.3) \doteq -0.4776682$$

which is nearly equal to the result in (5.29). Thus $c \doteq 0.3$ in (5.30).

Properties of Divided Differences

The divided differences (5.27) have a number of special properties that can simplify work with them. First, let $(i_0, i_1, \ldots, i_n)$ denote a permutation (or rearrangement) of the integers $(0, 1, \ldots, n)$. Then it can be shown that

$$f[x_{i_0}, x_{i_1}, \ldots, x_{i_n}] = f[x_0, x_1, \ldots, x_n] \tag{5.31}$$

The definition (5.27) seems to imply that the order of $x_0, x_1, \ldots, x_n$ will make a difference in the calculation of $f[x_0, \ldots, x_n]$; but (5.31) shows that this is not true. The proof of (5.31) is nontrivial, and we will consider only the cases $n = 1$ and $n = 2$. For $n = 1$,

$$f[x_1, x_0] = \frac{f(x_0) - f(x_1)}{x_0 - x_1} = \frac{f(x_1) - f(x_0)}{x_1 - x_0} = f[x_0, x_1]$$

For $n = 2$, we can expand (5.25) to obtain

$$f[x_0, x_1, x_2] = \frac{f(x_0)}{(x_0 - x_1)(x_0 - x_2)} + \frac{f(x_1)}{(x_1 - x_0)(x_1 - x_2)}$$

$$+ \frac{f(x_2)}{(x_2 - x_0)(x_2 - x_1)} \tag{5.32}$$

If we interchange values of x_0, x_1, and x_2, then the fractions on the right side will interchange their order, but the sum will remain the same. Try this with an interchange of x_0 and x_1 on the right-hand side of (5.31). We leave the derivation of (5.32) as problem 4.

A second useful property is that the definitions (5.21), (5.25) to (5.27) can be extended to the case where some or all of the node points x_i are coincident, provided that $f(x)$ is sufficiently differentiable. For example, define

$$f[x_0, x_0] = \lim_{x_1 \to x_0} f[x_0, x_1] = \lim_{x_1 \to x_0} \frac{f(x_1) - f(x_0)}{x_1 - x_0}$$

$$f[x_0, x_0] = f'(x_0)$$

For an arbitrary $n \geq 1$, let all of the nodes in (5.28) approach x_0. This leads to the definition

$$f[x_0, x_0, \ldots, x_0] = \frac{1}{n!} f^{(n)}(x_0) \tag{5.33}$$

where the left-hand side denotes an order n divided difference, all of whose nodes are x_0.

For cases where only some of the nodes are coincident, we can use (5.31), (5.33), and (5.27) to extend the definition of divided difference. For example,

$$\begin{aligned}
f[x_0, x_1, x_0] = f[x_0, x_0, x_1] &= \frac{f[x_0, x_1] - f[x_0, x_0]}{x_1 - x_0} \\
&= \frac{f[x_0, x_1] - f'(x_0)}{x_1 - x_0}
\end{aligned} \tag{5.34}$$

Further properties of divided differences are explored in the problems at the end of the section.

A Fortran Subroutine

Given a set of values $f(x_0), \ldots, f(x_n)$, we will often need to calculate the set of divided differences

$$f[x_0, x_1], f[x_0, x_1, x_2], \ldots, f[x_0, x_1, \ldots, x_n]$$

The following subroutine calculates these divided differences. In the program, note the use of arrays with zero subscripts.

```
            SUBROUTINE DIVDIF(N,X,F)
C
C           INPUT: (1) 'N' DENOTES A POSITIVE INTEGER.
C                  (2) 'X' DENOTES A VECTOR OF POINTS X(0),...,X(N).
C                  (3) 'F' DENOTES A VECTOR OF VALUES OF SOME FUNCTION
C                      EVALUATED AT THE NODES IN X.
C           OUTPUT: THE VECTOR F IS CONVERTED TO A VECTOR OF DIVIDED
C                   DIFFERENCES: F(I)=F[X(0),...,X(I)], I=0,...,N
C
            DIMENSION X(0:*),F(0:*)
C
            DO 1 I=1,N
            DO 1 J=N,I,-1
1             F(J)=(F(J)-F(J-1))/(X(J)-X(J-I))
            RETURN
            END
```

Newton's Divided Difference Interpolation Formula

From the results and examples in the first section of this chapter, it is clear that the Lagrange formula (5.17) is very inconvenient for actual calculations. Moreover, when computing with polynomials $P_n(x)$ of varying n, the calculation of $P_n(x)$ for a particular n is of little use in calculating a value with a larger n. These problems are avoided by using another formula for $P_n(x)$, one using the divided differences of the data being interpolated.

Let $P_n(x)$ denote the polynomial interpolating $f(x_i)$ at x_i, for $i = 0$, $1, \ldots, n$. Thus degree $(P_n) \leqq n$ and

$$P_n(x_i) = f(x_i), \qquad i = 0, 1, \ldots, n \tag{5.35}$$

Then

$$P_1(x) = f(x_0) + (x - x_0)f[x_0, x_1] \tag{5.36}$$

$$P_2(x) = f(x_0) + (x - x_0)f[x_0, x_1] + (x - x_0)(x - x_1)f[x_0, x_1, x_2] \tag{5.37}$$

$$\vdots$$

$$P_n(x) = f(x_0) + (x - x_0)f[x_0, x_1] + \ldots \\ + (x - x_0)(x - x_1) \ldots (x - x_{n-1})f[x_0, x_1, \ldots, x_n]. \tag{5.38}$$

This is called *Newton's divided difference formula* for the interpolating polynomial. Note that for $k \geqq 0$,

$$P_{k+1}(x) = P_k(x) + (x - x_0) \ldots (x - x_k)f[x_0, x_1, \ldots, x_k, x_{k+1}] \tag{5.39}$$

Thus we can go from degree k to degree $k + 1$ with a minimum of calculation, once the divided difference coefficients have been computed.

We will consider only the proof of (5.36) and (5.37). For the first case, consider $P_1(x_0)$ and $P_1(x_1)$. Easily, $P_1(x_0) = f(x_0)$; and

$$P_1(x_1) = f(x_0) + (x_1 - x_0)\left[\frac{f(x_1) - f(x_0)}{x_1 - x_0}\right]$$
$$= f(x_0) + [f(x_1) - f(x_0)] = f(x_1)$$

Thus degree $(P_1) \leq 1$, and it satisfies the interpolation conditions. By the uniqueness of polynomial interpolation (see Theorem 5.1), formula (5.36) is the linear interpolation polynomial to $f(x)$ at x_0, x_1. For (5.37), note that

$$P_2(x) = P_1(x) + (x - x_0)(x - x_1)f[x_0, x_1, x_2] \qquad (5.40)$$

Its degree is ≤ 2, and for x_0, x_1

$$P_2(x_i) = P_1(x_i) + 0 = f(x_i), \qquad i = 0, 1$$

Also,

$$
\begin{aligned}
P_2(x_2) &= f(x_0) + (x_2 - x_0)f[x_0, x_1] + (x_2 - x_0)(x_2 - x_1)f[x_0, x_1, x_2]\\
&= f(x_0) + (x_2 - x_0)f[x_0, x_1] + (x_2 - x_1)\{f[x_1, x_2] - f[x_0, x_1]\}\\
&= f(x_0) + (x_1 - x_0)f[x_0, x_1] + (x_2 - x_1)f[x_1, x_2]\\
&= f(x_0) + [f(x_1) - f(x_0)] + [f(x_2) - f(x_1)] = f(x_2)
\end{aligned}
$$

By the uniqueness of polynomial interpolation, this is the quadratic interpolating polynomial to $f(x)$ at x_0, x_1, x_2.

The general proof of (5.38) is more complicated than we wish to consider here. See Atkinson (1978, Chapter 3) for a detailed derivation of (5.38), along with further properties of divided differences.

Example

Let $f(x) = \cos(x)$. Table 5.2 contains a set of nodes x_i, the function values $f(x_i)$, and the divided differences

$$D_i = f[x_0, \ldots, x_i], \qquad i \geq 0$$

Table 5.3 contains the values of $P_n(x)$ for $x = 0.1, 0.3, 0.5$ for various

Table 5.2 Values and Divided Differences of $\cos(x)$

i	x_i	$\cos(x_i)$	D_i
0	0.0	1.000000	$0.1000000D+01$
1	0.2	0.980067	$-0.9966711D-01$
2	0.4	0.921061	$-0.4884020D+00$
3	0.6	0.825336	$0.4900763D-01$
4	0.8	0.696707	$0.3812246D-01$
5	1.0	0.540302	$-0.3962047D-02$
6	1.2	0.362358	$-0.1134890D-02$

values of n. The true values $f(x)$ are given in the last row of the table. The calculations were done using (5.38).

The above example used evenly spaced points x_i. But, in general, the interpolation node points need not be evenly spaced in order to use the divided difference interpolation formula (5.38); nor need they be arranged in any particular order. In the next chapter, we will use (5.38) with node points that are not evenly spaced.

To evaluate (5.38) in an efficient manner, we can use a variation on the nested multiplication algorithm of section 1.3 in Chapter 1. First, write (5.38) as

$$P_n(x) = D_0 + (x - x_0)D_1 + (x - x_0)(x - x_1)D_2$$
$$+ \ldots + (x - x_0) \ldots (x - x_{n-1})D_n \quad (5.41)$$

with

$$D_0 = f(x_0), \qquad D_i = f[x_0, \ldots, x_i] \qquad \text{for } i \geq 1$$

Table 5.3 Interpolation to $\cos(x)$

	$P_n(0.1)$	$P_n(0.3)$	$P_n(0.5)$
1	0.9900333	0.9700999	0.9501664
2	0.9949173	0.9554478	0.8769061
3	0.9950643	0.9553008	0.8776413
4	0.9950071	0.9553351	0.8775841
5	0.9950030	0.9553369	0.8775823
6	0.9950041	0.9553365	0.8775825
True	0.9950042	0.9553365	0.8775826

Next, rewrite (5.41) in the nested form

$$P_n(x) = D_0 + (x - x_0)[D_1 + (x - x_1)[D_2 + \ldots$$
$$+ (x - x_{n-2})[D_{n-1} + (x - x_{n-1})D_n] \ldots] \quad (5.42)$$

For example,

$$P_3(x) = D_0 + (x - x_0)[D_1 + (x - x_1)[D_2 + (x - x_2)D_3]]$$

Using (5.42), we need only n multiplications to evaluate $P_n(x)$.

To aid in constructing examples such as in Tables 5.2 and 5.3, we include the following interpolation program. It can be used to examine the error in polynomial interpolation, with varying x and n. Also, the program uses the Newton divided difference form of the interpolating polynomial, with the nested form given above. The program also uses subroutine DIVDIF from earlier in this section.

```
C       TITLE: NEWTON DIVIDED DIFFERENCE INTERPOLATING POLYNOMIAL
C
C       THIS ROUTINE WILL DO INTERPOLATION TO FCN(X), DEFINED BELOW.
C       THE PROGRAM INPUTS A DEGREE N AND AN INITIAL POINT X(0).
C       THEN THE DIVIDED DIFFERENCES
C           F[X(0),X(1),...,X(J)] ,   1 .LE. J .LE. N
C       ARE CONSTRUCTED FOR USE IN EVALUATING THE INTERPOLATING
C       POLYNOMIAL.  THE INTERPOLATION NODES X(J) ARE EVENLY SPACED.
C
C       FOLLOWING THIS, AN X AT WHICH THE INTERPOLATION IS TO BE
C       DONE IS REQUESTED FROM THE USER.  FOR THIS X, THE
C       INTERPOLATING POLYNOMIALS OF DEGREE .LE. N ARE
C       CONSTRUCTED, ALONG WITH THE ERROR IN THE INTERPOLATION.
C       THE INTERPOLATION IS DONE WITH THE NEWTON DIVIDED
C       DIFFERENCE FORM OF THE INTERPOLATING POLYNOMIAL.
C
        DIMENSION X(0:20),F(0:20),DF(0:20)
        DATA H/.2/
        FCN(X)=COS(X)
C
C       INPUT PROBLEM PARAMETERS.
        PRINT *, ' DEGREE, X(0)=?'
        READ *. N,X(0)
C       SET UP NODE AND FUNCTION VALUES.
        DO 10 I=0,N
          X(I)=X(0)+I*H
          F(I)=FCN(X(I))
10        DF(I)=F(I)
C
C       CALCULATE DIVIDED DIFFERENCES AND PRINT THEM.
        CALL DIVDIF(N,X,DF)
```

```
          PRINT 1000
1000      FORMAT('   I',4X,'X(I)',6X,'F(I)',8X,'DF(I)',/)
          DO 20 I=0,N
            PRINT 1001, I,X(I),F(I),DF(I)
20          CONTINUE
1001      FORMAT(I3,4X,F3.1,4X,F8.6,4X,E14.7)
C
C         BEGIN INTERPOLATION LOOP.
30        PRINT *, ' WHAT IS INTERPOLATION POINT X?'
          PRINT *, ' TO STOP, LET X=1000.'
          READ *, Z
          IF(Z .EQ. 1000.0) STOP
          PRINT 1002, Z
1002      FORMAT(///,' X=',F9.7,/)
C         CONSTRUCT INTERPOLATION POLYNOMIAL AT Z, FOR
C         ALL DEGREES I .LE. N.
          DO 50 I=1,N
            POLY=DF(I)
            DO 40 J=I-1,0,-1
40            POLY=(Z-X(J))*POLY+DF(J)
            ERROR=FCN(Z)-POLY
            PRINT 1003, I,POLY,ERROR
50          CONTINUE
1003      FORMAT(' DEGREE=',I2,4X,'INTERPOLANT=',F10.7,4X,
     *           'ERROR=',E10.3)
          GO TO 30
          END
```

PROBLEMS

1. Let $x_0 = 0.85, x_1 = 0.87, x_2 = 0.89$. Using Table 5.1, calculate $f[x_0, x_1]$, $f[x_1, x_2]$, and $f[x_0, x_1, x_2]$ for $f(x) = e^x$. Check the accuracy of the approximation (5.23).

2. Let $z = (x_0 + x_1)/2$, $h = (x_1 - x_0)/2$. Then to analyze the approximation (5.23), consider the error

$$E = f[x_0, x_1] - f'\left(\frac{x_0 + x_1}{2}\right) = \frac{f(z + h) - f(z - h)}{2h} - f'(z)$$

Expand $f(z + h)$ and $f(z - h)$ about z using Taylor's theorem from section 1.2, and include terms of degree ≤ 3 in the variable h. Use these expansions to show that

$$E \doteq \frac{h^2}{6} f'''(z)$$

for smaller values of h. Thus (5.23) is a good approximation when $x_1 - x_0$ is relatively small.

3. (a)   Let $f(x)$ be a polynomial of degree m. For $x \neq x_0$, define

$$g_1(x) = f[x_0, x] = \frac{f(x) - f(x_0)}{x - x_0}$$

Show $g(x)$ is a polynomial of degree $m - 1$. This is another justification for regarding the divided difference as an analog of the derivative.

Hint: The numerator $f(x) - f(x_0)$ is a polynomial of degree m with a zero at $x = x_0$. Use the fundamental theorem of algebra, noting that x_0 is a root of $f(x) - f(x_0)$.

(b)   For x_0, x_1, x distinct and for $f(x)$ a polynomial of degree m, define

$$g_2(x) = f[x_0, x_1, x]$$

Show $g_2(x)$ is a polynomial of degree $m - 2$.

4. Verify formula (5.32), showing the symmetry of $f[x_0, x_1, x_2]$ under permutations of the nodes.

5. Using (5.33) and the ideas used in obtaining (5.34), obtain a formula for calculating

(a) $f[x_0, x_0, x_1, x_1]$ (b) $f[x_0, x_0, x_0, x_1]$

6. Given the data below, find $f[x_0, x_1]$ and $f[x_0, x_1, x_2]$. Then calculate $P_1(0.15)$ and $P_2(0.15)$, the linear and quadratic interpolates at $x = 0.15$.

n	x_n	$f(x_n)$
0	.1	.2
1	.2	.24
2	.3	.30

7. Using subroutine DIVDIF, calculate the divided differences $D_0 = f(x_0)$, $D_1 = f[x_0, x_1]$, ..., $D_5 = f[x_0, x_1, x_2, x_3, x_4, x_5]$, for $f(x) = e^x$. Use $x_0 = 0$, $x_1 = 0.2$, $x_2 = 0.4$, ..., $x_5 = 1.0$.

8. Using the results of problem 7, calculate $P_j(x)$ for $x = 0.1, 0.3, 0.5$ and $j = 1, \ldots, 5$. Compare these results to the true values of e^x.

9. Produce a program to check the accuracy of higher order interpolation using Newton's formula (5.38). For some $f(x)$ and some node points $x_0, x_1, \ldots, x_6$, use DIVDIF to produce the divided differences

$$D_i = f[x_0, \ldots, x_i], \qquad i = 0, 1, \ldots, 6$$

Then evaluate $P_6(x)$ for a variety of values of x, and compare them to the true values of $f(x)$. Check the program by reproducing some of the results in Tables 5.2 and 5.3 for $f(x) = \cos(x)$. Then repeat the process with a nonpolynomial function $f(x)$ of your choice.

10. (a) In the linear formula (5.36), let $x_1 - x_0 = h$ and $\mu = (x - x_0)/(x_1 - x_0) = (x - x_0)/h$. Show (5.36) reduces to the earlier formula (5.6) of the first section.

 (b) In the quadratic formula (5.37), let $x_1 - x_0 = x_2 - x_1 = h$ and $\mu = (x - x_0)/(x_1 - x_0)$. Show (5.37) reduces to the earlier formula (5.14).

11. Let $f(x) = x^n$ for some integer $n \geq 0$. Let $x_0, x_1, \ldots, x_m$ be $m + 1$ distinct numbers. What is $f[x_0, x_1, \ldots, x_m]$ for $m = n$? For $m > n$?

5.3 ERROR IN POLYNOMIAL INTERPOLATION

For a given function $f(x)$ defined on an interval $[a, b]$, let $P_n(x)$ denote the polynomial of degree $\leq n$ interpolating $f(x)$ at $n + 1$ points $x_0, x_1, \ldots, x_n$ in $[a, b]$,

$$P_n(x) = \sum_{j=0}^{n} f(x_j)L_j(x) \tag{5.43}$$

In this section, we consider carefully the error in polynomial interpolation, giving more precise information on its behavior as x and n vary. We begin with a formula for the error.

Theorem 5.3

Let $n \geq 0$, let $f(x)$ have $n + 1$ continuous derivatives on $[a, b]$ and let $x_0, x_1, \ldots, x_n$ be distinct node points in $[a, b]$. Then

$$f(x) - P_n(x) = \frac{(x - x_0)(x - x_1) \ldots (x - x_n)}{(n + 1)!} f^{(n+1)}(c_x) \tag{5.44}$$

for $a \leq x \leq b$, where c_x is an unknown point between the minimum and maximum of $x_0, x_1, \ldots, x_n$, and x.

We omit the proof since it does not contain any useful additional information regarding interpolation.

Example

Take $f(x) = e^x$ on $[0, 1]$, and consider the error in linear interpolation to $f(x)$ at two points x_0 and x_1. Then from (5.44),

$$e^x - P_1(x) = \frac{(x - x_0)(x - x_1)}{2} e^{c_x} \qquad (5.45)$$

for some c_x between the minimum and maximum of x_0, x_1, and x. To apply this to table interpolation, assume $x_0 < x < x_1$. Then we note the interpolation error is negative, and we write

$$e^x - P_1(x) = -\frac{(x_1 - x)(x - x_0)}{2} e^{c_x}$$

This shows the error is like a quadratic polynomial with roots at x_0 and x_1. Since $x_0 \leqq c_x \leqq x_1$, we have the upper and lower bounds

$$\frac{(x_1 - x)(x - x_0)}{2} e^{x_0} \leqq |e^x - P_1(x)| \leqq \frac{(x_1 - x)(x - x_0)}{2} e^{x_1}$$

To obtain a bound independent of x, use

$$\max_{x_0 \leqq x \leqq x_1} \frac{(x_1 - x)(x - x_0)}{2} = \frac{h^2}{8}, \qquad h = x_1 - x_0 \qquad (5.46)$$

This follows easily by noting that $(x_1 - x)(x - x_0)$ is a quadratic with roots at x_0 and x_1 and, thus, its maximum value occurs midway between the roots. Substituting $x = (x_0 + x_1)/2$ yields the value $h^2/8$.

Noting that $e^{x_1} \leqq e$ on $[0, 1]$, we have the bound

$$|e^x - P_1(x)| \leqq \frac{h^2 e}{8}, \qquad 0 \leqslant x_0 \leqslant x \leqslant x_1 \leqslant 1 \quad (5.47)$$

independent of x, x_0, and x_1. Recall the example (5.3) to (5.5) of section 5.1. With $x = 0.826$ and $h = 0.01$, we have

$$|e^x - P_1(x)| \leqq \frac{(0.01)^2(2.72)}{8} = 0.0000340 \qquad (5.48)$$

The actual error was -0.0000276, which satisfies this bound.

Example

Again let $f(x) = e^x$ on $[0, 1]$, but consider the error in quadratic interpolation. Then

$$e^x - P_2(x) = \frac{(x - x_0)(x - x_1)(x - x_2)}{6} e^{c_x} \qquad (5.49)$$

for some c_x between the minimum and maximum of x_0, x_1, x_2, and x. For table interpolation, let $h = x_1 - x_0 = x_2 - x_1$, and assume that $x_0 < x < x_2$. Assuming all nodes are in $[0, 1]$, as before, we have

$$|e^x - P_2(x)| \leq \left| \frac{(x - x_0)(x - x_1)(x - x_2)}{6} \right| e^1 \qquad (5.50)$$

To proceed further, we use

$$\max_{x_0 \leq x \leq x_2} \left| \frac{(x - x_0)(x - x_1)(x - x_2)}{6} \right| = \frac{h^3}{9\sqrt{3}} \qquad (5.51)$$

This result is obtained using elementary calculus. To simplify the calculations, we consider the special case

$$w_2(x) = \frac{(x + h)x(x - h)}{6} = \frac{x^3 - xh^2}{6} \qquad (5.52)$$

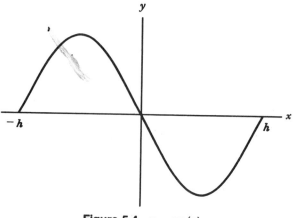

Figure 5.4 $y = w_2(x)$.

Its graph is shown in Figure 5.4. The cubic polynomial $w_2(x)$ is a translation along the x-axis of the polynomial in (5.51), and it has the same shape and size. With (5.52), we look for the points x at which

$$0 = w_2'(x) = \frac{3x^2 - h^2}{6}$$

Then $x = \pm h/\sqrt{3}$, and

$$\left| w_2\left(\pm \frac{h}{\sqrt{3}} \right) \right| = \frac{h^3}{9\sqrt{3}}$$

proving (5.51). Using this in (5.50), we get

$$|e^x - P_2(x)| \leq \frac{h^3 e}{9\sqrt{3}} \doteq 0.174 h^3 \tag{5.53}$$

For $h = 0.01$, $0 \leq x \leq 1$,

$$|e^x - P_2(x)| \leq 1.74 \times 10^{-7} \tag{5.54}$$

Compare this with the earlier result (5.48) for linear interpolation.

Example

Most students first encounter linear interpolation in connection with the use of tables of logarithms. With the use of calculators, the use of logarithms for arithmetic calculations is no longer needed. But it is still an interesting example to consider the linear interpolation error involved in using logarithm tables. We take $f(x) = \log_{10} x$ and $1 \leq x \leq 10$. To write out the error from (5.44), we need

$$f''(x) = \frac{-\log_{10} e}{x^2}$$

Take $x_0 < x < x_1$. Then

$$\log_{10} x - P_1(x) = \frac{(x - x_0)(x_1 - x)}{-2} \cdot \frac{\log_{10} e}{c^2}$$

with $x_0 \leq c \leq x_1$. Thus

$$\frac{(x - x_0)(x_1 - x)}{2} \cdot \frac{\log_{10} e}{x_1^2} \leq \log_{10} x - P_1(x)$$

$$\leq \frac{(x - x_0)(x_1 - x)}{2} \cdot \frac{\log_{10} e}{x_0^2} \qquad (5.55)$$

The error looks like a quadratic polynomial when x_0 and x_1 are close together. We also see that if the spacing $x_1 - x_0 = h$ remains constant over the interval, then the linear interpolation error decreases as we move to the right in [1, 10], that is as x_0, x_1 increase. Near $x = 10$, the error is about 100 times smaller than it is near $x = 1$.

To find a uniform error bound for the entire interval, note that $x_0 \geq 1$ and then use (5.46) to obtain

$$0 \leq \log_{10}(x) - P_1(x) \leq \frac{h^2 \log_{10} e}{8} \doteq 0.0542h^2$$

Recalling the tables used in high school algebra tests, these tables generally had $h = 0.01$. Thus

$$0 \leq \log_{10}(x) - P_1(x) \leq 5.42 \times 10^{-6} \qquad (5.56)$$

Since these tables gave only four digits of accuracy for $\log(x)$, the use of linear interpolation will give a correct answer to essentially four decimal places.

Another Error Formula

Construct the interpolation formula $P_{n+1}(x)$ that interpolates to $f(x)$ at $x_0, x_1, \ldots, x_n, x_{n+1}$. Using the Newton divided difference form gives us

$$P_{n+1}(x) = P_n(x)$$
$$+ (x - x_0)(x - x_1) \ldots (x - x_n)f[x_0, x_1, \ldots, x_n, x_{n+1}] \quad (5.57)$$

with $P_n(x)$ interpolating $f(x)$ at $x_0, x_1, \ldots, x_n$. By the interpolation property, letting $x = x_{n+1}$, we have

$$f(x_{n+1}) = P_n(x_{n+1}) + (x_{n+1} - x_0) \ldots (x_{n+1} - x_n)f[x_0, x_1, \ldots, x_n, x_{n+1}]$$

Regard x_{n+1} as a variable point, distinct from $x_0, \ldots, x_n$, and rename it t. Moving the P_n term to the left side of the equation, we have

$$f(t) - P_n(t) = (t - x_0)(t - x_1) \ldots (t - x_n) f[x_0, x_1, \ldots, x_n, t] \quad (5.58)$$

This is the interpolation error in $P_n(t)$. From what was said in the preceding section, the divided difference can also be extended to t a node point; and then the right side of (5.58) will be zero when $t = x_0, x_1, \ldots, x_n$.

Comparing (5.58) with (5.44), with $x = t$, we have two formulas for the interpolation error:

$$\frac{\psi_n(t)}{(n + 1)!} f^{(n+1)}(c_t) = \psi_n(t) f[x_0, x_1, \ldots, x_n, t]$$

with

$$\psi_n(t) = (t - x_0)(t - x_1) \ldots (t - x_n)$$

Cancelling $\psi_n(t)$ from both sides, we have

$$\frac{1}{(n + 1)!} f^{(n+1)}(c_t) = f[x_0, x_1, \ldots, x_n, t] \quad (5.59)$$

with c_t between the minimum and maximum of $x_0, x_1, \ldots, x_n$, and t. Let $m = n + 1$ and $t = x_m$. Then (5.59) becomes

$$\frac{f^{(m)}(c)}{m!} = f[x_0, x_1, \ldots, x_{m-1}, x_m]$$

where

$$\text{Minimum } \{x_0, \ldots, x_m\} \leq c \leq \text{Maximum } \{x_0, \ldots, x_m\}$$

This proves Theorem 5.2 of section 5.2.

Usually we use formula (5.44) to bound the interpolation error, but occasionally (5.58) is preferable. The derivation of error formulas for numerical integration methods often uses (5.58).

Behavior of the Error

When we consider the error formula (5.44), the polynomial

$$\psi_n(x) = (x - x_0) \ldots (x - x_n) \quad (5.60)$$

is the most important quantity in determining the behavior of the error. We will examine its behavior for $x_0 \leq x \leq x_n$ when the node points $x_0, \ldots, x_n$ are evenly spaced.

For larger values of n, say $n \geq 5$, the values of $\psi_n(x)$ change greatly through the interval $x_0 \leq x \leq x_n$. The values in $[x_0, x_1]$ and $[x_{n-1}, x_n]$ become much larger than the values in the middle of $[x_0, x_n]$. This can be proven theoretically, but we only suggest the result by looking at the graph of $\psi_n(x)$ when $n = 6$. This is given in Figure 5.5; note the relatively larger values in $[x_0, x_1]$ and $[x_5, x_6]$ as compared to the values in $[x_2, x_4]$. As n increases, this disparity also increases.

When considering $\psi_n(x)$ as a part of the error formula (5.44) for $f(x) - P_n(x)$, these remarks imply that the interpolation error at x is likely to be smaller when it is near the middle of the node points. In practical interpolation problems, high degree polynomial interpolation with evenly spaced nodes is seldom used because of these difficulties. However, we will see in Chapter 6 that high degree polynomial inter-polation with another set of node points can be very useful in obtaining polynomial approximations to functions.

Example

Let $f(x) = \cos(x)$, $h = 0.2$, $n = 8$, and interpolate at $x = 0.9$.

Case (i) $x_0 = 0.8$, $x_8 = 2.4$. Thus $x = 0.9$ is in the first subinterval $[x_0, x_1]$. By direct calculation of $P_8(0.9)$,

$$\cos(0.9) - P_8(0.9) = -5.51 \times 10^{-9}$$

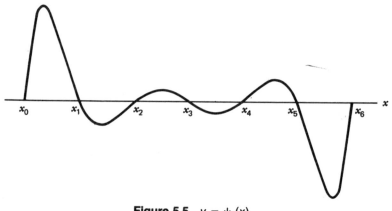

Figure 5.5 $y = \psi_6(x)$.

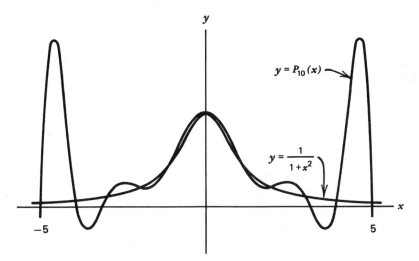

Figure 5.6 Interpolation to $1/(1 + x^2)$.

Case (ii) $x_0 = 0.2, x_8 = 1.8$. Thus $x = 0.9$ is in the subinterval $[x_3, x_4]$, where x_4 is the midpoint for the interpolation interval $[x_0, x_8]$. By direct calculation,

$$\cos(0.9) - P_8(0.9) = 2.26 \times 10^{-10}$$

a factor of 24 smaller than in case (i).

Example

Let $f(x) = 1/(1 + x^2)$ for $-5 \leq x \leq 5$. Let $n > 0$ be an even integer, and define $h = 10/n$,

$$x_j = -5 + jh, \qquad j = 0, 1, 2, \ldots, n$$

Then it can be shown that the interpolating polynomials $P_n(x)$, $n \geq 2$, do not converge to $f(x)$ as $h \to 0$, for many points x in the interval $[-5, 5]$. To illustrate the poor approximation of $P_n(x)$ to $f(x)$ in this case, we give the graphs of $f(x)$ and $P_{10}(x)$ in Figure 5.6. The divergence of $P_n(x)$ from $f(x)$, for $x \geq 4$, becomes worse as n increases.

PROBLEMS

1. Consider interpolating sin (x) from a table of values for $0 \leq x \leq 1.58$, with the x entries given in steps of $h = 0.01$. (a) Bound

the error of linear interpolation in this table. (b) Bound the error of quadratic interpolation.

2. Repeat problem 1 with $f(x) = \tan^{-1}(x)$ on $[0, .8]$

3. Repeat problem 1 with

$$f(x) = \int_0^x e^{-t^2}\, dt$$

for $0 \leq x \leq 1$.

4. Consider a table of values of $f(x) = \log(x)$, $1 \leq x \leq 7$, with a spacing of h for the values of x. Bound the error of linear interpolation in this table. How small should h be chosen if this interpolation error is to be less than 5×10^{-6}?

5. Suppose a table of values of $f(x) = \sin(x)$, $0 \leq x \leq 1.6$, is to be constructed, with the values of $\sin(x)$ given for $x = 0, h, 2h, 3h, \dots$. If quadratic interpolation is used in this table, how small should h be chosen in order to have the interpolation error be less than 10^{-6}?

6. Bound the error of cubic interpolation to $f(x) = e^x$ on $[0, 1]$ with evenly spaced node points.
   Hint: Replace the bounding of

$$\psi_3(x) = (x - x_0)(x - x_1)(x - x_2)(x - x_3), \qquad x_0 \leq x \leq x_3$$

with the bounding of its translate

$$\psi_3(x) = \left(x + \frac{3}{2}h\right)\left(x + \frac{1}{2}h\right)\left(x - \frac{1}{2}h\right)\left(x - \frac{3}{2}h\right)$$

on the interval $-\frac{3}{2}h \leq x \leq \frac{3}{2}h$. This generalizes the method used in obtaining (5.51).

7. To visualize the change in the values of

$$\psi_n(x) = (x - x_0)(x - x_1) \dots (x - x_n)$$

as n increases and as x varies over $[x_0, x_n]$, graph the special case

$$\psi_n(x) = x(x - 1) \dots (x - n)$$

for $0 \leq x \leq n$. Do this for $n = 3, 4$, and 8.

8. Analyze the error in quadratic interpolation to $f(x) = \log_{10} x$ on $1 \leq x \leq 10$. Compare it with the results for linear interpolation given in this section.

9. As noted following (5.55) the linear interpolation error for $f(x) = \log_{10} x$ varies greatly over the interval $1 \leq x \leq 10$. Consider

how to break [1, 10] into subintervals [1, a], [a, b], [b, 10], each having a different stepsize h for the table entries. The stepsize should be chosen to ensure the maximum error is about the same on each subinterval, say about 5×10^{-6} as in (5.56). The stepsize h should also be chosen as a simple number, for example, $h = 0.2$ rather than $h = 0.2152$.

10. For an interval [a, b], define $h = (b - a)/n$ for an integer $n > 0$. Define evenly spaced node points by

$$x_j = a + jh, \qquad j = 0, 1, \ldots, n$$

Thus $x_0 = a, x_1 = a + h, \ldots, x_n = a + nh = b$. Consider the polynomial

$$\psi_n(x) = (x - x_0)(x - x_1) \ldots (x - x_n)$$

and show

$$|\psi_n(x)| \leq n! h^{n+1}, \qquad a \leq x \leq b$$

Hint: Consider first a lower order case such as $n = 2$ or 3. Also, consider separately the cases $x_0 < x < x_1$, $x_1 < x < x_2$, $\ldots$, $x_{n-1} < x < x_n$. With each case, bound the various factors $x - x_j$ by multiples of h.

11. Let $P_n(x)$ be the degree n polynomial interpolating to $f(x) = e^x$ on $[a, b] = [0, 1]$. Using the result of problem 10, show that

$$\max_{0 \leq x \leq 1} |e^x - P_n(x)| \to 0$$

as $n \to \infty$.

5.4 INTERPOLATION USING SPLINE FUNCTIONS

To motivate the definition and use of spline functions, we begin with the problem of interpolating the data shown in Table 5.4. The simplest method of interpolation is to connect the node points by straight line segments; the resulting graph is shown in Figure 5.7. This is called *piecewise linear interpolation*, and the associated interpolating function is

Table 5.4 Interpolation Data Points

x	0	1	2	2.5	3	3.5	4
y	2.5	0.5	0.5	1.5	1.5	1.125	0

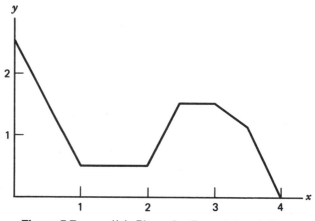

Figure 5.7 $y = l(x)$: Piecewise linear interpolation.

denoted by $l(x)$. It agrees with the data, but it has the disadvantage of not having a smooth graph. Most data will represent a smooth curved graph, one without the corners of $y = l(x)$.

The next choice of interpolation is to use polynomial interpolation. There are seven data points and, thus, we consider the degree six interpolating polynomial $P_6(x)$. Its graph is shown in Figure 5.8 (note the change in vertical scale), and it differs radically from that of $l(x)$. Although it is a smooth graph, it oscillates far too much between the interpolation node points. These comments presuppose that we wish a smooth curve that does not depart too radically from the graph of $y = l(x)$.

A third choice is to connect the data points of Table 5.4 by using a

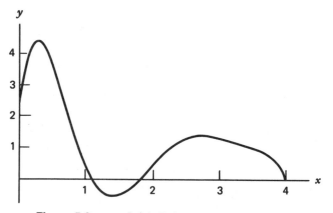

Figure 5.8 $y = P_6(x)$: Polynomial interpolation.

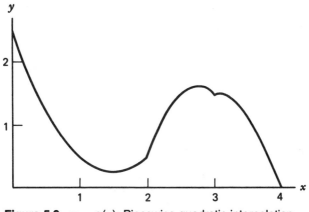

Figure 5.9 $y = q(x)$: Piecewise quadratic interpolation.

succession of quadratic interpolating polynomials. We denote this function on $0 \leq x \leq 4$ by $q(x)$, and we give its graph in Figure 5.9. On each of the subintervals $[0, 2], [2, 3]$, and $[3, 4]$, $q(x)$ is the quadratic polynomial interpolating the data on that subinterval. The graph of $q(x)$ looks quite a bit better than that of either $l(x)$ or $P_6(x)$. Nonetheless, there is still a problem at the points $x = 2$ and $x = 3$ where the graph has "corners"; the derivative $q'(x)$ is discontinuous at such points. We wish to find an interpolating function that is smooth and does not oscillate too much between the node points. For many applications, however, this interpolating function may be completely adequate as, for example, with the numerical integration of the data.

Spline Interpolation

To pose the problem more generally, suppose n data points (x_i, y_i) are given. For simplicity, assume

$$x_1 < x_2 < \ldots < x_n \qquad (5.61)$$

and let $a = x_1$, $b = x_n$. We seek a function $s(x)$ defined on $[a, b]$ that interpolates the data:

$$s(x_i) = y_i, \qquad i = 1, \ldots, n \qquad (5.62)$$

For smoothness of $s(x)$, we require that $s'(x)$ and $s''(x)$ be continuous. In addition, we would like to have the curve follow the general shape given by the piecewise linear function connecting the data points (x_i, y_i), as illustrated in Figure 5.7. The standard way in which this has been done has been to ask that the derivative $s'(x)$ not change too rapidly between node points. This has been carried out by requiring the second derivative $s''(x)$ to be as small as possible and, more precisely, by requiring that

$$\int_a^b [s''(x)]^2 \, dx \tag{5.63}$$

be made as small as possible. This may not be a perfect mathematical realization of the idea of a smooth shape-preserving interpolation function for the data $\{(x_i, y_i)\}$, but it usually gives a very good interpolating function.

There is a unique solution $s(x)$ to this problem, and it satisfies the following:

S1. $s(x)$ is a cubic polynomial on each subinterval $[x_{j-1}, x_j]$, for $j = 2,$ $3, \ldots, n$.

S2. $s''(x_1) = s''(x_n) = 0$.

The function $s(x)$ is called the *natural cubic spline function* that interpolates to the data $\{(x_i, y_i)\}$. We will give a method for constructing $s(x)$, and then apply it to the data in Table 5.4. Then, we will return to a more general discussion of cubic spline functions and interpolation with them.

Construction of s(x)

Introduce the variables $M_1, \ldots, M_n$ with

$$M_i \equiv s''(x_i), \qquad i = 1, 2, \ldots, n \tag{5.64}$$

We will express $s(x)$ in terms of the (unknown) values M_i; then we will produce a system of linear equations from which the values M_i can be calculated.

Since $s(x)$ is cubic on each interval $[x_{j-1}, x_j]$, the function $s''(x)$ is linear on the interval. A linear function is determined by two points, and we use

$$s''(x_{j-1}) = M_{j-1}, \qquad s''(x_j) = M_j \tag{5.65}$$

Then

$$s''(x) = \frac{(x_j - x)M_{j-1} + (x - x_{j-1})M_j}{x_j - x_{j-1}}, \qquad x_{j-1} \leqq x \leqq x_j \quad (5.66)$$

We will now form the second antiderivative of $s''(x)$ on $[x_{j-1}, x_j]$ and apply the interpolating conditions

$$s(x_{j-1}) = y_{j-1}, \qquad s(x_j) = y_j \qquad (5.67)$$

After quite a bit of manipulation, this results in the cubic polynomial

$$s(x) = \frac{(x_j - x)^3 M_{j-1} + (x - x_{j-1})^3 M_j}{6(x_j - x_{j-1})} + \frac{(x_j - x)y_{j-1} + (x - x_{j-1})y_j}{x_j - x_{j-1}}$$
$$- \frac{1}{6}(x_j - x_{j-1})[(x_j - x)M_{j-1} + (x - x_{j-1})M_j], \quad (5.68)$$

for $x_{j-1} \leqq x \leqq x_j$. It can be checked directly that the second derivative of this formula yields (5.66); by direct substitution, (5.68) satisfies the interpolating conditions (5.67).

Formula (5.68) applies to each of the intervals $[x_1, x_2], \ldots, [x_{n-1}, x_n]$. The formulas for adjacent intervals $[x_{j-1}, x_j]$ and $[x_j, x_{j+1}]$ will agree at their common point $x = x_j$ because of the interpolating condition $s(x_j) = y_j$, which is common to the definitions. This implies that $s(x)$ will be continuous over the entire interval $[a, b]$. Similarly, formula (5.66) for $s''(x)$ implies that it is continuous on $[a, b]$.

To ensure the continuity of $s'(x)$ over $[a, b]$, the formulas for $s'(x)$ on $[x_{j-1}, x_j]$ and $[x_j, x_{j+1}]$ are required to give the same value at their common point $x = x_j$, for $j = 2, 3, \ldots, n - 2$. After a great deal of simplification, this leads to the system of linear equations

$$\frac{x_j - x_{j-1}}{6} M_{j-1} + \frac{x_{j+1} - x_{j-1}}{3} M_j + \frac{x_{j+1} - x_j}{6} M_{j+1}$$

$$= \frac{y_{j+1} - y_j}{x_{j+1} - x_j} - \frac{y_j - y_{j-1}}{x_j - x_{j-1}}, \qquad j = 2, 3, \ldots, n - 1 \quad (5.69)$$

These $n - 2$ equations together with the earlier assumption (S2),

$$M_1 = M_n = 0 \qquad (5.70)$$

leads to the values $M_1, \ldots, M_n$ and thence to the interpolating function

$s(x)$. The linear system (5.69) is called a tridiagonal system, and there are special methods for its solution. One of these is given in Chapter 8, section 8.4; and most computer centers will have special routines for tridiagonal systems of equations.

Example

Calculate the natural cubic spline interpolating the data

$$\left\{ (1,1), \left(2, \frac{1}{2}\right), \left(3, \frac{1}{3}\right), \left(4, \frac{1}{4}\right) \right\} \tag{5.71}$$

The number of points is $n = 4$; and all $x_j - x_{j-1} = 1$. The system (5.69) becomes

$$\frac{1}{6} M_1 + \frac{2}{3} M_2 + \frac{1}{6} M_3 = \frac{1}{3}$$

$$\frac{1}{6} M_2 + \frac{2}{3} M_3 + \frac{1}{6} M_4 = \frac{1}{12} \tag{5.72}$$

Together with (5.70), this yields

$$M_2 = \frac{1}{2}, \qquad M_3 = 0$$

Substituting into (5.68), we obtain

$$s(x) = \begin{cases} \dfrac{1}{12} x^3 - \dfrac{1}{4} x^2 - \dfrac{1}{3} x + \dfrac{3}{2}, & 1 \leq x \leq 2 \\[2mm] -\dfrac{1}{12} x^3 + \dfrac{3}{4} x^2 - \dfrac{7}{3} x + \dfrac{17}{6}, & 2 \leq x \leq 3 \\[2mm] -\dfrac{1}{12} x + \dfrac{7}{12}, & 3 \leq x \leq 4 \end{cases} \tag{5.73}$$

It is left as an exercise to compute $s'(x)$ and $s''(x)$ and to check that they are continuous.

Example

Calculate the natural cubic spline interpolating the data in Table 5.4. Since there are $n = 7$ points, the system (5.69) will contain five equations.

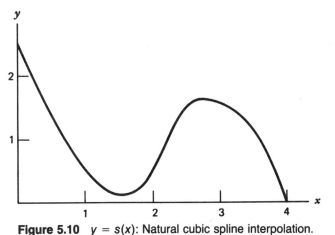

Figure 5.10 $y = s(x)$: Natural cubic spline interpolation.

The graph of the resulting function $s(x)$ is given in Figure 5.10, and it generally is quite similar to that of $q(x)$ in Figure 5.9. But with $s(x)$, the graph no longer contains the corners or discontinuous changes in slope that are present in the graph of $y = q(x)$ at $x = 2$ and 3. To the eye, the graph of $s(x)$ would seem a better interpolating function than $q(x)$ for the data of Table 5.4.

Other Interpolating Spline Functions

Up until this point, we have not considered the accuracy of the interpolating spline $s(x)$. This is satisfactory for cases where only the data points are known, and we only want a smooth curve that looks correct to the eye. But in many cases, we want the spline to interpolate a known function, and then we are also interested in accuracy.

Let $f(x)$ be given on $[a, b]$. We will consider the case where the interpolation of $f(x)$ is performed at evenly spaced values of x. Let $n > 1$,

$$h = \frac{b - a}{n - 1}, \quad x_j = a + (j - 1)h, \quad j = 1, 2, \ldots, n \quad (5.74)$$

Let $s_n(x)$ be the natural cubic spline interpolating $f(x)$ at $x_1, \ldots, x_n$. Then it can be shown that

$$\max_{a \le x \le b} |f(x) - s_n(x)| \le ch^2 \quad (5.75)$$

where c depends on $f''(x)$ and $f^{(4)}(x)$. The primary reason that the approximation $s_n(x)$ does not converge more rapidly (i.e., have a higher power of h) is that $f''(x)$ is generally nonzero at $x = a$ and b, whereas $s_n''(a) = s_n''(b) = 0$ by definition. For functions $f(x)$ with $f''(a) = f''(b) = 0$, the right-hand side of (5.75) can be replaced by ch^4.

To improve on $s_n(x)$, we can look at other cubic spline functions $s(x)$ that interpolate $f(x)$. Referring back to the general node point definition (5.61), we say $s(x)$ is a cubic spline on $[a, b]$ if (1) $s(x)$ is cubic on each subinterval $[x_{j-1}, x_j]$, and (2) if $s(x)$, $s'(x)$, and $s''(x)$ are all continuous on $[a, b]$. In general, cubic splines are fairly smooth functions that are convenient to work with, and they have come to be widely used in the past few years in many different areas of applied mathematics and statistics.

If $s(x)$ is again chosen to satisfy the earlier interpolating conditions of (5.62),

$$s(x_i) = y_i, \qquad i = 1, \ldots, n$$

then the representation formula (5.68) and the tridiagonal system (5.69) are still valid. This system contains $n - 2$ equations and the n unknowns $M_1, \ldots, M_n$. By replacing the endpoint conditions (5.70) with other conditions, we can obtain other interpolating cubic splines.

If the data are obtained by evaluating a function $f(x)$,

$$y_i = f(x_i), \qquad i = 1, \ldots, n$$

then we choose endpoint conditions for $s(x)$ that will result in a better approximation to $f(x)$. If possible, require

$$s'(x_1) = f'(x_1), \qquad s'(x_n) = f'(x_n) \tag{5.76}$$

or

$$s''(x_1) = f''(x_1), \qquad s''(x_n) = f''(x_n) \tag{5.77}$$

When combined with (5.68) and (5.69), either of these conditions leads to a unique interpolating spline $s(x)$, dependent on which of the conditions is chosen. In both cases, the right-hand side of (5.75) can be replaced by ch^4, where c depends on $f^{(4)}(x)$.

If the derivatives of $f(x)$ are not known, then extra interpolating end conditions can also be used to ensure that the error formula (5.75) is proportional to h^4. In particular, suppose

$$x_1 < z_1 < x_2, \qquad x_{n-1} < z_{n-1} < x_n$$

and suppose $f(z_1)$, $f(z_{n-1})$ are known. Then use the formula for $s(x)$ in (5.68) and

$$s(z_1) = f(z_1), \qquad s(z_{n-1}) = f(z_{n-1}) \tag{5.78}$$

This will add two new equations for M_1, M_2, M_{n-1}, M_n to the system (5.69). This form of spline would generally be preferable to the interpolating natural cubic spline, and it is almost equally easy to produce.

Most computer centers will have a number of programs involving cubic splines, for use in numerical integration, numerical differentiation, interpolation, and the curve-fitting of data. The interpolation programs will generally allow a variety of endpoint conditions, such as those given in (5.70), (5.76) to (5.78). The student is well-advised to use these packages rather than attempting to write his own programs.

Interpolating cubic spline functions have become a popular way to represent data analytically because they are relatively smooth (two continuous derivatives), they do not have the rapid oscillation that sometimes occurs with higher degree polynomial interpolation, and they are reasonably easy to work with on a computer. They do not replace polynomials, but they are a very useful extension of them.

PROBLEMS

1. Consider the data points $\{(0, 1), (1, 1), (2, 5)\}$.
   (a) Find the piecewise linear interpolating function for the data.
   (b) Find the quadratic interpolating polynomial.
   (c) Find the natural cubic spline that interpolates the data.
   In all three cases, graph the interpolating functions for $0 \le x \le 2$.

2. For the data

x	0	1	2	3	4
y	0	1	4	9	16

   consider constructing the cubic natural interpolating spline function $s(x)$. Set up a system of linear equations used in constructing $s(x)$, but do not attempt to solve it. Specify the meaning of the unknowns in your linear system.

3. Consider the data

x	0	1/2	1	2	3
y	0	1/4	1	-1	-1

(a) Find the piecewise linear interpolating function for the data.
(b) Find the piecewise quadratic interpolating function.
(c) Find the natural cubic spline that interpolates the data.
Graph all three cases for $0 \leq x \leq 3$.

4. Consider the data

x	0	1	2	2.5	3	4
y	1.4	0.6	1.0	0.65	0.6	1.0

Repeat problem 3. But note that the piecewise quadratic interpolation will require modification. Use quadratic interpolation on [0, 2] and [2, 3]. For $q(x)$ on [3, 4], construct the quadratic interpolating polynomial on [2.5, 4] and then use it on only [3, 4].
Warning: The linear system for the cubic spline function is best solved with a linear system solver such as those given in Chapter 8, section 8.4.

5. Show that the boundary conditions (5.76) lead to the respective equations

$$\frac{x_2 - x_1}{3} M_1 + \frac{x_2 - x_1}{6} M_2 = \frac{y_2 - y_1}{x_2 - x_1} - f'(x)$$

$$\frac{x_n - x_{n-1}}{6} M_{n-1} + \frac{x_n - x_{n-1}}{3} M_n = f'(x_n) - \frac{y_n - y_{n-1}}{x_n - x_{n-1}}$$

(5.79)

Hint: Differentiate (5.68) on $[x_1, x_2]$ and $[x_{n-1}, x_n]$, and then use (5.76).

6. (a) Solve for the cubic spline that interpolates the data in (5.71), with the addition of the boundary conditions (5.76),

$$s'(1) = -1, \qquad s'(4) = -\frac{1}{16}$$

Hint: Combine (5.79) in problem 5 with (5.72), and use a linear system solver to solve your linear system of four equations.

(b) Compare both the natural spline (5.73) and the present spline to the function $f(x) = 1/x$ from which the data were generated. Calculate $f(x) - s(x)$ for a sampling of points in $1 \leq x \leq 4$.

7. (a) Solve for the cubic spline that interpolates the data in (5.71), with the addition of the boundary conditions (5.77),

$$s''(1) = 2, \qquad s''(4) = 1/32$$

Hint: Combine $M_1 = 2$, $M_4 = \frac{1}{32}$ with the equations (5.72).

(b) Compare this spline and the natural cubic spline (5.73) to the function $f(x) = 1/x$ from which the data were generated.

8. Is the following function $s(x)$ a cubic spline for $1 \leq x \leq 3$? Is it a natural cubic spline function?

$$s(x) = \begin{cases} x^3 - 3x_2 + 2x + 1, & 1 \leq x \leq 2 \\ -x^3 + 9x^2 - 22x + 17, & 2 \leq x \leq 3 \end{cases}$$

9. (a) Formula (5.68) gives $s(x)$ on $x_{j-1} \leq x \leq x_j$. Write the corresponding formula when $x_j \leq x \leq x_{j+1}$.

(b) Form $s'(x)$ for $x_{j-1} \leq x \leq x_j$ and $x_j \leq x \leq x_{j+1}$.

(c) Set $x = x_j$ in both formulas for $s'(x)$, and require them to be equal. Derive (5.69).

10. (a) Define

$$s_a(x) = \begin{cases} 0, & x \leq a \\ (x - a)^3, & x \geq a \end{cases}$$

Show $s_a(x)$ is a cubic spline function.

(b) Let $x_1 < x_2 < \ldots < x_n$ and define

$$s(x) = \sum_{j=1}^{n} b_j s_{x_j}(x)$$

using the definition in part (a). Show $s(x)$ is a cubic spline. What are the points at which $s'''(x)$ is not continuous?

11. To study the accuracy of cubic spline interpolation, use a package program to construct interpolating splines to $y = e^x$ on $0 \leq x \leq 1$. Use evenly spaced nodes on $[0, 1]$ to generate the data, say with 10, 20, and 40 subdivisions. Then check the accuracy of the spline function at four times that many points. (a) Do this for the cubic natural interpolating spline function and (b) for the spline satisfying (5.76) or (5.77). Discuss your results. How does the error behave as the number of subdivisions is doubled?

SIX

APPROXIMATION OF FUNCTIONS

Most functions encountered in mathematics courses cannot be evaluated exactly, even though we usually handle them as if they were completely known quantities. The simplest and most important of these are $\sqrt{x}$, e^x, log (x), the trigonometric functions, and y^x; and there are many other functions that occur commonly in physics, engineering, and other areas.

In evaluating functions, by hand or using a computer, we are essentially limited to the elementary arithmetic operations $+$, $-$, $\times$, and $\div$. Combining these means we can evaluate polynomials and rational functions, polynomials divided by polynomials. All other functions must be evaluated using approximations based on polynomials or rational functions. In this chapter we discuss polynomial approximations of functions. Rational functions generally give slightly more efficient approximations, but polynomials are adequate for most problems and their theory is much easier to work with.

In Chapter One, we studied the Taylor polynomial as a means to evaluate a given function $f(x)$. This is a relatively easy way to approximate a function to any desired level of accuracy; and often the Taylor polynomial is the only direct method of approximating the function. Nonetheless, a Taylor polynomial for $f(x)$ is usually a very inefficient approximation; if the approximation is to be used many times, then it should be replaced by a formula requiring less evaluation time. If a

Taylor polynomial of some degree is being used, then there is usually another polynomial of much lower degree that will be of equal accuracy; and its lower degree will decrease both the evaluation time and the number of rounding errors in the evaluation process. The Taylor polynomial gives us an accurate, but inefficient way to approximate a function. Often it will be the only way to initially evaluate a function; then this approximation is used to find a more efficient approximation.

We begin this chapter by looking at the concept of best possible approximation. This is illustrated with improvements to the Taylor polynomials for $f(x) = e^x$. In section 6.2, we introduce a set of polynomials that are well-known in numerical analysis, the Chebyshev polynomials. They are used in section 6.3 to produce a much more efficient polynomial approximation method, one based on interpolation.

6.1 THE BEST APPROXIMATION PROBLEM

To understand how much it is possible to improve on the Taylor series approximation, we look at the concept of best possible approximation. Let $f(x)$ be a given function that is continuous on a given interval $a \leq x \leq b$. If $p(x)$ is a polynomial, then we are interested in measuring

$$E(p) = \underset{a \leq x \leq b}{\text{maximum}} |f(x) - p(x)| \qquad (6.1)$$

the worst possible error in the approximation of $f(x)$ by $p(x)$. For each degree $n \geq 0$, define

$$\begin{aligned} \rho_n(f) &= \underset{\deg(p) \leq n}{\text{minimum}} E(p) \\ &= \underset{\deg(p) \leq n}{\text{minimum}} \ \underset{a \leq x \leq b}{\text{maximum}} |f(x) - p(x)| \end{aligned} \qquad (6.2)$$

This is the smallest possible value for $E(p)$ that can be attained with a polynomial of degree $\leq n$. It is called the *minimax error*. The polynomial that gives this value is called the *minimax polynomial approximation* of order n, and it will be denoted by $m_n(x)$.

Example

Let $f(x) = e^x$ on $-1 \leq x \leq 1$, and consider linear polynomial approximations to $f(x)$. From section 1.2, the linear Taylor polynomial is

$$t_1(x) = 1 + x \qquad (6.3)$$

and by direct computation,

$$\max_{-1 \le x \le 1} |e^x - t_1(x)| = 0.718 \tag{6.4}$$

By using methods we will not discuss here, the linear minimax polynomial is

$$m_1(x) = 1.2643 + 1.1752x \tag{6.5}$$

and

$$\max_{-1 \le x \le 1} |e^x - m_1(x)| = 0.279 \tag{6.6}$$

The graphs of these two linear polynomials are given in Figure 6.1, along with that of e^x.

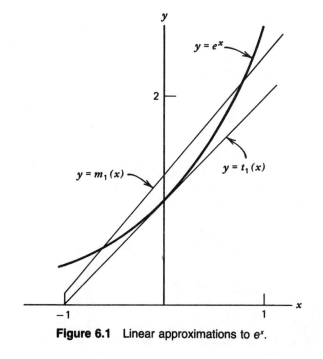

Figure 6.1 Linear approximations to e^x.

Example

Again let $f(x) = e^x$ for $-1 \leqq x \leqq 1$. We compare the maximum errors (6.1) for both the degree n Taylor polynomial $t_n(x)$ and the degree n minimax polynomial approximation $m_n(x)$. These results are shown in Table 6.1 for several values of n. Note that the accuracy of $m_n(x)$ relative to that of $t_n(x)$ becomes greater as n increases. This is a general characteristic of minimax approximations. This example is not as dramatic as for many functions $f(x)$ because the Taylor series for e^x converges very rapidly.

To further examine the differences in the Taylor and minimax approximations, we present graphs of the errors when $n = 3$. Then

$$t_3(x) = 1 + x + \frac{1}{2}x^2 + \frac{1}{6}x^3$$

$$m_3(x) = 0.994579 + 0.995668x + 0.542973x^2 + 0.179533x^3$$

(6.7)

The graphs are shown in Figures 6.2 and 6.3. Note that the vertical scales are quite different.

These examples illustrate several general properties of the minimax approximation $m_n(x)$ for approximating a function $f(x)$ on an interval $[a, b]$. First, $m_n(x)$ is usually a significant improvement on the Taylor polynomial $t_n(x)$. This means that to have an approximation with a given degree of accuracy, the minimax approximation will be of lower degree than the Taylor polynomials of equivalent accuracy, usually much lower.

A second characteristic of $m_n(x)$ is that the error $f(x) - m_n(x)$ has its error dispersed over the entire interval $[a, b]$. In comparison, the Taylor error $f(x) - t_n(x)$ is much smaller around the point of expansion than

Table 6.1   Taylor and Minimax Errors for e^x on $[-1, 1]$

n	Max Error in $t_n(x)$	Max Error in $m_n(x)$
1	7.18E-1	2.79E-1
2	2.18E-1	4.50E-2
3	5.16E-2	5.53E-3
4	9.95E-3	5.47E-4
5	1.62E-3	4.52E-5
6	2.26E-4	3.21E-6
7	2.79E-5	2.00E-7
8	3.06E-6	1.11E-8
9	3.01E-7	5.52E-10

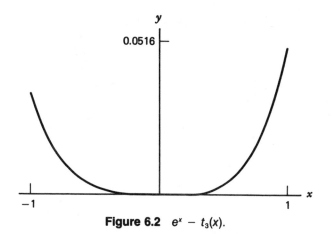

Figure 6.2 $e^x - t_3(x)$.

at other points of the interval $[a, b]$. In some sense, the smaller error in $m_n(x)$ is obtained by distributing the error evenly throughout the interval. An associated third characteristic is that the error $f(x) - m_n(x)$ is oscillatory on $[a, b]$, as illustrated in Figure 6.3. It can be shown that this error will change sign at least $n + 1$ times inside the interval $[a, b]$, and the sizes of the oscillations will be equal.

These properties can be used to construct $m_n(x)$, but it is still a process that is best left to an expert. Programs to do this are available in most computer center libraries. Our approach will be to use these properties to motivate the method given in section 6.3, and it will usually give a polynomial close to the minimax approximation.

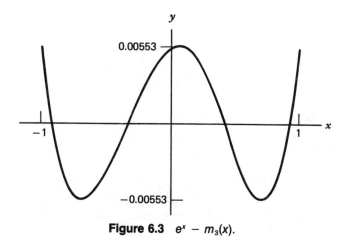

Figure 6.3 $e^x - m_3(x)$.

Accuracy of the Minimax Approximation

From the above examples for $f(x) = e^x$, it appears that $m_n(x)$ is very accurate for relatively small values of n. This can be made more precise in the case of some commonly occurring functions such as e^x, cos (x), and others. Assume $f(x)$ has an infinite number of continuous derivatives on an interval $[a, b]$, and let $m_n(x)$ be the minimax approximation of order n for $f(x)$ on $[a, b]$. Then the minimax error satisfies

$$\rho_n(f) \leq \frac{[(b - a)/2]^{n+1}}{(n + 1)! \, 2^n} \max_{a \leq x \leq b} |f^{(n+1)} (x)| \tag{6.8}$$

A derivation of this follows from the results of the next two sections. This error bound will not always become smaller with increasing n, but it will give a fairly accurate bound for most common functions $f(x)$.

Example

Let $f(x) = e^x$ for $-1 \leq x \leq 1$. Then (6.8) becomes

$$\rho_n(e^x) \leq \frac{e}{(n + 1)! 2^n} \tag{6.9}$$

Table 6.2 gives these values for various n, along with the corresponding exact value for $\rho_n(f)$ from Table 6.1. The computed value overestimates the true value by a factor of about 2.5, but this is still considered a fairly accurate estimator of $\rho_n(f)$.

Table 6.2 Bound on $\rho_n(e^x)$

n	1	2	3	4	5	6	7
Bound (6.9)	6.80E-1	1.13E-1	1.42E-2	1.42E-3	1.18E-4	8.43E-6	5.27E-7
$\rho_n(f)$	2.79E-1	4.50E-2	5.53E-3	5.47E-4	4.52E-5	3.21E-6	2.00E-7

PROBLEMS

1. Verify (6.4) and (6.6).

2. (a) For $f(x) = \tan^{-1} (x)$, calculate the Taylor approximations $t_1(x)$ and $t_3(x)$. Also find their maximum error relative to $\tan^{-1} (x)$ on $[-1, 1]$.

(b)   The linear and cubic minimax polynomials for $f(x) = \tan^{-1}(x)$ on $[-1, 1]$ are, respectively,

$$m_1(x) = 0.833278x$$
$$m_3(x) = 0.97238588x - 0.19193797x^3$$

Find their maximum errors on $[-1, 1]$.

(c)   Graph $f(x) - t_3(x)$ and $f(x) - m_3(x)$ on $[-1, 1]$.

3. With many functions $f(x)$, various identities can be used to reduce the interval of approximation from one of very large or infinite extent to a relatively small finite interval. For example, with $f(x) = e^x$, use

$$e^x = 1/e^{-x} \quad \text{if} \quad x < 0$$

Also, if $m \leq x < m + 1$ for some $m \geq 1$, then

$$e^x = e^m e^y, \quad y = x - m, \quad 0 \leq y < 1$$

These two identities reduce the evaluation of e^x to that of ordinary multiplication, $e^m = e \cdot e \cdot \cdots \cdot e$, and the evaluation of e^y with y between 0 and 1.

(a)   With this as motivation, compute the bound (6.8) with $f(x) = e^x$ on $[0, 1]$ and let $n = 1, 2, \ldots, 7$. Compare these results to those given in Table 6.2.

(b)   To check the accuracy of (a), compute the exact error in $m_3(x)$ for this interval, with

$$m_3(x) = 0.9994552 + 1.0166023x$$
$$+ 0.4217030x^2 + 0.2799765x^3$$

Plot the graph of $e^x - m_3(x)$ on $[0, 1]$.

4. Based on the interval reduction discussion in problem 3, write a section of Fortran code to reduce the evaluation of $\cos(x)$ to that of $\cos(y)$ where $0 \leq y \leq \pi/2$ for some y related to x.

5. (a)   Compute the bound (6.8) for $f(x) = \cos(x)$, $0 \leq x \leq \pi/2$, for $n = 1, 2, \ldots, 7$. DONE / CRESFICAL PS.

(b)   To check the accuracy of (a), compute the exact error in $m_3(x) \approx \cos(x)$ on $[0, \pi/2]$, where   SOLUTIONS IN $[0, \pi/2]$

$$m_3(x) = 0.9986329 + 0.0296140x$$

ROOT FINDING METHODS

$$- 0.6008616x^2 + 0.1125060x^3$$

(c)   Graph $\cos(x) - m_3(x)$ on $[0, \pi/2]$.

6. For $f(x) = \log x$, bound the minimax approximation error $\rho_n(f)$ on $1 \leq x \leq 2$. Find a bound for each case $n \geq 1$, and have the bound converge to zero as $n \to \infty$.

7. (a) Verify the identities

$$\tan^{-1}(-x) = -\tan^{-1}(x)$$

$$\tan^{-1}(x) = \frac{\pi}{2} - \tan^{-1}\left(\frac{1}{x}\right), \qquad x > 0$$

(b) With these, what smaller interval can be used to approximate $f(x) = \tan^{-1}(x)$, $-\infty < x < \infty$?

6.2 CHEBYSHEV POLYNOMIALS

We introduce a family of polynomials, the Chebyshev polynomials, which are used in many parts of numerical analysis, and, more generally, in mathematics and physics. A few of their properties will be given in this section, and then they will be used in section 6.3 to produce a polynomial approximation close to the minimax approximation.

For an integer $n \geq 0$, define the function

$$T_n(x) = \cos[n \cos^{-1} x], \qquad -1 \leq x \leq 1 \tag{6.10}$$

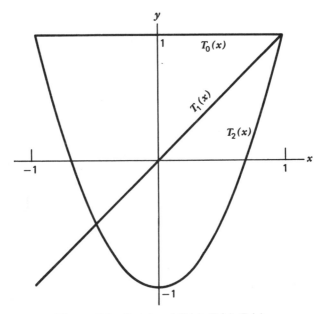

Figure 6.4 Graphs of $T_0(x)$, $T_1(x)$, $T_2(x)$.

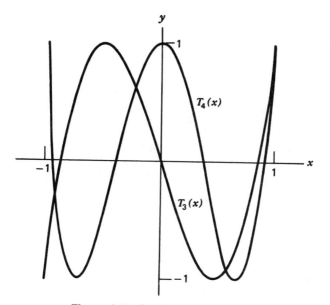

Figure 6.5 Graphs of $T_3(x)$, $T_4(x)$.

This may not appear to be a polynomial, but we will show it is a polynomial of degree n. To simplify the manipulation of (6.10), we introduce

$$\theta = \cos^{-1}(x) \quad \text{or} \quad x = \cos(\theta), \quad 0 \leq \theta \leq \pi \quad (6.11)$$

Then

$$T_n(x) = \cos(n\theta) \quad (6.12)$$

Graphs of $T_0(x)$, $T_1(x)$, ..., $T_4(x)$ are given in Figures 6.4 and 6.5.

Example

(a) Let $n = 0$. Then

$$T_0(x) = \cos(0 \cdot \theta) = 1$$

(b) Let $n = 1$. Then

$$T_1(x) = \cos(\theta) = x$$

(c) Let $n = 2$. Then using a common trigonometric identity, we obtain

$$T_2(x) = \cos(2\theta) = 2\cos^2(\theta) - 1 = 2x^2 - 1$$

The Triple Recursion Relation

Recall the trigonometric addition formulas,

$$\cos(\alpha \pm \beta) = \cos(\alpha)\cos(\beta) \mp \sin(\alpha)\sin(\beta)$$

For any $n \geq 1$, apply these identities to get

$$T_{n+1}(x) = \cos[(n+1)\theta] = \cos(n\theta + \theta)$$
$$= \cos(n\theta)\cos(\theta) - \sin(n\theta)\sin(\theta)$$
$$T_{n-1}(x) = \cos[(n-1)\theta] = \cos(n\theta - \theta)$$
$$= \cos(n\theta)\cos(\theta) + \sin(n\theta)\sin(\theta)$$

Add these two equations, and then use (6.11) and (6.12) to obtain

$$T_{n+1}(x) + T_{n-1} = 2\cos(n\theta)\cos(\theta) = 2x\,T_n(x)$$
$$T_{n+1}(x) = 2x T_n(x) - T_{n-1}(x), \qquad n \geq 1 \qquad (6.13)$$

This is called the triple recursion relation for the Chebyshev polynomials. It is often used in evaluating them, rather than using the explicit formula (6.10).

Example

(a) Let $n = 2$ in (6.13). Using the previously computed values of $T_1(x)$ and $T_2(x)$,

$$T_3(x) = 2x T_2(x) - T_1(x) = 2x(2x^2 - 1) - x$$
$$= 4x^3 - 3x$$

(b) Let $n = 3$. Then

$$T_4(x) = 2x T_3(x) - T_2(x) = 2x(4x^3 - 3x) - (2x^2 - 1)$$
$$= 8x^4 - 8x^2 + 1$$

The Minimum Size Property

Before stating the main result, note that

$$|T_n(x)| \leq 1, \qquad -1 \leq x \leq 1 \tag{6.14}$$

for all $n \geq 0$. Also,

$$T_n(x) = 2^{n-1} x^n + \text{lower degree terms}, \qquad n \geq 1. \tag{6.15}$$

The first result follows directly from the definition (6.10). The second can be proven using mathematical induction and the recursion relation (6.13). Note that the earlier computations of T_1, T_2, T_3, T_4 are also examples of (6.15).

Introduce a modified version of $T_n(x)$,

$$\tilde{T}_n(x) = \frac{1}{2^{n-1}} T_n(x) = x^n + \text{lower degree terms} \tag{6.16}$$

From (6.14) and (6.15),

$$|\tilde{T}_n(x)| \leq \frac{1}{2^{n-1}}, \qquad -1 \leq x \leq 1, \qquad n \geq 1 \tag{6.17}$$

A polynomial whose highest degree term has a coefficient of 1 is called a monic polynomial. Formula (6.17) says the monic polynomial $\tilde{T}_n(x)$ has size $1/2^{n-1}$ on $-1 \leq x \leq 1$, and this becomes smaller as the degree n increases. In comparison,

$$\max_{-1 \leq x \leq 1} |x^n| = 1$$

Thus x^n is a monic polynomial whose size does not change with increasing n.

Theorem 6.1

Let $n \geq 1$ be an integer, and consider all possible monic polynomials of degree n. Then the degree n monic polynomial with the smallest maximum on $[-1, 1]$ is the modified Chebyshev polynomial $\tilde{T}_n(x)$, and its maximum value on $[-1, 1]$ is $1/2^{n-1}$.

This theorem leads to a number of useful applications for Chebyshev polynomials, one of which we consider in the next section. A proof of the theorem is suggested in problem 9.

PROBLEMS

1. Find $T_5(x)$ explicitly in polynomial form, and then graph it on $-1 \leq x \leq 1$.

2. Demonstrate as best you can why (6.15) is true.

3. From (6.10), we know that $|T_n(x)| \leq 1$ on $[-1, 1]$ and $T_n(1) = 1$. Find a general formula for the points x at which $T_n(x) = \pm 1$. How many such points are there on $[-1, 1]$?

4. Evaluate $T_n(0.5)$ for $2 \leq n \leq 10$, without using the definitions (6.10) to (6.12). Use $T_0(0.5) = 1$, $T_1(0.5) = 0.5$.

5. Suppose that we require $T_0(x), T_1(x), \ldots, T_n(x)$ for a particular value of x. Do a count of the number of multiplications that are needed to produce these values if we use the triple recursion formula (6.13).

6. Let $q(x)$ be a polynomial of degree $\leq n - 1$, and consider

$$\max_{-1 \leq x \leq 1} |x^n - q(x)|$$

What is the smallest possible value for this quantity? Solve for the $q(x)$ for which the smallest value is attained.

7. For $n, m \geq 0$, and $n \neq m$, show

$$\int_{-1}^{1} \frac{T_n(x) T_m(x) \, dx}{\sqrt{1 - x^2}} = 0$$

This is called the orthogonality relation for Chebyshev polynomials. *Hint:* Use (6.10) and the change of variable $x = \cos \theta$.

8. The functions

$$S_n(x) = \frac{1}{n + 1} T'_{n+1}(x), \qquad n \geq 0$$

are called Chebyshev polynomials of the second kind.
(a) Calculate $S_0(x)$, $S_1(x)$, $S_2(x)$, $S_3(x)$.
(b) Show

$$S_n(x) = \frac{\sin (n + 1)\theta}{\sin \theta} \qquad \text{with } x = \cos \theta$$

for $0 \leq \theta \leq \pi$.

(c)   Use addition formulas for sin $(\alpha \pm \beta)$ to produce a triple recursion formula for $S_{n+1}(x)$.

9. To indicate the proof of Theorem 6.1, consider the case of degree $n = 3$.

(a)   Using problem 3, find the values of x for which

$$T_3(x) = \pm\frac{1}{4}$$

Call these values $x_4 < x_3 < x_2 < x_1$.

(b)   Assume there is another monic polynomial $q(x)$ for which

(i) $\displaystyle \max_{-1 \leq x \leq 1} |q(x)| < \frac{1}{2^{n-1}} = \frac{1}{4}$

(ii)   Degree $(q) \leq 3$.

(c)   Evaluate the polynomial

$$R(x) = \tilde{T}_3(x) - q(x)$$

at x_1, x_2, x_3, x_4. Show that

$$R(x_1) > 0, \qquad R(x_2) < 0, \qquad R(x_3) > 0, \qquad R(x_4) < 0$$

(d)   Show $R(x)$ has degree ≤ 2 and that it must have at least three roots.

(e)   Show $T_3(x) \equiv q(x)$, contrary to the assumption (i) in (b). This proves there is no smaller maximum on $[-1, 1]$ than $\frac{1}{4}$ for a monic polynomial of degree 3.

6.3   A NEAR-MINIMAX APPROXIMATION METHOD

Since we are looking for polynomial approximations to a given function $f(x)$, it would seem reasonable to consider using an interpolating polynomial. The most obvious choice is to choose an evenly spaced set of interpolation node points on the interval $a \leq x \leq b$ of interest. Unfortunately, this often gives an interpolating polynomial that is a very poor approximation to $f(x)$, for reasons we will not go into here. This was illustrated in the example in Figure 5.6 of Chapter 5. To consider interpolation in a more methodical way, we will examine it by means of the error formula (5.44) from Chapter 5.

To simplify the presentation, we choose the special interval $-1 \leq x \leq 1$ as the approximation interval for $f(x)$, and we limit the degree of the approximating polynomial to $n = 3$. Let x_0, x_1, x_2, x_3 be the interpolation node points in $[-1, 1]$, and let $P_3(x)$ denote the polynomial of degree

≤ 3 that interpolates to $f(x)$ at x_0, x_1, x_2, and x_3. Then from (5.44), the interpolation error is given by

$$f(x) - P_3(x) = \frac{(x - x_0)(x - x_1)(x - x_2)(x - x_3)}{4!} f^{(4)}(c_x), \quad (6.18)$$

for $-1 \leq x \leq 1$ and for some c_x in $[-1, 1]$. The nodes x_0, x_1, x_2, x_3 are to be chosen so that the maximum value of $|f(x) - P_3(x)|$ on $[-1, 1]$ is made as small as possible.

Looking at the right side of (6.18), we see that the only quantity we can use to influence the size of the error is the degree 4 polynomial

$$\omega(x) = (x - x_0)(x - x_1)(x - x_2)(x - x_3) \quad (6.19)$$

We want to choose the interpolation points x_0, x_1, x_2, x_3 so that

$$\max_{-1 \leq x \leq 1} |\omega(x)| \quad (6.20)$$

is made as small as possible.

If $\omega(x)$ is multiplied out, it is fairly easy to see that

$$\omega(x) = x^4 + \text{lower degree terms}$$

This is a monic polynomial of degree 4. From Theorem 6.1 in the preceding section, the smallest possible value for (6.20) is obtained with

$$\omega(x) = \frac{T_4(x)}{2^3} = \frac{1}{8}(8x^4 - 8x^2 + 1) \quad (6.21)$$

and the smallest value of (6.20) is $1/2^3$ in this case.

The choice (6.21) defines implicitly the interpolation node points. From (6.19), the node points are the zeros of $\omega(x)$, and from (6.21), they must therefore be the zeros of $T_4(x)$. The node points could be calculated by finding the roots of the right side of (6.21); but there is a simpler procedure, which is also much better to use when the polynomial degree becomes larger.

Look at the definition (6.10), (6.11). In our case,

$$T_4(x) = \cos(4\theta), \qquad x = \cos(\theta)$$

This is zero when

$$4\theta = \pm \frac{\pi}{2}, \pm \frac{3\pi}{2}, \pm \frac{5\pi}{2}, \pm \frac{7\pi}{2}, \ldots$$

$$\theta = \pm \frac{\pi}{8}, \pm \frac{3\pi}{8}, \pm \frac{5\pi}{8}, \pm \frac{7\pi}{8}, \ldots$$

$$x = \cos\left(\frac{\pi}{8}\right), \quad \cos\left(\frac{3\pi}{8}\right), \quad \cos\left(\frac{5\pi}{8}\right),$$

$$\cos\left(\frac{7\pi}{8}\right), \quad \cos\left(\frac{9\pi}{8}\right), \ldots \tag{6.22}$$

using $\cos(-\theta) = \cos(\theta)$. The first four values of x are distinct, but the following values repeat the first four values. Thus evaluating (6.22), the nodes are

$$\pm 0.382683, \pm 0.923880 \tag{6.23}$$

Example

Let $f(x) = e^x$ on $[-1, 1]$, and use the nodes (6.23) to produce the interpolating polynomials $P_3(x)$ of degree 3. Table 6.3 contains the nodes, function values, and divided differences needed for Newton's divided difference formula for the interpolating polynomial; see (5.38) in Chapter 5.

By evaluating $P_3(x)$ at a large number of points, we find

$$\max_{-1 \leq x \leq 1} |e^x - P_3(x)| \doteq 0.00666 \tag{6.24}$$

Table 6.3 Interpolation Data for $P_3(x)$

i	x_i	$f(x_i)$	$f[x_0, \ldots, x_i]$
0	0.923880	2.5190442	2.5190442
1	0.382683	1.4662138	1.9453769
2	-0.382683	0.6820288	0.7047420
3	-0.923880	0.3969760	0.1751757

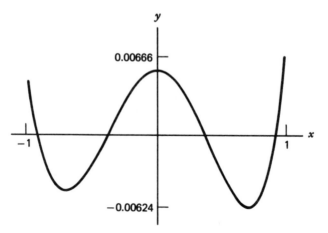

Figure 6.6 $e^x - P_3(x)$ with Chebyshev nodes.

The graph of $e^x - P_3(x)$ is given in Figure 6.6. Compare this error to the corresponding value from Table 6.1,

$$\rho_3(e^x) = 0.00553$$

and compare the graph to that given in Figure 6.3 for $m_3(x) \approx e^x$.

The above construction of $P_3(x)$ generalizes to finding a degree n near-minimax approximation to $f(x)$ on $[-1, 1]$. The interpolation error is given by

$$f(x) - P_n(x) = \frac{(x - x_0) \dots (x - x_n)}{(n + 1)!} f^{(n+1)}(c_x), \qquad -1 \le x \le 1 \quad (6.25)$$

and we seek to minimize

$$\max_{-1 \le x \le 1} |(x - x_0) \dots (x - x_n)| \qquad (6.26)$$

The polynomial being minimized is monic of degree $n + 1$. And from Theorem 6.1, this minimum is attained by the monic polynomial

$$\frac{1}{2^n} T_{n+1}(x)$$

Thus the interpolation nodes are the zeros of $T_{n+1}(x)$; and by the procedure leading to (6.22), they are given by

$$x_j = \cos\left(\frac{2j+1}{2n+2}\pi\right), \quad j = 0, 1, \ldots, n \qquad (6.27)$$

The near-minimax approximation $P_n(x)$ of degree n is obtained by interpolating to $f(x)$ at these $n+1$ nodes on $[-1, 1]$.

Example

Let $f(x) = e^x$. Then the maximum error in $P_n(x)$ on $[-1, 1]$ is given in Table 6.4. For comparison, we also include the corresponding minimax errors. These figures show that for practical purposes, $P_n(x)$ is a satisfactory replacement for the minimax approximation $m_n(x)$.

An Illustrative Program

A program is given below for constructing and evaluating the near-minimax approximation $P_n(x)$ for a given function $f(x)$ on $[-1, 1]$. The polynomial $P_n(x)$ is written in the Newton divided difference form of the interpolating polynomial; and it is evaluated using the nested multiplication method described in (5.42) of Chapter 5. Note that subroutine DIVDIF from section 5.2 is used by the program.

The program uses $f(x) = e^x$, but this is easily changed to some other function. The maximum error

$$\max_{-1 \leq x \leq 1} |e^x - P_n(x)|$$

is estimated by evaluating the error at 101 evenly spaced points x in $[-1, 1]$. The program can be modified to print out the nodes and divided differences, so that $P_n(x)$ could be evaluated in a separately constructed

Table 6.4 Near-Minimax Errors for e^x on $[-1, 1]$

n	1	2	3	4	5	6
Max $|e^x - p_n(x)|$	0.372	5.65E-2	6.66E-3	6.40E-4	5.18E-5	3.80E-6
$\rho_n(e^x)$	0.279	4.50E-2	5.53E-3	5.47E-4	4.52E-5	3.21E-6

function subprogram. Also, the program is written to be interactive, but it can be changed without difficulty to operate in a batch mode.

For a practical problem, we would have $f(x)$ evaluated by some accurate, but inefficient method, often a Taylor polynomial. This approximation would be given in the function subprogram $f(x)$, to be used by the main program to produce a more efficient approximation. One other difficulty is that most functions do not have $[-1, 1]$ as the interval on which we wish to approximate them. This limitation is removed in problem 3.

```
C       TITLE: FIND A NEAR-MINIMAX APPROXIMATION
C
C       THIS IS THE PROGRAM IN SECTION 6.3.
C
C       THIS PROGRAM CALCULATES A NEAR-MINIMAX APPROXIMATION
C       TO A GIVEN FUNCTION 'FCN' ON THE INTERVAL [-1,1].
C       THE APPROXIMATION IS BASED ON INTERPOLATION AT THE
C       CHEBYSHEV ZEROS ON [-1,1]. THE PROGRAM THEN
C       ESTIMATES THE MAXIMUM ERROR BY CALCULATING THE ERROR
C       AT A LARGE SAMPLE OF EVENLY SPACED POINTS ON [-1,1].
C       THE PROGRAM CAN BE EASILY MODIFIED TO PRINT THE
C       DIVIDED DIFFERENCES AND THE ERRORS.
C
        DIMENSION FX(0:20),X(0:20)
        PARAMETER(ZERO=0.0,ONE=1.0,P02=.02)
C
C       THE FOLLOWING STATEMENT DEFINES THE FUNCTION TO BE
C       APPROXIMATED. IT CAN ALSO BE SPECIFIED USING A
C       FUNCTION SUBPROGRAM.
        FCN(X)=EXP(X)
C
C       INITIALIZE VARIABLES.
        PI=4*ATAN(ONE)
        PRINT *,' WHAT IS THE DEGREE N?'
        READ *, N
        H=PI/(2*(N+1))
C
C       SET UP INTERPOLATION NODES AND CORRESPONDING
C       FUNCTION VALUES.
        DO 10 I=0,N
          X(I)=COS((2*I+1)*H)
          FX(I)=FCN(X(I))
10        CONTINUE
C
C       CALCULATE THE DIVIDED DIFFERENCES FOR THE
C       INTERPOLATION POLYNOMIAL.
        CALL DIVDIF(N,X,FX)
C
C       CALCULATE THE MAXIMUM INTERPOLATION ERROR AT 101
```

```
C       EVENLY SPACED POINTS ON [-1,1].
        ERRMAX=ZERO
        DO 20 I=0,100
           Z=-ONE+PO2*I
C       USE NESTED MULTIPLICATION TO EVALUATE THE
C       INTERPOLATION POLYNOMIAL AT Z.
        P=FX(N)
        DO 30 J=N-1,0,-1
30         P=FX(J)+(Z-X(J))*P
        ERROR=FCN(Z)-P
        ERRMAX=MAX(ABS(ERROR),ERRMAX)
20         CONTINUE
C
C       PRINT MAXIMUM ERROR.
        PRINT *,' FOR N=',N
        PRINT *,' MAXIMUM ERROR=',ERRMAX
        STOP
        END
```

Odd and Even Functions

A word of warning in using this algorithm. If $f(x)$ is what is called an even or odd function on $[-1, 1]$, then n should be chosen in a more restricted way. We say $f(x)$ is *even* if

$$f(-x) = f(x), \quad \text{all } x \qquad (6.28)$$

An example is $f(x) = \cos(x)$. Such functions have graphs that are symmetric about the y-axis. We say $f(x)$ is *odd* if

$$f(-x) = -f(x), \quad \text{all } x \qquad (6.29)$$

An example is $f(x) = \sin(x)$. Such functions are said to be symmetric about the origin.

For these two cases, choose n in the above algorithm as follows.

$$\text{If } f(x) \text{ is odd, then choose } n \text{ even.} \qquad (6.30)$$
$$\text{If } f(x) \text{ is even, then choose } n \text{ odd.}$$

This will result in $P_n(x)$ having degree only $n - 1$, but it will give the correct formula. The difficulty arises from the following. If a polynomial is odd as a function, as in (6.29), then $P_n(x)$ will have only odd degree terms; and if $P_n(x)$ is even, then it will have only even degree terms. The restriction (6.30) will take this difficulty into account in using the program.

PROBLEMS

1. (a) Demonstrate that there are only four distinct values in (6.22).
   (b) Show that the zeros of $T_{n+1}(x)$ are given by (6.27).

2. Give the interpolation nodes for the linear near-minimax approximation of this section, for the interval $[-1, 1]$. Give the linear near-minimax approximation for $f(x) = e^x$ on $[-1, 1]$.

3. (a) Implement the program given in this section. Include a print-out of either the divided differences and nodes or the errors; these can be used to generate a graph of the error function.
   (b) Calculate the cubic near-minimax for $f(x) = \tan^{-1}(x)$ on $[-1, 1]$. Compare it to the cubic minimax approximation given in problem 2(b) of section 6.1.
   Note: Use $n = 4$ in the program, based on (6.30), because $\tan^{-1}(x)$ is an odd function on $[-1, 1]$.

4. Most functions do not have $[-1, 1]$ as the interval on which they are to be approximated. Suppose $g(t)$ is to be evaluated for $a \leq t \leq b$. Then define a new function $f(x)$ on $[-1, 1]$ by

$$f(x) = g\left(\frac{(b + a) + x(b - a)}{2}\right), \qquad -1 \leq x \leq 1$$

where

$$t = \frac{1}{2}[(b + a) + x(b - a)]$$

represents a linear change of variable. We now approximate $f(x)$ on $[-1, 1]$.

As a specific example, produce the cubic near-minimax approximation for $g(t) = e^t$, $0 \leq t \leq 1$. Compare this to the minimax approximation given in problem 3(b) of section 6.1.

5. Find the cubic near-minimax approximation to $g(t) = \cos(t)$, $0 \leq t \leq \pi/2$. Compare it to the minimax approximation given in problem 5(b) of section 6.1.

6. Find the maximum error in the degree n near-minimax approximation to $g(t) = \tan^{-1}(t)$, $0 \leq t \leq 1$. Do this for $n = 1, 2, \ldots, 6$.

7. Repeat problem 6 with $g(t) = \log(t)$, $\frac{1}{2} \leq t \leq 1$. Why is this a natural interval for approximating $\log(t)$ on a binary computer?

8. Repeat problem 6 for the function

$$g(t) = \frac{1}{t} \int_0^t \frac{e^u - 1}{u} \, du, \qquad -1 \leq t \leq 1$$

To initially evaluate $g(t)$, use a Taylor polynomial approximation, say with an accuracy of 5×10^{-12} on the interval $-1 \leq t \leq 1$. Write a FUNCTION subprogram to evaluate this polynomial, following the example of FUNCTION SF(X) in section 1.3 of Chapter one.

9. Show that if a polynomial $p(x)$ is an even function, satisfying (6.28), then all odd degree terms in $p(x)$ will be zero. Generalize this result to polynomials that are odd functions.

10. Using (6.25) and (6.26), along with Theorem 6.1, derive the bound (6.8) of section 6.1 in the case $[a, b] = [-1, 1]$.

SEVEN

NUMERICAL INTEGRATION
AND DIFFERENTIATION

The definite integral

$$I(f) = \int_a^b f(x)\, dx \tag{7.1}$$

is defined in the calculus as a limit of what are called Riemann sums. It is then proven that

$$I(f) = F(b) - F(a) \tag{7.2}$$

where $F(x)$ is any antiderivative of $f(x)$; this is the Fundamental Theorem of calculus. Many integrals can be evaluated using this formula, and a significant portion of most calculus textbooks is devoted to this approach. Nonetheless, most integrals cannot be evaluated using (7.2) because most integrands $f(x)$ do not have antiderivatives expressible in terms of elementary functions. Examples of such integrals are

$$\int_0^1 e^{-x^2}\, dx, \qquad \int_0^\pi x^\pi \sin(\sqrt{x})\, dx \tag{7.3}$$

Other methods are needed for evaluating such integrals.

In the first section of this chapter, we define two of the oldest and most popular numerical methods for approximating (7.1): the trapezoidal rule and Simpson's rule. Section 7.2 analyzes the error in using these methods and then obtains improvements on them. Section 7.3 gives another approach to numerical integration, Gaussian quadrature. It is more complicated than the Simpson and trapezoidal rules, but it is almost always much superior in accuracy for equivalent amounts of computation. The last section discusses numerical differentiation. Some simple methods are derived and their sensitivity to rounding errors is analyzed.

7.1 THE TRAPEZOIDAL AND SIMPSON RULES

The central idea behind most ideas for approximating

$$I(f) = \int_a^b f(x)\, dx$$

is to replace $f(x)$ by an approximating function whose integral can be evaluated. In this section, we look at methods based on using linear and quadratic interpolation.

Approximate $f(x)$ by the linear polynomial

$$P_1(x) = \frac{(b - x)f(a) + (x - a)f(b)}{b - a}$$

which interpolates $f(x)$ at a and b (Figure 7.1). The integral of $P_1(x)$ over $[a, b]$ is the area of the shaded trapezoid shown in Figure 7.1; it is given by

$$T_1(f) = (b - a)\left[\frac{f(a) + f(b)}{2}\right] \tag{7.4}$$

This approximates the integral $I(f)$ if $f(x)$ is almost linear on $[a,b]$.

Example

Approximate the integral

$$I = \int_0^1 \frac{dx}{1 + x} \tag{7.5}$$

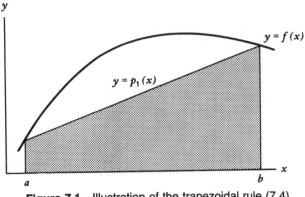

Figure 7.1   Illustration of the trapezoidal rule (7.4).

The true value is $I = \log (2) \doteq 0.693147$. Using (7.4), we obtain

$$T_1 = \frac{1}{2}\left[1 + \frac{1}{2}\right] = \frac{3}{4} = 0.75 \tag{7.6}$$

This is in error by

$$I - T_1 \doteq -0.0569 \tag{7.7}$$

To improve on the approximation $T_1(f)$ in (7.4) when $f(x)$ is not a nearly linear function on $[a, b]$, break the interval $[a, b]$ into smaller subintervals, and apply (7.4) on each subinterval. If the subintervals are small enough, then $f(x)$ will be nearly linear on each one. This idea is illustrated in Figure 7.2.

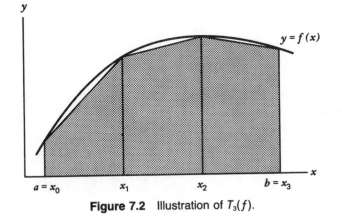

Figure 7.2   Illustration of $T_3(f)$.

Example

Evaluate the preceding example using $T_1(f)$ on two subintervals of equal length. For two subintervals,

$$I = \int_0^{1/2} \frac{dx}{1 + x} + \int_{1/2}^1 \frac{dx}{1 + x}$$

$$\doteq \frac{1}{2}\left[\frac{1 + \frac{2}{3}}{2}\right] + \frac{1}{2}\left[\frac{\frac{2}{3} + \frac{1}{2}}{2}\right]$$

$$T_2 = \frac{17}{24} = 0.70833 \tag{7.8}$$

$$I - T_2 \doteq -0.0152 \tag{7.9}$$

The error in T_2 is about 1/4 of that given for T_1 in (7.7).

We will derive a general formula in order to simplify the calculations when using several subintervals of equal length. Let the number of subintervals be denoted by n, and let

$$h = (b - a)/n$$

be the length of each subinterval. The endpoints of the subintervals are given by

$$x_j = a + jh, j = 0,1,\dots,n$$

Then break the integral into n subintegrals,

$$
\begin{aligned}
I(f) &= \int_a^b f(x)\ dx = \int_{x_0}^{x_n} f(x)\ dx \\
&= \int_{x_0}^{x_1} f(x)\ dx + \int_{x_1}^{x_2} f(x)\ dx + \dots + \int_{x_{n-1}}^{x_n} f(x)\ dx
\end{aligned}
\tag{7.10}
$$

Approximate each subintegral using (7.4), noting that each subinterval $[x_{j-1}, x_j]$ has width h. Then

$$I(f) \doteq h\left[\frac{f(x_0) + f(x_1)}{2}\right] + h\left[\frac{f(x_1) + f(x_2)}{2}\right] + h\left[\frac{f(x_2) + f(x_3)}{2}\right]$$

$$+ \dots + h\left[\frac{f(x_{n-1}) + f(x_n)}{2}\right]$$

The terms on the right can be combined to give the simpler formula

$$T_n(f) = h \left[\frac{f(x_0)}{2} + f(x_1) + f(x_2) + \ldots + f(x_{n-1}) + \frac{f(x_n)}{2} \right] \quad (7.11)$$

This is called the *trapezoidal numerical integration rule*. The subscript n gives the number of subintervals being used. And the points $x_0, x_1, \ldots, x_n$ are called the numerical integration *node points*.

Before giving some numerical examples of $T_n(f)$, we would like to discuss the choice of n. With a sequence of increasing values of n, $T_n(f)$ will usually be an increasingly accurate approximation of $I(f)$. But which sequence of values of n should be used? If n is doubled repeatedly, then the function values used in each $T_{2n}(f)$ will include all of the earlier function values used in the preceding $T_n(f)$. Thus the doubling of n will insure that all previously computed information is used in the new calculation, making the trapezoidal rule more efficient than it would be otherwise. To illustrate how function values are reused when n is doubled, consider $T_2(f)$ and $T_4(f)$.

$$T_2(f) = h \left[\frac{f(x_0)}{2} + f(x_1) + \frac{f(x_2)}{2} \right] \quad (7.12)$$

with

$$h = \frac{b-a}{2}, \qquad x_0 = a, \qquad x_1 = \frac{a+b}{2}, \qquad x_2 = b$$

Also,

$$T_4(x) = h \left[\frac{f(x_0)}{2} + f(x_1) + f(x_2) + f(x_3) + \frac{f(x_4)}{2} \right] \quad (7.13)$$

with

$$h = \frac{b-a}{4}, \qquad x_0 = a, \qquad x_1 = \frac{3a+b}{4},$$

$$x_2 = \frac{a+b}{2}, \qquad x_3 = \frac{a+3b}{4}, \qquad x_4 = b$$

In (7.13), only $f(x_1)$ and $f(x_3)$ need be evaluated, as the other function values are known from (7.12). For this and other reasons, all of our examples of $T_n(f)$ are based on doubling n.

Example

We give calculations of $T_n(f)$ for three integrals:

$$I^{(1)} = \int_0^1 e^{-x^2} \, dx \doteq 0.74682413281234 \qquad (7.14)$$

$$I^{(2)} = \int_0^4 \frac{dx}{1 + x^2} = \tan^{-1}(4) \doteq 1.3258176636680 \qquad (7.15)$$

$$I^{(3)} = \int_0^{2\pi} \frac{dx}{2 + \cos(x)} = \frac{2\pi}{\sqrt{3}} \doteq 3.6275987284684 \qquad (7.16)$$

The results are shown in Table 7.1. Only the errors

$$I(f) - T_n(f)$$

are given since this is the main quantity of interest in considering the speed with which $T_n(f)$ approaches $I(f)$. The column labeled "Ratio" gives the ratio of successive errors, the factor by which the error decreases when n is doubled.

From the table, the error in calculating $I^{(1)}$ and $I^{(2)}$ decreases by a factor of about 4 when n is doubled. The third example of $I^{(3)}$ converges much more rapidly. The answers for $n = 32, 64, 128$ were correct up to the limits due to rounding error on the computer, and this is denoted by * in the table. The explanation of all of these results will be given in the next section.

Simpson's Rule

To improve on $T_1(f)$ in (7.4), use quadratic interpolation to approximate $f(x)$ on $[a, b]$. Let $P_2(x)$ be the quadratic polynomial that interpolates $f(x)$ at $a, c = (a + b)/2$ and b. Using this to approximate $I(f)$, we get

$$I(f) \doteq \int_a^b p_2(x) \, dx$$

$$= \int_a^b \left[\frac{(x - c)(x - b)}{(a - c)(a - b)} f(a) + \frac{(x - a)(x - b)}{(c - a)(c - b)} f(c) \right.$$

$$\left. + \frac{(x - a)(x - c)}{(b - a)(b - c)} f(b) \right] dx \qquad (7.17)$$

This integral can be evaluated directly, but it is easier to first introduce

Table 7.1 Examples of the Trapezoidal Rule

	$I^{(1)}$		$I^{(2)}$		$I^{(3)}$	
n	Error	Ratio	Error	Ratio	Error	Ratio
2	1.55E-2		−1.33E-1		−5.61E-1	
4	3.84E-3	4.02	−3.59E-3	37	−3.76E-2	14.9
8	9.59E-4	4.01	5.64E-4	−6.37	−1.93E-4	195
16	2.40E-4	4.00	1.44E-4	3.92	−5.19E-9	37600
32	5.99E-5	4.00	3.60E-5	4.00	*	
64	1.50E-5	4.00	9.01E-6	4.00	*	
128	3.74E-6	4.00	2.25E-6	4.00	*	

$h = (b - a)/2$ and then to change the variable of integration. We will evaluate the first term to illustrate the general procedure. Let $u = x - a$. Then

$$\int_a^b \frac{(x - c)(x - b)}{(a - c)(a - b)} \, dx = \frac{1}{2h^2} \int_a^{a+2h} (x - c)(x - b) \, dx$$

$$= \frac{1}{2h^2} \int_0^{2h} (u - h)(u - 2h) \, du = \frac{1}{2h^2} \left[\frac{u^3}{3} - \frac{3}{2} u^2 h + 2h^2 u \right]_0^{2h} = \frac{h}{3}$$

The complete evaluation of (7.17) yields

$$S_2(f) = \frac{h}{3} \left[f(a) + 4f\left(\frac{a + b}{2} \right) + f(b) \right] \qquad (7.18)$$

The method is illustrated in Figure 7.3.

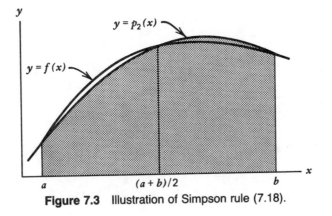

Figure 7.3 Illustration of Simpson rule (7.18).

Example

Use the earlier integral from (7.5),

$$I = \int_0^1 \frac{dx}{1 + x}$$

Then $h = (b - a)/2 = 1/2$, and

$$S_2 = \frac{1/2}{3}\left[1 + 4\left(\frac{2}{3}\right) + \frac{1}{2}\right] = \frac{25}{36} \doteq 0.69444 \qquad (7.19)$$

The error is

$$I - S_2 = \log(2) - S_2 \doteq -0.00130 \qquad (7.20)$$

To compare with the trapezoidal rule, use T_2 from (7.7), since the number of function evaluations is the same for both S_2 and T_2. The error in S_2 is smaller than that in (7.9) for T_2 by a factor of about 12, a significant increase in accuracy.

The rule $S_2(f)$ will be an accurate approximation to $I(f)$ if $f(x)$ is nearly quadratic on $[a, b]$. For other cases, proceed in the same manner as for the trapezoidal rule. Let n be an even integer, $h = (b - a)/n$, and define the evaluation points for $f(x)$ by

$$x_j = a + jh, \qquad j = 0, 1, \ldots, n \qquad (7.21)$$

Follow the idea of (7.10), but break $[a, b] = [x_0, x_n]$ into larger subintervals, each containing three interpolation node points. Thus

$$I(f) = \int_a^b f(x)\, dx = \int_{x_0}^{x_n} f(x)\, dx$$

$$= \int_{x_0}^{x_2} f(x)\, dx + \int_{x_2}^{x_4} f(x)\, dx + \ldots + \int_{x_{n-2}}^{x_n} f(x)\, dx$$

Approximate each subintegral by (7.18). This yields

$$I(f) \doteq \frac{h}{3}[f(x_0) + 4f(x_1) + f(x_2)] + \frac{h}{3}[f(x_2) + 4f(x_3) + f(x_4)]$$

$$+ \ldots + \frac{h}{3}[f(x_{n-2}) + 4f(x_{n-1}) + f(x_n)]$$

Table 7.2 Examples of Simpson's Rule

n	$I^{(1)}$ Error	Ratio	$I^{(2)}$ Error	Ratio	$I^{(3)}$ Error	Ratio
2	−3.56E-4		8.66E-2		−1.26	
4	−3.12E-5	11.4	3.95E-2	2.2	1.37E-1	−9.2
8	−1.99E-6	15.7	1.95E-3	20.3	1.23E-2	11.2
16	−1.25E-7	15.9	4.02E-6	485	6.43E-5	191
32	−7.79E-9	16.0	2.33E-8	172	1.71E-9	37600
64	−4.87E-10	16.0	1.46E-9	16.0	*	
128	−3.04E-11	16.0	9.15E-11	16.0	*	

If these terms are combined and simplified, we obtain the formula

$$S_n(f) = \frac{h}{3} \left[f(x_0) + 4f(x_1) + 2f(x_2) + 4f(x_3) + 2f(x_4) \right.$$
$$\left. + \ldots + 2f(x_{n-2}) + 4f(x_{n-1}) + f(x_n) \right] \quad (7.22)$$

This is called *Simpson's rule*, and it has been among the most popular numerical integration methods for over two centuries. The index n gives the number of subdivisions used in defining the integration node points $x_0, \ldots, x_n$.

Example

We evaluate the integrals (7.14) to (7.16), which were used previously to illustrate the trapezoidal rule. The results of using Simpson's rule are given in Table 7.2. For integrals $I^{(1)}$ and $I^{(2)}$, the ratio by which the error decreases approaches 16. For integral $I^{(3)}$, the errors converge to zero much more rapidly. An explanation of these results is given in the next section.

A Simpson's Rule Program

We give a program for Simpson's rule, for $n = 2, 4, 8, 16, \ldots, 512$ subdivisions. When n is doubled to $2n$, all of the function values occurring in $S_n(f)$ are also used in computing $S_{2n}(f)$. We also allow for a variety of integrands in the function F, with the user specifying the integrand through the variable *NUMF*. Note that double-precision constants should always be terminated by $D0$, as a matter of good (and safe) programming practice when writing double-precision programs. The program comments should explain the organization of the program.

```
C       TITLE: EXAMPLE OF SIMPSON'S NUMERICAL INTEGRATION METHOD.
C
C          THE INTEGRANDS ARE GIVEN IN FUNCTION 'F'. THE VARIABLE
C       'NUMF' IS AN INDEX AS TO WHICH INTEGRAND IS TO BE USED.
C       THE PROGRAM IS WRITTEN TO BE USED INTERACTIVELY.
C       FOR ILLUSTRATIVE PURPOSES, THE PROGRAM WILL COMPUTE THE
C       ERROR WHEN THE TRUE INTEGRAL IS KNOWN.
C
        IMPLICIT DOUBLE PRECISION(A-H,O-Z)
        COMMON/PARAM/NUMF
C
C       INPUT THE PROBLEM PARAMETERS.
10      PRINT *,' GIVE NUMBER OF INTEGRAND.'
        READ *, NUMF
        IF(NUMF .EQ. 0) STOP
        PRINT *,' GIVE INTEGRATION LIMITS A AND B.'
        READ *, A,B
        PRINT *,' GIVE TRUE ANSWER, IF KNOWN.'
        PRINT *,' IF UNKNOWN, GIVE ZERO.'
        READ *, TRUE
C
C       INITIALIZE FOR NUMERICAL INTEGRATION.
        SUMEND=F(A)+F(B)
        SUMODD=0.0D0
        SUMEVN=0.0D0
C
C       BEGIN MAIN LOOP. NINE NUMERICAL INTEGRATIONS WILL BE
C       PERFORMED, FOR N=2,4,8,...,512.
        DO 30 J=1,9
C         SET INTEGRATION PARAMETERS.
          N=2**J
          H=(B-A)/N
          SUMEVN=SUMEVN+SUMODD
          SUMODD=0.0D0
C         CALCULATE NEEDED ADDITIONAL FUNCTION VALUES.
          DO 20 K=1,N-1,2
            X=A+K*H
20          SUMODD=SUMODD+F(X)
C
C         CALCULATE SIMPSON RULE. THEN PRINT THE RESULTS,
C         INCLUDING THE ERROR.
          SIMPSN=H*(SUMEND+4.0D0*SUMODD+2.0D0*SUMEVN)/3.0D0
          ERROR=TRUE-SIMPSN
          PRINT *,' N=',N,'    INTEGRAL=',SIMPSN
          PRINT *,' ERROR=',ERROR
30        CONTINUE
        GO TO 10
        END

        FUNCTION F(X)
C
C       INTEGRANDS FOR NUMERICAL INTEGRATION.
C
        IMPLICIT DOUBLE PRECISION(A-H,O-Z)
        COMMON/PARAM/NUMF
C
        GO TO (10,20,30), NUMF
```

```
10       F=EXP(-X*X)
         RETURN
20       F=1.0D0/(1.0D0+X*X)
         RETURN
30       F=1.0D0/(2.0D0+COS(X))
         RETURN
         END
```

PROBLEMS

1. Compute $T_4(f)$ and $S_4(f)$ for the integral I in (7.5). Compute the errors $I - T_4$ and $I - S_4$; compare them to the errors $I - T_2$ and $I - S_2$, respectively.

2. Using the general form of the Simpson program, prepare a program for the trapezoidal rule. With it, prepare a table of values of $T_n(f)$ for $n = 2, 4, 8, \ldots, 512$ for the following integrals. Also find the errors and the ratios by which the errors decrease.

(a) $\displaystyle\int_0^\pi e^x \cos(4x)\, dx = \frac{e^\pi - 1}{17}$

(b) $\displaystyle\int_0^1 x^{5/2}\, dx = \frac{2}{7}$

(c) $\displaystyle\int_0^5 \frac{dx}{1 + (x - \pi)^2} = \tan^{-1}(5 - \pi) + \tan^{-1}(\pi)$

(d) $\displaystyle\int_{-\pi}^\pi e^{\cos(x)}\, dx \doteq 7.95492652101$

(e) $\displaystyle\int_0^1 \sqrt{x}\, dx = \frac{2}{3}.$

3. Repeat problem 2 using Simpson's rule.

4. Calculate the last two integrals in (7.17), verifying (7.18).

5. (a) As another approximation to $I(f) = \int_a^b f(x)\, dx$, replace $f(x)$ by the constant $f((a + b)/2)$ on the entire interval $a \leq x \leq b$. Show that this leads to the numerical integration formula

$$M_1(f) = (b - a)f\left(\frac{a + b}{2}\right)$$

Graphically illustrate this approximation.

(b) In analogy with the derivation of the trapezoidal rule (7.11) and Simpson's rule (7.22), generalize (a) to the numerical integration formula

$$M_n(f) = h[f(x_1) + f(x_2) + \ldots + f(x_n)]$$

where $h = (b - a)/n$ and

$$x_j = a + \left(j - \frac{1}{2}\right)h, \quad j = 1, \ldots, n$$

This is called the *midpoint rule;* it is a popular alternative to the trapezoidal rule.

(c) For the integral,

$$I = \int_0^1 \frac{dx}{1 + x}$$

of (7.5), calculate $M_1(f)$ and $M_2(f)$. Compare the errors to those for T_1 and T_2, given in (7.7) and (7.9).

6 (a) As another approximation to $I(f) = \int_a^b f(x)\,dx$, replace $f(x)$ by the degree four polynomial $P_4(x)$ interpolating $f(x)$ at the five points

$$x_j = a + jh, \quad j = 0, 1, 2, 3, 4, \quad h = \frac{b - a}{4}$$

Show that this leads to the approximating formula

$$B_4(f) = \frac{2h}{45} [7f(x_0) + 32f(x_1) + 12f(x_2) + 32f(x_3) + 7f(x_4)]$$

This is called *Boole's rule;* and it can be generalized to a larger number of subintervals in the same manner as was done for the Simpson and trapezoidal rules.

(b) Compute $B_4(f)$ for $I = \int_0^1 dx/(1 + x)$, and compare the results to those obtained in problem 1.

7 The *degree of precision* of a numerical integration formula is defined as follows: If the formula has zero error when integrating any polynomial of degree $\leq r$, and if the error is nonzero for some polynomial of degree $r + 1$, then we say the formula has degree of precision equal to r. Show the following rules have the indicated degree of precision r.

(a) $M_1(f)$: $r = 1$ (b) $T_1(f)$: $r = 1$
(c) $S_2(f)$: $r = 3$ (d) $B_4(f)$: $r = 5$

Are these results still valid when the various subscripts are replaced by n?

Hint: In your work, consider only $f(x) = 1, x, x^2$, etc.

8. Let $I_h(f) = (3h/4) [f(0) + 3f(2h)]$. What is the degree of precision of the approximation $I_h(f) \doteq \int_0^{3h} f(x) \, dx$?

Hint: Consider $f(x) = 1, x, x^2, x^3$, etc.

7.2 ERROR FORMULAS

In the preceding section, all but one of the numerical examples showed a regular behavior in the error for both the trapezoidal and Simpson rules. To explain this behavior, we consider error formulas for these integration methods. These formulas will lead to a better understanding of the methods, showing both their weaknesses and strengths. We begin by examining the error for the trapezoidal rule.

Theorem 7.1

Let $f(x)$ have two continuous derivatives on $[a, b]$, and let n be a positive integer. Then for the error in integrating

$$I(f) = \int_a^b f(x) \, dx$$

using the trapezoidal rule $T_n(f)$ of (7.11), we have

$$E_n^T(f) \equiv I(f) - T_n(f) = \frac{-h^2(b - a)}{12} f''(c_n) \qquad (7.23)$$

The number c_n is some unknown point in $[a, b]$; and $h = (b - a)/n$.

The proof is omitted, although part of it is given later. The formula (7.23) can be used to bound the error in $T_n(f)$, generally by bounding the term $|f''(c_n)|$ by its largest possible value on the interval $[a, b]$. This will be illustrated in the following example. Also note that the formula for $E_n^T(f)$ is consistent with the behavior of the errors observed in Table 7.1 for the integrals $I^{(1)}$ and $I^{(2)}$. When n is doubled, h is halved, and the term h^2 decreases by a factor of 4. This is exactly the factor observed in Table 7.1 for the decrease in the trapezoidal error.

Example

Recall the examples involving (7.5) in the preceding section, with

$$I = \int_0^1 \frac{dx}{1 + x} = \log (2)$$

Here $f(x) = 1/(1 + x)$, $[a, b] = [0, 1]$, and $f''(x) = 2/(1 + x)^3$. Substituting into (7.23), we obtain

$$E_n^T(f) = \frac{-h^2}{12} f''(c_n), \qquad 0 \le c_n \le 1, \qquad h = \frac{1}{n} \qquad (7.24)$$

This formula cannot be computed exactly because c_n is not known. But we can bound the error by looking at the largest possible value for $|f''(c_n)|$; bound $|f''(x)|$ on $[a, b] = [0, 1]$.

$$\max_{0 \le x \le 1} \frac{2}{(1 + x)^3} = 2$$

Then

$$|E_n^T(f)| \le \frac{h^2}{12} (2) = \frac{h^2}{6} \qquad (7.25)$$

For $n = 1$ and $n = 2$, we have

$$|E_1^T(f)| \le \frac{1}{6} \doteq 0.167, \qquad |E_2^T(f)| \le \frac{(1/2)^2}{6} \doteq 0.0417$$

Comparing with the true errors given in (7.7) and (7.9), these bounds are two to three times the actual errors.

A possible weakness in the trapezoidal rule can be inferred from the assumptions of Theorem 7.1. If $f(x)$ does not have two continuous derivatives on $[a, b]$, then does $T_n(f)$ converge more slowly? The answer is yes for some functions, especially if the first derivative is not continuous. This is explored experimentally in problem 10.

An Estimate of the Trapezoidal Error

The error formula (7.23) can only be used to bound the error, because $f''(c_n)$ is unknown. This will be improved upon by a more careful consideration of the error formula.

The essence of the proof of (7.23) lies in being able to demonstrate the $n = 1$ case on an interval $[\alpha, \alpha + h]$,

$$\int_{\alpha}^{\alpha+h} f(x)\, dx - h\left[\frac{f(\alpha) + f(\alpha + h)}{2}\right] = -\frac{h^3}{12} f''(c) \qquad (7.26)$$

for some c in $[\alpha, \alpha + h]$. We will give the part of the proof that this then leads to the general formula (7.23) in the theorem. Recall the derivation of the trapezoidal rule $T_n(f)$ as given in and following (7.10) in section 7.1. Then

$$E_n^T(f) = \int_a^b f(x)\, dx - T_n(f) = \int_{x_0}^{x_n} f(x)\, dx - T_n(f)$$

$$= \int_{x_0}^{x_1} f(x)\, dx - h\left[\frac{f(x_0) + f(x_1)}{2}\right]$$

$$+ \int_{x_1}^{x_2} f(x)\, dx - h\left[\frac{f(x_1) + f(x_2)}{2}\right] \qquad (7.27)$$

$$+ \ldots$$

$$+ \int_{x_{n-1}}^{x_n} f(x)\, dx - h\left[\frac{f(x_{n-1}) + f(x_n)}{2}\right]$$

Apply (7.26) to each of the terms on the right side of (7.27), to obtain

$$E_n^T(f) = -\frac{h^3}{12} f''(\gamma_1) - \frac{h^3}{12} f''(\gamma_2) - \ldots - \frac{h^3}{12} f''(\gamma_n) \qquad (7.28)$$

The unknown constants $\gamma_1, \ldots, \gamma_n$ are located in the respective subintervals

$$[x_0, x_1],\ [x_1, x_2],\ \ldots,\ [x_{n-1}, x_n] \qquad (7.29)$$

Factoring (7.28), we obtain

$$E_n^T(f) = -\frac{h^2}{12}[hf''(\gamma_1) + \ldots + hf''(\gamma_n)] \qquad (7.30)$$

It is left to problem 8 to show that the quantity in brackets equals $(b - a)f''(c_n)$ for some c_n in $[a, b]$, thus obtaining the general case of (7.23).

To estimate the trapezoidal error, observe that the term in brackets in (7.30) is a Riemann sum for the integral

$$\int_a^b f''(x) \, dx = f'(b) - f'(a) \qquad (7.31)$$

The Riemann sum is based on the partition (7.29) of $[a, b]$; as $n \to \infty$, this sum will approach the integral (7.31). Using (7.31) to estimate the right side of (7.30), we find that

$$E_n^T(f) \doteq \frac{-h^2}{12}[f'(b) - f'(a)] \qquad (7.32)$$

This error estimate will be denoted by $\tilde{E}_n^T(f)$. It is called an *asymptotic estimate* of the error because it improves as n increases. As long as $f'(x)$ is computable, $\tilde{E}_n^T(f)$ will be very easy to compute.

Example

Again consider the case $I = \int_0^1 dx/(1 + x)$. Then $f'(x) = -1/(1 + x)^2$, and (7.32) yields the estimate

$$\tilde{E}_n^T(f) = \frac{-h^2}{12}\left[\frac{-1}{(1 + 1)^2} - \frac{-1}{(1 + 0)^2}\right] = \frac{-h^2}{16}, \qquad h = \frac{1}{n} \quad (7.33)$$

For $n = 1$ and $n = 2$,

$$\tilde{E}_1^T(f) = -\frac{1}{16} = -0.0625, \quad \tilde{E}_2^T(f) = -0.0156$$

These compare quite closely to the true errors given in (7.7) and (7.9).

The estimate $\tilde{E}_n^T(f)$ has several practical advantages over the earlier error formula (7.23). First, it confirms that when n is doubled (or h is halved), the error decreases by a factor of about 4, *provided* that $f'(b) - f'(a) \neq 0$. This agrees with the results for $I^{(1)}$ and $I^{(2)}$ in Table 7.1. Second, (7.32) implies that the convergence of $T_n(f)$ will be more rapid when $f'(b) - f'(a) = 0$. This is a partial explanation of the very rapid convergence observed with $I^{(3)}$ in Table 7.1; a further explanation is given at the end of this section. Finally, (7.32) leads to a more accurate numerical integration formula by taking $\tilde{E}_n^T(f)$ into account:

$$I(f) - T_n(f) \doteq \frac{-h^2}{12} [f'(b) - f'(a)]$$

$$I(f) \doteq T_n(f) - \frac{h^2}{12} [f'(b) - f'(a)] \tag{7.34}$$

This is called the *corrected trapezoidal rule*, and it will be denoted by $CT_n(f)$.

Example

Recall the integral $I^{(1)}$ used in Table 7.1,

$$I = \int_0^1 e^{-x^2} \, dx \doteq 0.74682413281234$$

In Table 7.3, we give the results of using $T_n(f)$ and $CT_n(f)$, including their errors and the estimate

$$\tilde{E}_n^T(f) = \frac{h^2 e^{-1}}{6}, \qquad h = \frac{1}{n}$$

Note that $\tilde{E}_n^T(f)$ is a very accurate estimator of the true error. Also, the error in $CT_n(f)$ converges to zero at a more rapid rate than does the error for $T_n(f)$. When n is doubled, the error in $CT_n(f)$ decreases by a factor of about 16.

Table 7.3 Example of $CT_n(f)$ and $\tilde{E}_n^T(f)$

n	$I - T_n(f)$	$\tilde{E}_n^T(f)$	$CT_n(f)$	$I - CT_n(f)$	Ratio
2	1.545E-2	1.533E-2	0.746698561877	1.26E-4	
4	3.840E-3	3.832E-3	0.746816175313	7.96E-6	15.8
8	9.585E-4	9.580E-4	0.746823634224	4.99E-7	16.0
16	2.395E-4	2.395E-4	0.746824101633	3.12E-8	16.0
32	5.988E-5	5.988E-5	0.746824130863	1.95E-9	16.0
64	1.497E-5	1.497E-5	0.746824132690	2.22E-10	16.0

Error Formulas for Simpson's Rule

The type of analysis used above can also be used to derive corresponding error formulas for Simpson's rule. These are stated in the following theorem, with the proof omitted.

Theorem 7.2

Assume $f(x)$ has four continuous derivatives on $[a, b]$, and let n be an even positive integer. Then the error in using Simpson's rule is given by

$$E_n^S(f) = I(f) - S_n(f) = -\frac{h^4(b - a)}{180} f^{(4)}(c_n) \qquad (7.35)$$

with c_n an unknown point in $[a, b]$ and $h = (b - a)/n$. Moreover, this error can be estimated with the asymptotic error formula

$$\bar{E}_n^S(f) = -\frac{h^4}{180} [f'''(b) - f'''(a)] \qquad (7.36)$$

Example

Recall (7.19) where $S_2(f)$ was applied to $\int_0^1 dx/(1 + x)$. Then

$$f(x) = \frac{1}{1 + x}, \qquad f^{(3)}(x) = \frac{-6}{(1 + x)^4}, \qquad f^{(4)}(x) = \frac{24}{(1 + x)^5}$$

The exact error is given by

$$E_n^S(f) = -\frac{h^4}{180} f^{(4)}(c_n), \qquad h = \frac{1}{n}$$

for some $0 \leq c_n \leq 1$. We can bound it by

$$|E_n^S(f)| \leq \frac{h^4}{180} (24) = \frac{2h^4}{15}$$

The asymptotic error is given by

$$\bar{E}_n^S(f) = \frac{-h^4}{180} \left[\frac{-6}{(1 + 1)^4} - \frac{-6}{(1 + 0)^4} \right] = \frac{-h^4}{32}$$

For $n = 2$, $\tilde{E}^S(f) \doteq -0.00195$; for comparison from (7.20), the actual error is -0.00130.

From (7.36), the behavior in $I(f) - S_n(f)$ can be derived. When n is doubled, h is halved, and h^4 decreases by a factor of 16; thus the error $E_n^S(f)$ should decrease by the same factor, provided that $f'''(b) \neq f'''(a)$. This is the error behavior observed in Table 7.2 with integrals $I^{(1)}$ and $I^{(2)}$. When $f'''(b) = f'''(a)$, the error will decrease more rapidly, which is a partial explanation of the rapid convergence for $I^{(3)}$ in Table 7.2.

The theory of asymptotic error formulas

$$E_n(f) \doteq \tilde{E}_n(f) \tag{7.37}$$

such as for $E_n^T(f)$ and $E_n^S(f)$, says that (7.37) is valid provided that n is large enough. The needed size of n will vary with the integrand f, which is illustrated with the two cases $I^{(1)}$ and $I^{(2)}$ in Table 7.2. For $I^{(2)}$, the behavior (7.37) is not valid until n becomes larger, $n \geq 64$.

From (7.35) and (7.36), we also are lead to infer that Simpson's rule will not perform as well if $f(x)$ is not four times continuously differentiable on $[a, b]$. This is correct for most such functions, and other numerical methods are often necessary for integrating them.

Example

Use Simpson's rule to approximate

$$I = \int_0^1 \sqrt{x}\, dx = \tfrac{2}{3}$$

the results are shown in Table 7.4. The column "Ratio" shows the convergence is much slower.

Table 7.4 Simpson's Rule for $\sqrt{x}$

n	Error	Ratio
2	2.860E-2	
4	1.012E-2	2.82
8	3.587E-3	2.83
16	1.268E-3	2.83
32	4.485E-4	2.83

As was done for the trapezoidal rule, a corrected Simpson rule can be defined,

$$CS_n(f) = S_n(f) - \frac{h^4}{180} [f'''(b) - f'''(a)] \tag{7.38}$$

This will usually be a more accurate approximation than $S_n(f)$.

Richardson Extrapolation

The error estimates (7.32) and (7.36) are both of the form

$$I - I_n \doteq \frac{c}{n^p} \tag{7.39}$$

where I_n denotes the numerical integral and h has been replaced by $(b - a)/n$. The constants c and p vary with the method and the function. With most integrands $f(x)$, $p = 2$ for the trapezoidal rule and $p = 4$ for Simpson's rule. There are other numerical methods that satisfy (7.39), with other values of p and c. We will use (7.39) to obtain a computable estimate of the error $I - I_n$, without needing to know c explicitly.

In (7.39), replace n by $2n$ to obtain

$$I - I_{2n} \doteq \frac{c}{2^p n^p} \tag{7.40}$$

Comparing this to (7.39), we see that

$$2^p[I - I_{2n}] \doteq \frac{c}{n^p} \doteq I - I_n$$

Solving for I gives us

$$(2^p - 1)I \doteq 2^p I_{2n} - I_n$$

$$I \doteq \frac{1}{2^p - 1} [2^p I_{2n} - I_n] \equiv R_{2n} \tag{7.41}$$

R_{2n} is an improved estimate of I, based on using I_n, I_{2n}, p, and the assumption (7.39). It is called *Richardson's extrapolation formula*, and generally it is a more accurate approximation to I than is I_{2n}. How much more accurate it is depends on the validity of (7.39), (7.40).

To estimate the error in I_{2n}, compare it to the more accurate value R_{2n}.

$$I - I_{2n} \doteq R_{2n} - I_{2n} = \frac{1}{2^p - 1} [2^p I_{2n} - I_n] - I_{2n}$$

$$I - I_{2n} \doteq \frac{1}{2^p - 1} [I_{2n} - I_n] \qquad (7.42)$$

This is *Richardson's error estimate.*

Example

In using the trapezoidal rule to approximate

$$I = \int_0^1 e^{-x^2} \, dx \doteq 0.74682413281234$$

we have

$$T_2 = 0.7313702518, \qquad T_4 = 0.7429840978$$

Using (7.41) with $p = 2$ and $n = 2$, we obtain

$$I \doteq R_4 = \tfrac{1}{3} [4I_4 - I_2] = \tfrac{1}{3} [4T_4 - T_2]$$
$$= 0.7468553797$$

The error in R_4 is -0.0000312; from Table 7.1, R_4 is more accurate than T_{32}. To estimate the error in T_4, use (7.42) to get

$$I - T_4 \doteq \tfrac{1}{3} [T_4 - T_2] = 0.00387$$

The actual error in T_4 is 0.00384; thus (7.42) is a very accurate error estimate.

Richardson extrapolation and error estimation is not always as accurate as this example might suggest, but it is usually a fairly accurate procedure. The main assumption that must be satisfied is (7.39); and problem 9(b) gives a way of testing whether this assumption is valid for the actual values of I_n being used.

Most computer program libraries contain one or more programs for *automatic numerical integration.* This means the user of such a program presents an integral to it along with a desired error tolerance. The program then attempts to find a numerical integral within that error limit. There are many such programs, based on many different numerical integration rules. But most of them use error estimation techniques related to the ideas of Richardson extrapolation and estimation. We will not consider such programs here, as they are very complex in structure; but they are often the best way to integrate a function if there are not too many such integrals. These programs usually have high *program overhead* costs, but these are generally more than offset by the programming time that they save the user.

Periodic Integrands

A function $f(x)$ is periodic with period τ if

$$f(x) = f(x + \tau), \qquad -\infty < x < \infty \qquad (7.43)$$

and this equation should not be true with any smaller value of τ. For example,

$$f(x) = e^{\cos(\pi x)}$$

is periodic with period $\tau = 2$. If $f(x)$ is also differentiable, then its derivatives are also periodic with period τ.

Consider integrating

$$I = \int_a^b f(x)\,dx$$

with the trapezoidal or Simpson rule, and assume that $(b - a)$ is an integer multiple of the period τ. Then for all derivatives of $f(x)$, the periodicity of $f(x)$ implies that

$$f^{(r)}(a) = f^{(r)}(b), \qquad r \geq 0 \qquad (7.44)$$

If we now look at the asymptotic error formulas for the trapezoidal and Simpson rules, they become zero because of (7.44). Thus the error formulas $E_n^T(f)$ and $E_n^S(f)$ should converge to zero more rapidly for $f(x)$ a periodic function, if $b - a$ is a multiple of the period of f.

The asymptotic error formulas $\bar{E}_n^T(f)$ and $\bar{E}_n^S(f)$ can be extended to higher degree terms in h, using what is called the *Euler–MacLaurin expansion*. Using it, we can prove the errors $E_n^T(f)$ and $E_n^S(f)$ converge to zero even more rapidly than implied by the earlier comments, for $f(x)$ periodic. This work is omitted, but note that the trapezoidal rule is the preferred integration rule when dealing with periodic integrands. The earlier results for integral $I^{(3)}$ in Tables 7.1 and 7.2 are illustrations of these comments.

PROBLEMS

1. Using the error formula (7.23), bound the error in $T_n(f)$ applied to the following integrals.
   (a) $\int_0^{\pi/2} \cos(x)\, dx$ (b) $\int_0^1 e^{-x^2}\, dx$ (c) $\int_0^{\sqrt{\pi}} \cos(x^2)\, dx$

2. Repeat problem 1 using the integrals (a), (b), and (c) of problem 2 in section 7.1. Compare your bounds to the actual errors.

3. Apply the trapezoidal error estimate (7.32) to the integrals (a), (b), and (c) of problem 2 in section 7.1. Compare with the actual errors.

4. Repeat problem 3 using the integral $I^{(2)}$ of Table 7.1 in section 7.1.

5. Estimate the number of integration points needed to numerically integrate $\int_1^3 \log(x)\, dx$ with an accuracy of 10^{-8}, using the trapezoidal rule. Repeat using Simpson's rule.

6. The error $E_n(f)$ for both the trapezoidal and Simpson rules has some useful properties that simplify its calculation.
   (a) Show $E_n(f + g) = E_n(f) + E_n(g)$, for all continuous functions $f(x)$ and $g(x)$.
   (b) Show that $E_n(cf) = cE_n(f)$, for all continuous functions $f(x)$ and constants c.
   (c) If $p(x)$ is a linear polynomial, what is $E_n^T(p)$? If $p(x)$ is a cubic polynomial, what is $E_n^S(p)$?
   (d) For a twice continuously differentiable function $f(x)$ on $[a, b]$, use the linear Taylor polynomial

   $$p_1(x) = f(a) + (x - a)f'(a)$$

   to write $f(x) = p_1(x) + R_1(x)$. Show that

   $$E_n^T(f) = E_n^T(R_1)$$

   This can sometimes be used to simplify the calculation of $E_n^T(f)$.

7. The proof of the basic trapezoidal error formula (7.26) is based on using Taylor's formula to expand $f(x)$ about α. To give a heuristic proof of (7.26), write

$$f(x) \doteq f(\alpha) + (x - \alpha)f'(\alpha) + \frac{(x - \alpha)^2}{2} f''(\alpha)$$

Substitute this into the left side of (7.26) and obtain something quite close to its right side. Problem 6(d) can be used to simplify these calculations.

8. Recall the formula (7.30), an intermediate step in obtaining the error formula (7.23) for $E_n^T(f)$. To complete the proof, apply the ideas embodied in formula (A1.1) and problem 9 of Appendix A1, with $f''(x)$ playing the role of $f(x)$ in those statements.
   Hint: Write the term in brackets in (7.30) as

$$(b - a) \left[\frac{f''(\gamma_1) + f''(\gamma_2) + \ldots + f''(\gamma_n)}{n} \right]$$

9. (a) From (7.39) derive that

$$\frac{I - I_n}{I - I_{2n}} \doteq 2^p$$

for all n for which (7.39) is valid.
   Hint: Consider (7.39) and (7.40).

   (b) From (7.39), derive the computable estimate

$$\frac{I_{2n} - I_n}{I_{4n} - I_{2n}} \doteq 2^p$$

This gives a practical means of checking the value of p, using three successive values I_n, I_{2n} and I_{4n}. Using the log function,

$$p = \log \left(\frac{I_{2n} - I_n}{I_{4n} - I_{2n}} \right) / \log 2$$

   Hint: Write $I_{2n} - I_n = (I - I_n) - (I - I_{2n})$, and do similarly for the denominator.

10. (a) Use the formula in 9(a) with the results in Table 7.4 to derive the appropriate p in (7.39) for the Simpson rule applied to $\int_0^1 \sqrt{x}\, dx$.

   (b) Apply $T_n(f)$ to $\int_0^1 \sqrt{x}\, dx$ for $n = 2, 4, 8, \ldots, 128$. Compute the value of p in (7.39) for these numerical integrals. Use 9(a) or (b).

11. Use Richardson extrapolation to estimate the errors in problems 2(a), (b), (c) of section 7.1.

12. Use Richardson extrapolation to estimate the errors in problems 3(a), (b), (c) of section 7.1.

13. In the following table of numerical integrals and their differences, give the likely value of p if we assume the error behaves like $I - I_n \doteq \dfrac{c}{n^p}$. Also, estimate the error in I_{64}.

   Hint: Use problem 9(b).

n	I_n	$I_n - I_{\frac{n}{2}}$
2	0.702877396	
4	0.781978959	0.07910
8	0.804500932	0.02252
16	0.810303086	0.005802
32	0.811764354	0.001461
64	0.812130341	0.0003660

14. Using the Simpson error estimate (7.36), calculate how large n should be chosen in order to evaluate $I = \int_0^4 dx/(1 + x^2)$ with an accuracy of 10^{-12}.

15. (a) Apply Simpson's rule to $I = \int_0^1 \sin(\sqrt{x})\, dx$ with $n = 2, 4, 8, \ldots, 128$. Use problem 9(b) to calculate the rate of convergence.

   (b) Transform I using the change of variable $x = t^2$. Then $I = 2\int_0^1 t \sin(t)\, dt$. Apply Simpson's rule to this new integral, and again compute the rate of convergence using 9(b). Explain the difference in the results for parts (a) and (b).

7.3 GAUSSIAN NUMERICAL INTEGRATION

The numerical methods studied in the first two sections were based on integrating linear and quadratic interpolating polynomials, and the resulting formulas were applied on subdivisions of ever smaller subintervals. In this section, we look at a numerical method based on the exact integration of polynomials of increasing degree; no subdivision of the integration interval is used. To motivate this approach, recall from Chapter 6 the material on approximation of functions.

Let $f(x)$ be continuous on $[a, b]$. Then $\rho_n(f)$ denotes the smallest error bound that can be attained in approximating $f(x)$ with a polynomial $p(x)$

of degree $\leq n$ on a given interval $a \leq x \leq b$. The polynomial $q(x)$ that yields this approximation is called the minimax approximation of order n for $f(x)$,

$$\max_{a \leq x \leq b} |f(x) - q(x)| = \rho_n(f) \tag{7.45}$$

and $\rho_n(f)$ is called the minimax error. From formula (6.8), it can be seen that $\rho_n(f)$ will often converge to zero quite rapidly.

Example

Let $f(x) = e^{-x^2}$ for $0 \leq x \leq 1$. Table 7.5 contains the minimax errors $\rho_n(f)$ for $n = 1, 2, \ldots, 10$. They converge to zero rapidly, although not at a uniform rate.

If we have a numerical integration formula to integrate low to moderate degree polynomials exactly, then the hope is that the same formula will integrate other functions $f(x)$ almost exactly, if $f(x)$ is well approximable by such polynomials. To illustrate the derivation of such integration formulas, we restrict our attention to the integral

$$I(f) = \int_{-1}^{1} f(x)\, dx \tag{7.46}$$

Its relation to integrals over other intervals $[a, b]$ will be discussed later. The integration formula is to have the general form

$$I_n(f) = \sum_{j=1}^{n} w_j f(x_j) \tag{7.47}$$

Table 7.5 Minimax Errors for e^{-x^2}, $0 \leq x \leq 1$

n	$\rho_n(f)$	n	$\rho_n(f)$
1	5.30E-2	6	7.82E-6
2	1.79E-2	7	4.62E-7
3	6.63E-4	8	9.64E-8
4	4.63E-4	9	8.05E-9
5	1.62E-5	10	9.16E-10

and we require that the nodes $\{x_1, \ldots, x_n\}$ and weights $\{w_1, \ldots, w_n\}$ be so chosen that $I_n(f) = I(f)$ for all polynomials $f(x)$ of as large a degree as possible.

Case $n = 1$ The integration formula has the form

$$\int_{-1}^{1} f(x) \, dx \doteq w_1 f(x_1) \tag{7.48}$$

It is to be exact for polynomials of as large a degree as possible.

Using $f(x) \equiv 1$ and forcing equality in (7.48) gives us

$$2 = w_1$$

Now use $f(x) = x$ and again force equality in (7.48). Then

$$0 = w_1 x_1$$

which implies $x_1 = 0$. Thus (7.48) becomes

$$\int_{-1}^{1} f(x) \, dx \doteq 2f(0) \equiv I_1(f) \tag{7.49}$$

This is the midpoint formula of problem 5(a) in section 7.1. The formula (7.49) is exact for all linear polynomials; the proof of that is left as problem 4 for the reader.

To see that (7.49) is not exact for quadratics, let $f(x) = x^2$. Then the error in (7.49) is given by

$$\int_{-1}^{1} x^2 \, dx - 2(0)^2 = \tfrac{2}{3} \neq 0$$

Case $n = 2$ The integration formula is

$$\int_{-1}^{1} f(x) \, dx \doteq w_1 f(x_1) + w_2 f(x_2) \tag{7.50}$$

and it has four unspecified coefficients, x_1, x_2, w_1, and w_2. To determine these, we require it to be exact for the four monomials

$$f(x) = 1, x, x^2, x^3 \tag{7.51}$$

This leads to the four equations

$$
\begin{aligned}
2 &= w_1 + w_2 \\
0 &= w_1 x_1 + w_2 x_2 \\
\tfrac{2}{3} &= w_1 x_1^2 + w_2 x_2^2 \\
0 &= w_1 x_1^3 + w_2 x_2^3
\end{aligned}
\tag{7.52}
$$

This is a nonlinear system in four unknowns; its solution can be shown to be

$$
w_1 = w_2 = 1, \qquad x_1 = -\frac{\sqrt{3}}{3}, \qquad x_2 = \frac{\sqrt{3}}{3} \tag{7.53}
$$

This yields the integration formula

$$
\int_{-1}^{1} f(x)\, dx \doteq f\left(-\frac{\sqrt{3}}{3}\right) + f\left(\frac{\sqrt{3}}{3}\right) \equiv I_2(f) \tag{7.54}
$$

From being exact for the monomials in (7.51), one can show this formula will be exact for all polynomials of degree ≤ 3. It also can be shown by direct calculation to not be exact for the degree 4 polynomial $f(x) = x^4$.

Example

Approximate

$$
I = \int_{-1}^{1} e^x\, dx = e - e^{-1} \doteq 2.3504024
$$

Using (7.54), we get

$$
I_2 = e^{-\sqrt{3}/3} + e^{\sqrt{3}/3} = 2.3426961
$$
$$
I - I_2 \doteq 0.00771
$$

The error is quite small for using such a small number of node points.

Case $n > 2$ We seek the formula (7.47), which has $2n$ unspecified coefficients, $x_1, \ldots, x_n, w_1, \ldots, w_n$, by forcing the integration formula to be exact for the $2n$ monomials

$$
f(x) = 1, x, x^2, \ldots, x^{2n-1} \tag{7.55}
$$

This forces $I_n(f) = I(f)$ for all polynomials f of degree $\leq 2n - 1$. This leads to the following system of $2n$ nonlinear equations in $2n$ unknowns:

$$2 = w_1 + w_2 + \ldots + w_n$$
$$0 = w_1 x_1 + w_2 x_2 + \ldots + w_n x_n$$
$$\frac{2}{3} = w_1 x_1^2 + w_2 x_2^2 + \ldots + w_n x_n^2$$
$$0 = w_1 x_1^3 + w_2 x_2^3 + \ldots + w_n x_n^3 \qquad (7.56)$$
$$\vdots$$
$$\frac{2}{2n - 1} = w_1 x_1^{2n-2} + \ldots + w_n x_n^{2n-2}$$
$$0 = w_1 x_1^{2n-1} + \ldots + w_n x_n^{2n-1}$$

Solving this system is a formidable problem. Thankfully, this has already been done and the solutions have been collected in tables for the most commonly used values of n. Table 7.6 contains the solutions for

Table 7.6 Gaussian Integration Nodes and Weights

n	x_i	w_i
2	±.5773502692	1.0
3	±.7745966692	0.5555555556
	0.0	0.8888888889
4	±.8611363116	0.3478548451
	±.3399810436	0.6521451549
5	±.9061798459	0.2369268851
	±.5384693101	0.4786286705
	0.0	0.5688888889
6	±.9324695142	0.1713244924
	±.6612093865	0.3607615730
	±.2386191861	0.4679139346
7	±.9491079123	0.1294849662
	±.7415311856	0.2797053915
	±.4058451514	0.3818300505
	0.0	0.4179591837
8	±.9602898565	0.1012285363
	±.7966664774	0.2223810345
	±.5255324099	0.3137066459
	±.1834346425	0.3626837834

$n = 2, 3, \ldots, 8$. For more complete tables, see A. Stroud and D. Secrest (1966). Most computer centers will have programs to produce these nodes and weights or to directly perform the numerical integration.

There is also another approach to the development of the numerical integration formula (7.47), using the *theory of orthogonal polynomials*. From that theory, it can be shown that the nodes $\{x_1, \ldots, x_n\}$ are the zeros of the Legendre polynomial of degree n on the interval $[-1, 1]$. Since these polynomials are well known, the nodes $\{x_j\}$ can be found without any recourse to the nonlinear system (7.56). For an introduction to this theory, see Atkinson (1978, page 231).

The sequence of formulas (7.47) is called the *Gaussian numerical integration* method. From its definition, $I_n(f)$ uses n nodes, and it is exact for all polynomials of degree $\leq 2n - 1$. $I_n(f)$ is limited to (7.46), an integral over $[-1, 1]$; but this limitation is easily removed. Given an integral over $[a, b]$,

$$I(f) = \int_a^b f(x)\, dx \qquad (7.57)$$

introduce the linear change of variable

$$x = \frac{b + a + t(b - a)}{2}, \qquad -1 \leq t \leq 1 \qquad (7.58)$$

transforming the integral to

$$I(f) = \frac{b - a}{2} \int_{-1}^{1} f\left(\frac{b + a + t(b - a)}{2}\right) dt \qquad (7.59)$$

Now apply $I_n(f)$ to this new integral.

Example

Apply Gaussian numerical integration to the three integrals $I^{(1)}$, $I^{(2)}$, and $I^{(3)}$ of (7.14)–(7.16), which were used as examples for the trapezoidal and Simpson rules in section 7.1. All are reformulated as integrals over $[-1, 1]$, in the manner described above. The error results are shown in Table 7.7. The entry (*) means the error was zero relative to the accuracy possible on the computer being used.

If these results are compared to those in Tables 7.1 and 7.2, then Gaussian integration of $I^{(1)}$ and $I^{(2)}$ is much more efficient than are the trapezoidal and Simpson rules. But the integration of the periodic in-

Table 7.7 Gaussian Numerical Integration Examples

n	Error in $I^{(1)}$	Error in $I^{(2)}$	Error in $I^{(3)}$
2	2.29E-4	−2.33E-2	8.23E-1
3	9.55E-6	−3.49E-2	−4.30E-1
4	−3.35E-7	−1.90E-3	1.77E-1
5	6.05E-9	1.70E-3	−8.12E-2
6	−7.77E-11	2.74E-4	3.55E-2
7	8.60E-13	−6.45E-5	−1.58E-2
10	*	1.27E-6	1.37E-3
15	*	7.40E-10	−2.33E-5
20	*	*	3.96E-7

tegral $I^{(3)}$ is not as efficient as with the trapezoidal rule. These results are also true for most other integrals. Except for periodic integrals, Gaussian numerical integration is usually much more accurate than the trapezoidal and Simpson rules. This is even true with many integrals in which the integrand does not have a continuous derivative.

Example

Use Gaussian integration on

$$I = \int_0^1 \sqrt{x} \, dx = \tfrac{2}{3} \qquad (7.60)$$

The results are shown in Table 7.8, with n the number of node points. The ratio column shows that the error behaves like

$$I - I_n \doteq \frac{c}{n^3} \qquad (7.61)$$

Table 7.8 Gaussian Integration of (7.60)

n	$I - I_n$	Ratio
2	−7.22E-3	
4	−1.16E-3	6.2
8	−1.69E-4	6.9
16	−2.30E-5	7.4
32	−3.00E-6	7.6
64	−3.84E-7	7.8

for some c. Compare this with Table 7.4, which gives the results of Simpson's rule for (7.60). There the empirical rate of convergence is proportional to only $1/n^{3/2}$, a much slower rate than in (7.61).

This section concludes by relating the minimax error to the Gaussian numerical integration error.

Theorem 7.3

Let $f(x)$ be continuous for $a \leq x \leq b$, and let $n \geq 1$. Then applying Gaussian numerical integration to $I = \int_a^b f(x)\,dx$, the error in I_n satisfies

$$|I - I_n| \leq 2(b - a)\rho_{2n-1}(f) \tag{7.62}$$

where $\rho_{2n-1}(f)$ is the minimax error of order $2n - 1$ for $f(x)$ on $[a, b]$.

Example

Using Table 7.5, apply (7.62) to

$$I = \int_0^1 e^{-x^2}\,dx \tag{7.63}$$

For $n = 3$, the above bound implies

$$|I - I_3| \leq 2\rho_5(e^{-x^2}) = 3.24 \times 10^{-5}$$

The actual error is 9.55E-6, from Table 7.7.

Gaussian numerical integration is not as simple to use as are the trapezoidal and Simpson rules, partly because the Gaussian nodes and weights do not have simple formulas and also because the error is harder to predict. Nonetheless, the increase in the speed of convergence is so rapid and dramatic in most cases that the method should always be considered seriously when doing many integrations. Estimating the error is quite difficult, and most people satisfy themselves by looking at two or more successive values. If n is doubled, repeatedly then comparing two successive values, I_n and I_{2n}, is probably adequate for estimating the error in I_n,

$$I - I_n \doteq I_{2n} - I_n \tag{7.64}$$

This is somewhat inefficient, but the speed of convergence in I_n is so rapid that this will still not diminish its advantage over most other methods.

PROBLEMS

1. Recalling the example following (7.54), apply I_3 and I_4 to $\int_{-1}^{1} e^x \, dx$. Use the nodes and weights given in Table 7.6.

2. Apply $I_2, I_3,$ and I_4 to the integrals 2(a) to (e) of section 7.1. Calculate the errors and compare to the earlier results for the trapezoidal and Simpson rules. If your computer center has a Gaussian numerical integration program, then do these same calculations for $I_5, \ldots, I_{10}$.

3. Repeat the example following (7.63) with $n = 2, 4,$ and 5.

4. Show that if an integration formula of the form (7.47) is exact when integrating $1, x, x^2, \ldots, x^m$, then it is exact for all polynomials of degree $\leq m$.
   Hint: Recall problem 6 in section 7.2.

5. Consider integrals

   $$I(f) = \int_0^1 f(x) \log (x) \, dx$$

   with $f(x)$ a function with several continuous derivatives on $0 \leq x \leq 1$. We want to imitate the ideas of this section to develop an approximating formula for this $I(f)$.
   (a) Find a formula

   $$\int_0^1 f(x) \log (x) \, dx \doteq w_1 f(x_1)$$

   which is exact if $f(x)$ is any linear polynomial.

   Hint: $\displaystyle \int_0^1 x^m \log (x) \, dx = \frac{-1}{(m + 1)^2}, \quad m \geq 0$

   (b) To find a formula

   $$\int_0^1 f(x) \log (x) \, dx \doteq w_1 f(x_1) + w_2 f(x_2)$$

   which is exact for all polynomials of degree ≤ 3, set up a system of four equations with unknowns w_1, w_2, x_1, x_2. Do not attempt to solve them.

(c) Use I_1 of (a) to approximate

$$I = \int_0^1 \cos(x) \log(x)\, dx \doteq -0.9460829$$

7.4 NUMERICAL DIFFERENTIATION

To numerically calculate the derivative of $f(x)$, begin by recalling the definition of derivative:

$$f'(x) = \lim_{h \to 0} \frac{f(x+h) - f(x)}{h}$$

This justifies using

$$f'(x) \doteq \frac{f(x+h) - f(x)}{h} \equiv D_h f(x) \qquad (7.65)$$

for small values of h. $D_h f(x)$ is called a *numerical derivative* of $f(x)$ with stepsize h.

Example

Use $D_h f(x)$ to approximate the derivative of $f(x) = \cos(x)$ at $x = \pi/6$. Table 7.9 contains the results for various values of h. Looking at the error column, the error is proportional to h; when h is halved, the error is also halved.

Table 7.9 Numerical Differentiation of $f(x) = \cos(x)$ Using (7.65)

h	$D_h f$	Error	Ratio
0.1	−0.54243	0.04243	
0.05	−0.52144	0.02144	1.98
0.025	−0.51077	0.01077	1.99
0.0125	−0.50540	0.00540	1.99
0.0625	−0.50270	0.00270	2.00
0.03125	−0.50135	0.00135	2.00

To explain the behavior in this example, Taylor's theorem can be used to find an error formula. Expanding $f(x + h)$ about x,

$$f(x + h) = f(x) + hf'(x) + \frac{h^2}{2} f''(c)$$

for some c between x and $x + h$. Substituting in the right side of (7.65), we obtain

$$D_h f(x) = \frac{1}{h} \left\{ \left[f(x) + hf'(x) + \frac{h^2}{2} f''(c) \right] - f(x) \right\}$$

$$= f'(x) + \frac{h}{2} f''(c)$$

$$f'(x) - D_h f(x) = -\frac{h}{2} f''(c) \tag{7.66}$$

The error is proportional to h, agreeing with the results in Table 7.9 above. For that example,

$$f'(\pi/6) - D_h f(\pi/6) = \frac{h}{2} \cos(c) \tag{7.67}$$

where c is between $\pi/6$ and $\pi/6 + h$. The reader should check that if c is replaced by $\pi/6$, then the right side of (7.67) agrees with the error column in Table 7.9.

Differentiation Using Interpolation

Let $P_n(x)$ denote the degree n polynomial that interpolates $f(x)$ at $n + 1$ node points $x_0, x_1, \ldots, x_n$. To calculate $f'(x)$ at some point $x = t$, use

$$f'(t) \doteq p_n'(t) \tag{7.68}$$

Many different formulas can be obtained by varying n and by varying the placement of the nodes $x_0, \ldots, x_n$ relative to the point t of interest

As an especially useful example of (7.68), take $n = 2, t = x_1, x_0 = x_1 - h$, $x_2 = x_1 + h$. Then

$$P_2(x) = \frac{(x - x_1)(x - x_2)}{2h^2} f(x_0) + \frac{(x - x_0)(x - x_2)}{-h^2} f(x_1)$$

$$+ \frac{(x - x_0)(x - x_1)}{2h^2} f(x_2)$$

$$P_2'(x) = \left(\frac{2x - x_1 - x_2}{2h^2}\right) f(x_0) + \left(\frac{2x - x_0 - x_2}{-h^2}\right) f(x_1)$$

$$+ \left(\frac{2x - x_0 - x_1}{2h^2}\right) f(x_2) \quad (7.69)$$

$$P_2'(x_1) = \left(\frac{x_1 - x_2}{2h^2}\right) f(x_0) + \left(\frac{2x_1 - x_0 - x_2}{-h^2}\right) f(x_1)$$

$$+ \left(\frac{x_1 - x_0}{2h^2}\right) f(x_2)$$

$$= \frac{f(x_2) - f(x_0)}{2h}$$

Replacing x_0 and x_2 by $x_1 - h$ and $x_1 + h$, (7.68) becomes

$$f'(x_1) \doteq \frac{f(x_1 + h) - f(x_1 - h)}{2h} \equiv D_h f(x_1) \qquad (7.70)$$

It will be shown below that this is a more accurate approximation to $f'(x)$ than is (7.65).

The error for the general procedure (7.68) can be obtained from the interpolation error formula (5.58) in Chapter 5. The main result is given in the following theorem; the proof is omitted.

Theorem 7.4

Assume $f(x)$ has $n + 2$ continuous derivatives on an interval $[a, b]$. Let $x_0, x_1, \ldots, x_n$ be $n + 1$ distinct interpolation nodes in $[a, b]$, and let t be an arbitrary given point in $[a, b]$. Then

$$f'(t) - P_n'(t) = \psi_n(t) \frac{f^{(n+2)}(c_1)}{(n + 2)!} + \psi_n'(t) \frac{f^{(n+1)}(c_2)}{(n + 1)!} \qquad (7.71)$$

with

$$\psi_n(t) = (t - x_0)(t - x_1) \ldots (t - x_n)$$

The constants c_1 and c_2 are unknown points located between the maximum and minimum of $x_0, x_1, \ldots, x_n$ and t.

To illustrate this result, an error formula can be derived for (7.70). Since $t = x_1$ in deriving (7.70), we find that the first term on the right side of (7.71) is zero. Also, $n = 2$ and

$$\psi_2(x) = (x - x_0)(x - x_1)(x - x_2)$$
$$\psi_2'(x) = (x - x_1)(x - x_2) + (x - x_0)(x - x_2) + (x - x_0)(x - x_1)$$
$$\psi_2'(x_1) = (x_1 - x_0)(x_1 - x_2) = -h^2$$

Using this in (7.71), we get

$$f'(x_1) - \frac{f(x_1 + h) - f(x_1 - h)}{2h} = -\frac{h^2}{6} f'''(c_2) \qquad (7.72)$$

with $x_1 - h \leq c_2 \leq x_1 + h$. This says that for small values of h, the formula (7.70) should be more accurate than the earlier approximation (7.65), because the error term of (7.70) decreases more rapidly with h.

Example
The earlier example in Table 7.9 is repeated using (7.70). Recall that $f(x) = \cos(x)$ and $x_1 = \pi/6$. The results are shown in Table 7.10, with (7.70) given in the column labeled $D_h f$. The results confirm the rate of convergence given in (7.72), and they illustrate that (7.70) will usually be superior to the earlier approximation (7.65).

Table 7.10 Numerical Differentiation Using (7.70)

h	$D_h f$	Error	Ratio
0.1	−0.49916708	−0.0008329	
0.05	−0.49979169	−0.0002083	4.00
0.025	−0.49994792	−0.00005208	4.00
0.0125	−0.49998698	−0.00001302	4.00
0.00625	−0.49999674	−0.000003255	4.00

Undetermined Coefficients

The method of undetermined coefficients is a procedure used in deriving formulas for numerical differentiation, interpolation, and integration. We will explain the method by using it to derive an approximation for $f''(x)$.

To approximate $f''(x)$ at some point $x = t$, write

$$f''(t) \doteq D_h^{(2)} f(t) \equiv Af(t + h) + Bf(t) + Cf(t - h) \qquad (7.73)$$

with A, B, and C unspecified constants. Replace $f(t - h)$ and $f(t + h)$ by the Taylor polynomial approximations

$$f(t - h) \doteq f(t) - hf'(t) + \frac{h^2}{2} f''(t) - \frac{h^3}{6} f'''(t) + \frac{h^4}{24} f^{(4)}(t)$$

$$f(t + h) \doteq f(t) + hf'(t) + \frac{h^2}{2} f''(t) + \frac{h^3}{6} f'''(t) + \frac{h^4}{24} f^{(4)}(t) \qquad (7.74)$$

Including more terms would give higher powers of h, and for small values of h, these should be much smaller than the terms included in (7.74). Substituting these approximations into the formula for $D_h^{(2)} f(t)$ and collecting together common powers of h gives us

$$D_h^{(2)} f(t) \doteq (A + B + C)f(t) + h(A - C)f'(t) + \frac{h^2}{2} (A + C)f''(t)$$

$$+ \frac{h^3}{6} (A - C)f'''(t) + \frac{h^4}{24} (A + C)f^{(4)}(t) \qquad (7.75)$$

To have

$$D_h^{(2)} f(t) \doteq f''(t)$$

for arbitrary functions $f(x)$, it is necessary to require

$$\begin{aligned}
A + B + C &= 0 : \text{coefficient of } f(t) \\
h(A - C) &= 0 : \text{coefficient of } f'(t) \\
\frac{h^2}{2} (A + C) &= 1 : \text{coefficient of } f''(t)
\end{aligned} \qquad (7.76)$$

These have the solution

$$A = C = \frac{1}{h^2}, \qquad B = -\frac{2}{h^2} \tag{7.77}$$

This determines

$$D_h^{(2)} f(t) = \frac{f(t + h) - 2f(t) + f(t - h)}{h^2} \tag{7.78}$$

To determine an error formula for $D_h^{(2)} f(t)$, substitute (7.77) into (7.75) to obtain

$$D_h^{(2)} f(t) \doteq f''(t) + \frac{h^2}{12} f^{(4)}(t)$$

The approximation in this arises from not including terms in the Taylor polynomials (7.74) corresponding to higher powers of h. Thus

$$f''(t) - \frac{f(t + h) - 2f(t) + f(t - h)}{h^2} \doteq \frac{-h^2}{12} f^{(4)}(t) \tag{7.79}$$

This is an accurate estimate of the error for small values of h. Of course, in a practical situation we would not know $f^{(4)}(t)$. But the error formula shows that the error decreases by a factor of about 4 when h is halved. This can be used to justify Richardson's extrapolation to obtain an even more accurate estimate of the error and of $f''(t)$; see problem 5.

Example

Let $f(x) = \cos(x)$, $t = \pi/6$, and use (7.78) to calculate $f''(t) = -\cos(\pi/6)$. The results are shown in Table 7.11. Note that the ratio column is consistent with the error formula (7.79).

In the derivation of (7.78), the form (7.73) was assumed for the approximate derivative. We could equally well have chosen to evaluate $f(x)$ at points other than those used there, for example,

$$f''(t) \doteq Af(t + 2h) + Bf(t + h) + Cf(t) \tag{7.80}$$

Table 7.11 Numerical Differentiation Using $D_h^{(2)}f$

h	$D_h^{(2)}f$	Error	Ratio
0.5	−0.84813289	−1.789E-2	
0.25	−0.86152424	−4.501E-3	3.97
0.125	−0.86489835	−1.127E-3	3.99
0.0625	−0.86574353	−2.819E-4	4.00
0.03125	−0.86595493	−7.048E-5	4.00

Or we could have chosen more evaluation points, as in

$$f''(t) \doteq Af(t + 3h) + Bf(t + 2h) + Cf(t + h) + Df(t) \quad (7.81)$$

The extra degree of freedom could have been used to obtain a more accurate approximation to $f''(t)$, by forcing the error term to be proportional to a higher power of h. Both of the possibilities (7.80) and (7.81) are explored in problem 9.

Many of the formulas derived by the method of undetermined coefficients can also be derived by differentiating and evaluating a suitably chosen interpolation polynomial. But often, it is easier to visualize the desired formula as a combination of certain function values and to then derive the proper combination, as was done above for (7.78).

Effects of Error in Function Values

The formulas derived above are useful for differentiating functions that are known analytically and for setting up numerical methods for solving differential equations. Nonetheless, they are very sensitive to errors in the function values, especially if these errors are not sufficiently small compared to the stepsize h used in the differentiation formula. To explore this, we analyze the effect of such errors in the formula $D_h^{(2)}f(t)$ approximating $f''(t)$.

Rewrite (7.78) as

$$D_h^{(2)}f(x_1) = \frac{f(x_2) - 2f(x_1) + f(x_0)}{h^2} \doteq f''(x_1) \quad (7.82)$$

where $x_2 = x_1 + h$, $x_0 = x_1 - h$. Let the actual function values used in the computation be denoted by $\bar{f}_0$, $\bar{f}_1$, and $\bar{f}_2$ with

$$f(x_i) - \bar{f}_i = \epsilon_i, \quad i = 0, 1, 2 \quad (7.83)$$

the errors in the function values. Thus the actual quantity calculated is

$$\tilde{D}_h^{(2)} f(x_1) = \frac{\tilde{f}_2 - 2\tilde{f}_1 + \tilde{f}_0}{h^2} \tag{7.84}$$

For the error in this quantity, replace $\tilde{f}_j$ by $f(x_j) - \epsilon_j$, $j = 0, 1, 2$, to obtain

$$f''(x_1) - \tilde{D}_h^{(2)} f(x_1)$$

$$= f''(x_1) - \frac{[f(x_2) - \epsilon_2] - 2[f(x_1) - \epsilon_1] + [f(x_0) - \epsilon_0]}{h^2}$$

$$= \left[f''(x_1) - \frac{f(x_2) - 2f(x_1) + f(x_0)}{h^2} \right] + \frac{\epsilon_2 - 2\epsilon_1 + \epsilon_0}{h^2}$$

$$\doteq \frac{-h^2}{12} f^{(4)}(x_1) + \frac{\epsilon_2 - 2\epsilon_1 + \epsilon_0}{h^2} \tag{7.85}$$

The last step used (7.79).

The errors ϵ_0, ϵ_1, ϵ_2 are generally random in some interval $[-\delta, \delta]$. If the values $\tilde{f}_0$, $\tilde{f}_1$, $\tilde{f}_2$ are experimental data, then δ is a bound on the experimental error. Also, if these function values $\tilde{f}_j$ are obtained from computing $f(x)$ in a computer, then the errors ϵ_j are the combination of rounding or chopping errors and δ is a bound on these errors. In either case, (7.85) yields

$$|f''(x_1) - \tilde{D}_h^{(2)} f(x_1)| \le \frac{h^2}{12} |f^{(4)}(x_1)| + \frac{4\delta}{h^2} \tag{7.86}$$

This error bound suggests that as $h \to 0$, the error will eventually increase, because of the final term $4\delta/h^2$.

Example

Calculate $\tilde{D}_h^{(2)} f(x_1)$ for $f(x) = \cos(x)$ at $x_1 = \pi/6$. To show the effect of rounding errors, the values $\tilde{f}_i$ are obtained by rounding $f(x_i)$ to six significant digits; and the errors satisfy

$$|\epsilon_i| \le 5.0 \times 10^{-7} = \delta, \qquad i = 0, 1, 2$$

Other than these rounding errors, the formula $\tilde{D}_h^{(2)} f(x_1)$ is calculated exactly. The results are shown in Table 7.12. In this example, the bound (7.86) becomes

$$|f''(x_1) - \tilde{D}_h^{(2)}f(x_1)| \leq \frac{h^2}{12} \cos{(\pi/6)} + \left(\frac{4}{h^2}\right) (5 \times 10^{-7})$$

$$= 0.0722h^2 + \frac{2 \times 10^{-6}}{h^2} \equiv E(h) \quad (7.87)$$

For $h = 0.0625$, the bound $E(h) = 0.000794$, which is not too far off from the actual error given in the table.

The bound $E(h)$ indicates that there is a smallest value of h, call it h^*, below which the error bound will begin to increase. To find it, let $E'(h) = 0$, with its root being h^*. This leads to $h^* = 0.0726$, which is consistent with the behavior of the errors in Table 7.12.

One must be very cautious in using numerical differentiation, because of the sensitivity to errors in the function values. This is especially true if the function values are obtained empirically with relatively large experimental errors, as is common in practice. In this latter case, you should probably use a carefully prepared package program for numerical differentiation. Such programs take into account the error in the data, finding numerical derivatives that are as accurate as can be justified by the data. In the absence of such a program, you should consider producing a cubic spline function that approximates the data, and then use its derivative as a numerical derivative for the data. The cubic spline function could be based on interpolation; or better for data with relatively large errors, construct a cubic spline that is a *least squares approximation* to the data. The concept of least squares approximation is introduced in section 8.6 of Chapter 8.

Table 7.12 Calculation of $\tilde{D}_h^{(2)}f(x_1)$, Involving Rounding Errors

h	$\tilde{D}_h^{(2)}f(x_1)$	Error
0.5	−0.848128	−0.017897
0.25	−0.861504	−0.004521
0.125	−0.864832	−0.001193
0.0625	−0.865536	−0.000489
0.03125	−0.865280	−0.000745
0.015625	−0.860160	−0.005865
0.0078125	−0.851968	−0.014057
0.00390625	−0.786432	−0.079593

PROBLEMS

1. In the following cases, find the numerical derivative $D_h f(x)$ at the indicated point, using formula (7.65). Use $h = 0.1, 0.05, 0.025, 0.0125, 0.00625$. As in Table 7.9, calculate the error and the ratio with which the error decreases. Also estimate the error by using (7.66) with c replaced by x.
   (a) $f(x) = e^x$ at $x = 0$
   (b) $f(x) = \tan^{-1}(x^2 - x + 1)$ at $x = 1$
   (c) $f(x) = \tan^{-1}(100x^2 - 199x + 100)$ at $x = 1$

2. Repeat problem 1, but use the numerical derivative $D_h f(x_1)$ from (7.70) and estimate the error using (7.72) with c_2 replaced by x_1.

3. Let $h > 0$, $x_j = x_0 + jh$ for $j = 0, 1, 2, \ldots, n$, and let $P_n(x)$ be the degree n polynomial interpolating $f(x)$ at $x_0, \ldots, x_n$. Use this polynomial to estimate $f'(x_0)$. Produce the actual formulas involving $f(x_0), \ldots, f(x_n)$ for $n = 1, 2, 3, 4$. Also produce their error formulas.

4. Use the degree 4 polynomial $P_4(x)$, defined in problem 3, to find an approximation to $f'(x_2)$. Also find its error formula.

5. The error formulas (7.66), (7.72), and (7.79) all are proportional to a power of h. These justify the use of Richardson extrapolation, as in section 7.2. For (7.65), derive the extrapolation formula

$$f'(x) \doteq 2D_h f(x) - D_{2h} f(x)$$

   Derive corresponding extrapolation formulas for (7.70) and (7.78), based on the error formulas (7.72) and (7.79).

6. Use the extrapolation formula for (7.65), given in problem 5, to improve the answers given in Table 7.9. Produce a table of extrapolated values from Table 7.9; include the errors in the extrapolated values and the ratios by which they decrease.

7. Use the polynomial $P_2(x)$ preceding (7.69) to obtain an estimate for $f''(x_1)$.

8. Use $D_h^{(2)} f(x)$ from (7.78) to estimate $f''(x)$ for the functions and points in problem 1. Also calculate the errors and the ratios by which they decrease. Use $h = 0.5, 0.25, 0.125, 0.0625, 0.03125$.

9. (a) Use the method of undetermined coefficients to derive the formula (7.80).
   (b) Use the method of undetermined coefficients to derive the formula (7.81), with the error as small as possible.

10. Repeat the example summarized in Table 7.12, but use $f(x) = e^x$ and $x_1 = 0$.

11. Repeat the rounding error analysis that led to (7.85), but do it for $D_h f(t)$ in (7.70).

12. On your computer with single-precision arithmetic, perform the numerical differentiation $D_h^{(2)} f(t)$ of (7.78). Successively halve h, calculating $D_h^{(2)} f(t)$ and its error; continue this until the error begins to increase.

13. Using the following table of rounded values of $f(x)$, estimate $f''(0.5)$ numerically with stepsizes $h = 0.2, 0.1$. Also estimate the possible size of that part of the error in your answer that is due to the rounding errors in the table entries. Is this a serious source of error in this case?

x	$f(x)$	x	$f(x)$
0.3	7.3891	0.6	7.6141
0.4	7.4633	0.7	7.6906
0.5	7.5383		

EIGHT

SOLUTION OF SYSTEMS OF LINEAR EQUATIONS

Systems of simultaneous linear equations occur in solving problems in a wide variety of areas, including mathematics, statistics, the physical, biological, and social sciences, engineering, and business. They arise directly in solving real-world problems, and they also occur as part of the solution process for other problems, for example, solving systems of simultaneous nonlinear equations. In this chapter, we will look at the solution of some of the more commonly occurring systems of linear equations.

The notation and theory for linear systems is given in the first section, and a general method of solution is given in section 8.2. Most work with systems of linear equations is simpler to express and understand when matrix algebra is used, and this is introduced in section 8.3. Using matrix algebra, sections 8.4 and 8.5 extend the ideas of section 8.2 for solving linear systems, considering special types of linear systems and the effects of rounding errors. Section 8.6 defines the least squares data fitting method, which leads to some well-known linear systems. Section 8.7 introduces the matrix eigenvalue problem, along with one numerical method for its solution.

8.1   SYSTEMS OF LINEAR EQUATIONS

One of the topics studied in elementary algebra is the solution of pairs of linear equations,

$$\begin{aligned} ax + by &= c \\ dx + ey &= f \end{aligned} \tag{8.1}$$

The coefficients a, b, ..., f are given constants, and the task is to find the unknown values x, y. In this chapter, we look at the more general problem of finding solutions to larger systems of equations, containing more equations and unknowns.

To write the most general system of linear equations that we will study, we must change the notation used in (8.1) to something more convenient. Letting n denote the number of equations, the general form for a system of n linear equations in the n unknowns $x_1, x_2, \ldots, x_n$ is

$$\begin{aligned} a_{1,1}x_1 + a_{1,2}x_2 + a_{1,3}x_3 + \ldots + a_{1,n}x_n &= b_1 \\ a_{2,1}x_1 + a_{2,2}x_2 + a_{2,3}x_3 + \ldots + a_{2,n}x_n &= b_2 \\ &\vdots \\ a_{n,1}x_1 + a_{n,2}x_2 + a_{n,3}x_3 + \ldots + a_{n,n}x_n &= b_n \end{aligned} \tag{8.2}$$

The coefficients are given symbolically by a_{ij}, with i the number of the equation and j the number of the associated unknown. The right-hand sides $b_1, \ldots, b_n$ are given constants; one is to calculate the unknowns $x_1, \ldots, x_n$. The linear system is said to be of *order n*.

Example

Define the coefficients of (8.2) by

$$a_{ij} = \max\{i, j\} \tag{8.3}$$

for $1 \le i \le n$, $1 \le j \le n$. Also define the right sides by

$$b_1 = b_2 = \ldots = b_n = 1 \tag{8.4}$$

The solution of this linear system (8.2) to (8.4) is

$$x_1 = x_2 = \ldots = x_{n-1} = 0, \qquad x_n = \frac{1}{n} \tag{8.5}$$

This can also be shown to be the only solution to this linear system. To make the example more concrete, let $n = 3$. Then (8.2) to (8.4) is

$$
\begin{aligned}
x_1 + 2x_2 + 3x_3 &= 1 \\
2x_1 + 2x_2 + 3x_3 &= 1 \\
3x_1 + 3x_2 + 3x_3 &= 1
\end{aligned}
\qquad (8.6)
$$

and its solution is

$$
x_1 = x_2 = 0, \qquad x_3 = \frac{1}{3} \qquad (8.7)
$$

The linear system of equations (8.2) is completely specified by knowing the coefficients a_{ij} and the right-hand constants b_i. Partly as a notational convenience, these coefficients are given as elements of an array of numbers called a *matrix:*

$$
A = \begin{bmatrix}
a_{11} & a_{12} & a_{13} & \cdots & a_{1n} \\
a_{21} & a_{22} & a_{23} & \cdots & a_{2n} \\
& & \vdots & & \\
a_{n1} & a_{n2} & a_{n3} & \cdots & a_{nn}
\end{bmatrix}, \qquad
b = \begin{bmatrix}
b_1 \\
b_2 \\
\vdots \\
b_n
\end{bmatrix}
\qquad (8.8)
$$

The letters A and b are the names given to these matrices. The indices of a_{ij} now give the numbers of the row and column of A that contain a_{ij}. The solution $x_1, \ldots, x_n$ is written similarly,

$$
x = \begin{bmatrix}
x_1 \\
x_2 \\
\vdots \\
x_n
\end{bmatrix}
\qquad (8.9)
$$

In using a computer program to solve the linear system (8.2), the user specifies the system by defining the two matrices in (8.8).

Example

For (8.6) to (8.7),

$$
A = \begin{bmatrix}
1 & 2 & 3 \\
2 & 2 & 3 \\
3 & 3 & 3
\end{bmatrix}, \qquad
b = \begin{bmatrix}
1 \\
1 \\
1
\end{bmatrix}, \qquad
x = \begin{bmatrix}
0 \\
0 \\
\frac{1}{3}
\end{bmatrix}
\qquad (8.10)
$$

In the theory of the solvability of the linear system (8.2), an important role is played by the special system obtained by letting all of the constants $b_i = 0$. With this definition of b, the system is called *homogeneous*; all other systems are called nonhomogeneous. The following theorem summarizes some important theoretical results about linear systems. Its proof can be based on the numerical method to be presented in section 8.2, but it will be omitted here.

Theorem 8.1

Let n be a positive integer, and let A be given as in (8.8). Then the following are equivalent statements about the linear system (8.2) of order n.

S1. For each right side b, the system (8.2) has exactly one solution x.

S2. For each right side b, the system (8.2) has at least one solution x.

S3. The homogeneous form of (8.2) has exactly one solution,

$$x_1 = x_2 = \ldots = x_n = 0$$

S4. Det $(A) \neq 0$.

Det (A) denotes the determinant of A. Many people have studied the determinant in beginning algebra, in association with solving (8.1), but we omit any theoretical discussion of it. Matrices A for which det $(A) = 0$ are called *singular*, and those with det $(A) \neq 0$ are called *nonsingular*.

For A of order $n = 2$,

$$A = \begin{bmatrix} a_{11} & a_{12} \\ a_{21} & a_{22} \end{bmatrix}$$

and the determinant of A is

$$\det (A) = a_{11}a_{22} - a_{12}a_{21} \tag{8.11}$$

Thus the linear system of order 2 associated with this A will be uniquely solvable for all right-hand sides b if and only if

$$a_{11}a_{22} - a_{12}a_{21} \neq 0$$

which is an easy condition to check.

This chapter considers the solution of nonsingular systems of linear equations. From (S1) in the above theorem, such systems have a unique solution $x_1, \ldots, x_n$ for each set of right-hand constants $b_1, \ldots, b_n$. The remaining parts of the theorem, (S2) to (S4), are equivalent to (S1); but it is often easier to show that (S1) is true for your system by instead showing (S2), (S3), or (S4). For example, questions involving the existence and uniqueness of interpolating polynomials are often treated by first reformulating them as questions involving linear systems of the form (8.2); then, these systems are often treated using (S3). We omit such examples here, but they occur commonly in higher level numerical analysis.

Computing with Matrices

The matrices in (8.8) were used to organize or represent the coefficients needed to define the linear system (8.2). If the linear system is to be solved using a computer program, then these matrices must be stored in arrays (indexed variables) in the computer. Such arrays are first introduced into a program by using a DIMENSION statement to specify the number of rows and columns in the array, implicitly specifying the total amount of memory that will be needed to store the array. When writing a program that uses arrays, the DIMENSION statement should reserve the largest amount of memory that might be needed (subject to reasonable upper limits, of course).

To illustrate the storage of matrices in a computer program, we give a short section of computer code, creating the matrices associated with (8.3), (8.4).

In this code, the matrix of coefficients (8.8) has order N, and there is an upper limit of $N \le 20$ based on the space reserved in the DIMENSION statement. Other examples will be given in the following sections.

```
      DIMENSION A(20,20),B(20)
C
      PRINT *,' WHAT IS ORDER N?'
      PRINT *,' IT SHOULD BE .LE. 20.'
      READ *, N
C
      DO 10 I=1,N
        B(I)=1.0
        DO 10 J=1,N
          A(I,J)=MAX(I,J)
10          CONTINUE
```

PROBLEMS

1. (a) Consider the problem of cubic polynomial interpolation,

$$p(x_i) = y_i, \quad i = 0, 1, 2, 3$$

   with degree $(p) \leq 3$ and x_0, x_1, x_2, x_3 distinct. Convert the problem of finding $p(x)$ to another problem involving the solution of a system of linear equations.
   Hint: Write

$$p(x) = a_0 + a_1 x + a_2 x^2 + a_3 x^3$$

   and determine $a_0, a_1, a_2,$ and a_3.
   (b) Write the matrices of (8.8) for the system in (a).

2. Recall the equations (5.69) to (5.70) for the interpolating cubic spline function. Assuming the interpolation node points are evenly spaced, with a spacing of h, set up the matrices of (8.8) for this linear system.

3. Consider the matrices

$$A = \begin{bmatrix} 2 & 1 & 0 & 0 & \cdots\cdots\cdots & 0 \\ 1 & 2 & 1 & 0 & & 0 \\ 0 & 1 & 2 & 1 & 0 & 0 \\ \vdots & & & \ddots & & \vdots \\ \vdots & & & & & 0 \\ \vdots & & & & 1 & 2 & 1 \\ 0 & \cdots\cdots\cdots & & 0 & 1 & 2 \end{bmatrix}, \quad b = \begin{bmatrix} 3 \\ 4 \\ 4 \\ \vdots \\ \vdots \\ 4 \\ 3 \end{bmatrix}, \quad x = \begin{bmatrix} 1 \\ 1 \\ \vdots \\ \vdots \\ \vdots \\ \vdots \\ 1 \end{bmatrix}$$

   Set up the linear system (8.2) of order n associated with A and b. Verify that the given x solves this linear system.

4. In imitation of the Fortran code of this section, write a section of code to define the following matrices A and b.

$$a_{ij} = \begin{cases} \dfrac{i}{j}, & 1 \leq i \leq j \leq n \\[2mm] \dfrac{j}{i}, & 1 \leq j \leq i \leq n \end{cases} \qquad b_i = \begin{cases} -1, & i \text{ even} \\ 1, & i \text{ odd} \end{cases}$$

5. Repeat problem 4 with the following matrices A and b.

$$a_{1j} = a_{j1} = \beta, \quad j = 1, 2, \ldots, n$$
$$a_{ij} = a_{i-1,j} + a_{i,j-1}, \quad i, j = 2, 3, \ldots, n$$
$$b_i = (-1)^{i-1}/i, \quad i = 1, \ldots, n$$

   Let β be an input parameter to the program, $\beta \neq 0$. Write out a specific example when $\beta = 1$.

6. For linear systems of order 2, verify that (S1) is equivalent to (S4) in Theorem 8.1.

7. Recall the reformulation of cubic polynomial interpolation given in problem 1. Show that the linear system for a_0, a_1, a_2, a_3 always has a unique solution by proving the equivalent statement (S3) in Theorem 8.1.

    Hint: Statement (S3) can be interpreted as a statement about zeros of a cubic polynomial. How many zeros can a cubic polynomial possess?

8.2 GAUSSIAN ELIMINATION

To solve large systems of linear equations, we use a structured form of the method taught in beginning algebra courses. It is introduced here by first using it to solve a particular system of order 3. Consider the system

$$
\begin{array}{lll}
x_1 + 2x_2 + x_3 \ \ \ = 0 & \quad E(1) & \\
2x_1 + 2x_2 + 3x_3 = 3 & \quad E(2) & \quad (8.12) \\
-x_1 - 3x_2 \ \ \ \ \ \ \ \ = 2 & \quad E(3) &
\end{array}
$$

The equations have been labeled for easier reference.

Step 1 Eliminate x_1 from E(2) and E(3). Subtract 2 times E(1) from E(2); and subtract -1 times E(1) from E(3). This results in the new system

$$
\begin{array}{lll}
x_1 + 2x_2 + x_3 = 0 & \quad E(1) & \\
-2x_2 + x_3 = 3 & \quad E(2) & \quad (8.13) \\
-x_2 + x_3 = 2 & \quad E(3) &
\end{array}
$$

Step 2 Eliminate x_2 from E(3). Subtract $\tfrac{1}{2}$ times E(2) from E(3). This yields

$$
\begin{array}{lll}
x_1 + 2x_2 + x_3 = 0 & \quad E(1) & \\
-2x_2 + x_3 = 3 & \quad E(2) & \quad (8.14) \\
\tfrac{1}{2}x_3 = \tfrac{1}{2} & \quad E(3) &
\end{array}
$$

Step 3 In succession, solve for x_3, x_2, and x_1.

$$x_3 = 1$$
$$-2x_2 + 1 = 3$$
$$x_2 = -1$$
$$x_1 + 2(-1) + 1 = 0$$
$$x_1 = 1$$

Steps 1 and 2 are the *elimination* steps, resulting in (8.14), which is called an upper triangular system of linear equations. This system has exactly the same solutions as (8.12), but (8.14) is in a form that is easier to solve. Step 3 is called solution by *back substitution*. The entire process is called *Gaussian elimination*, and it is generally the most efficient means of solving a system of linear equations.

In algebra courses, variables are often eliminated in a somewhat ad hoc manner. One is taught to look for simple integer coefficients to use in the elimination, in order to minimize calculations with fractions. But on a computer, dealing with fractions is no more difficult than using integers. To write a computer program, the algorithm must have a precisely stated set of directions, ones that do not depend on whether or not a coefficient happens to be an integer.

To see the formal structure of Gaussian elimination, we again look at the $n = 3$ case, but with general coefficients. The system to be solved is

$$a_{11}x_1 + a_{12}x_2 + a_{13}x_3 = b_1 \qquad \text{E}(1)$$
$$a_{21}x_1 + a_{22}x_2 + a_{23}x_3 = b_2 \qquad \text{E}(2) \qquad (8.15)$$
$$a_{31}x_1 + a_{32}x_2 + a_{33}x_3 = b_3 \qquad \text{E}(3)$$

Step 1 Eliminate x_1 from E(2) and E(3). To simplify the presentation, assume $a_{11} \neq 0$; this assumption will be removed later in the section. Define

$$m_{21} = \frac{a_{21}}{a_{11}}, \qquad m_{31} = \frac{a_{31}}{a_{11}} \qquad (8.16)$$

Subtract m_{21} times E(1) from E(2), and subtract m_{31} times E(1) from E(3). This changes (8.15) to the equivalent system

$$
\begin{aligned}
a_{11}x_1 + a_{12}x_2 + a_{13}x_3 &= b_1 & \text{E(1)} \\
a_{22}^{(2)}x_2 + a_{23}^{(2)}x_3 &= b_2^{(2)} & \text{E(2)} \\
a_{32}^{(2)}x_2 + a_{33}^{(2)}x_3 &= b_3^{(2)} & \text{E(3)}
\end{aligned} \tag{8.17}
$$

The coefficients $a_{ij}^{(2)}$ are defined by

$$
\begin{aligned}
a_{ij}^{(2)} &= a_{ij} - m_{i1}a_{1j}, & i, j &= 2, 3 \\
b_i^{(2)} &= b_i - m_{i1}b_1, & i &= 2, 3
\end{aligned} \tag{8.18}
$$

Step 2 Eliminate x_2 from E(3). Again assume temporarily that $a_{22}^{(2)} \neq 0$. Define

$$
m_{32} = \frac{a_{32}^{(2)}}{a_{22}^{(2)}} \tag{8.19}
$$

Subtract m_{32} times E(2) from E(3). This yields

$$
\begin{aligned}
a_{11}x_1 + a_{12}x_2 + a_{13}x_3 &= b_1 & \text{E(1)} \\
a_{22}^{(2)}x_2 + a_{23}^{(2)}x_3 &= b_2^{(2)} & \text{E(2)} \\
a_{33}^{(3)}x_3 &= b_3^{(3)} & \text{E(3)}
\end{aligned} \tag{8.20}
$$

The new coefficients are defined by

$$
a_{33}^{(3)} = a_{33}^{(2)} - m_{32}a_{23}^{(2)}, \qquad b_3^{(3)} = b_3^{(2)} - m_{32}b_2^{(2)}
$$

Step 3 Use back substitution to solve successively for x_3, x_2, and x_1.

$$
\begin{aligned}
x_3 &= b_3^{(3)}/a_{33}^{(3)} \\
x_2 &= (b_2^{(2)} - a_{23}^{(2)}x_3)/a_{22}^{(2)} \\
x_1 &= (b_1 - a_{12}x_2 - a_{13}x_3)/a_{11}
\end{aligned} \tag{8.21}
$$

This algorithm for $n = 3$ is easily extended to one for a general nonsingular system of n linear equations. We will again assume that certain elements will be nonzero, but this assumption will be removed later in the section. Also, for uniformity in the presentation, denote the original system by

$$a_{11}^{(1)}x_1 + \ldots + a_{1n}^{(1)}x_n = b_1^{(1)} \qquad E(1)$$

$$\vdots \qquad\qquad \vdots \qquad\qquad (8.22)$$

$$a_{n1}^{(1)}x_1 + \ldots + a_{nn}^{(1)}x_n = b_n^{(1)} \qquad E(n)$$

For $k = 1, 2, \ldots, n - 1$ carry out the following elimination step.

Step k   Eliminate x_k from $E(k + 1)$ through $E(n)$. The results of the preceding steps $1, \ldots, k - 1$ will have yielded a system of the form

$$a_{11}^{(1)}x_1 + a_{12}^{(1)}x_2 + \ldots\ldots\ldots\ldots + a_{1n}^{(1)}x_n = b_1^{(1)} \qquad E(1)$$

$$a_{22}^{(2)}x_2 + \ldots\ldots\ldots\ldots + a_{2n}^{(2)}x_n = b_2^{(2)} \qquad E(2)$$

$$\vdots \qquad\qquad\qquad (8.23)$$

$$a_{kk}^{(k)}x_k + \ldots + a_{kn}^{(k)}x_n = b_k^{(k)} \qquad E(k)$$

$$\vdots \qquad\qquad \vdots$$

$$a_{nk}^{(k)}x_k + \ldots + a_{nn}^{(k)}x_n = b_n^{(k)} \qquad E(n)$$

Assume $a_{kk}^{(k)} \neq 0$, and define the *multipliers*

$$m_{ik} = a_{ik}^{(k)}/a_{kk}^{(k)}, \qquad i = k + 1, \ldots, n \qquad (8.24)$$

For equations $i = k + 1, \ldots, n$, subtract m_{ik} times $E(k)$ from $E(i)$, eliminating x_k from $E(i)$. The new coefficients in $E(k + 1)$ through $E(n)$ are defined by

$$a_{ij}^{(k+1)} = a_{ij}^{(k)} - m_{ik}a_{kj}^{(k)}, \qquad i, j = k + 1, \ldots, n \qquad (8.25)$$

$$b_i^{(k+1)} = b_i^{(k)} - m_{ik}b_k^{(k)}, \qquad i = k + 1, \ldots, n$$

When step $n - 1$ is completed, the linear system will be in upper triangular form. We will denote it by

$$u_{11}x_1 + \ldots + u_{1n}x_n = g_1$$

$$\vdots \qquad\qquad (8.26)$$

$$u_{nn}x_n = g_n$$

These coefficients are related to the earlier ones by

$$u_{ij} = a_{ij}^{(i)}, \qquad g_i = b_i^{(i)}$$

Step n Solve (8.26) using back substitution. Solve successively for x_n, $x_{n-1}, \ldots, x_1$ in (8.26).

$$x_n = g_n/u_{nn}$$

$$x_i = \left(g_i - \sum_{j=i+1}^{n} u_{ij}x_j \right)/u_{ii}, \qquad i = n-1, \ldots, 1 \tag{8.27}$$

This completes the definition of Gaussian elimination.

Before giving another example of Gaussian elimination, note that at every step, only the coefficients in the linear system are changing. The unknowns remain, except when the coefficient is zero, and the unknowns serve primarily as position markers for the coefficients. For that reason, we could equally well define Gaussian elimination using just the coefficient matrices A and b (8.22), modifying their rows by the operations given in steps $1, \ldots, n-1$.

Example

Solve

$$
\begin{aligned}
4x_1 + 3x_2 + 2x_3 + x_4 &= 1 \\
3x_1 + 4x_2 + 3x_3 + 2x_4 &= 1 \\
2x_1 + 3x_2 + 4x_3 + 3x_4 &= -1 \\
x_1 + 2x_2 + 3x_3 + 4x_4 &= -1
\end{aligned}
\tag{8.28}
$$

For convenience in showing the solution of this system, matrices are written in a combined form called the *augmented matrix*. We first give the results of steps 1, 2, 3 of Gaussian elimination for this system.

$$
\left[\begin{array}{cccc|c}
4 & 3 & 2 & 1 & 1 \\
3 & 4 & 3 & 2 & 1 \\
2 & 3 & 4 & 3 & -1 \\
1 & 2 & 3 & 4 & -1
\end{array}\right]
\quad
\begin{array}{l}
m_{21} = \dfrac{3}{4} \\[6pt]
\longrightarrow \\[6pt]
m_{31} = \dfrac{1}{2} \\[6pt]
m_{41} = \dfrac{1}{4}
\end{array}
\quad
\left[\begin{array}{cccc|c}
4 & 3 & 2 & 1 & 1 \\[4pt]
0 & \dfrac{7}{4} & \dfrac{3}{2} & \dfrac{5}{4} & \dfrac{1}{4} \\[6pt]
0 & \dfrac{3}{2} & 3 & \dfrac{5}{2} & \dfrac{-3}{2} \\[6pt]
0 & \dfrac{5}{4} & \dfrac{5}{2} & \dfrac{15}{4} & \dfrac{-5}{4}
\end{array}\right]
$$

$$
m_{32} = \dfrac{6}{7} \quad\downarrow\quad m_{42} = \dfrac{5}{7}
$$

$$
\left[\begin{array}{cccc|c}
4 & 3 & 2 & 1 & 1 \\[4pt]
0 & \dfrac{7}{4} & \dfrac{3}{2} & \dfrac{5}{4} & \dfrac{1}{4} \\[6pt]
0 & 0 & \dfrac{12}{7} & \dfrac{10}{7} & \dfrac{-12}{7} \\[6pt]
0 & 0 & 0 & \dfrac{5}{3} & 0
\end{array}\right]
\quad
\begin{array}{l}
\longleftarrow \\[6pt]
m_{43} = \dfrac{5}{6}
\end{array}
\quad
\left[\begin{array}{cccc|c}
4 & 3 & 2 & 1 & 1 \\[4pt]
0 & \dfrac{7}{4} & \dfrac{3}{2} & \dfrac{5}{4} & \dfrac{1}{4} \\[6pt]
0 & 0 & \dfrac{12}{7} & \dfrac{10}{7} & \dfrac{-12}{7} \\[6pt]
0 & 0 & \dfrac{10}{7} & \dfrac{20}{7} & \dfrac{-10}{7}
\end{array}\right]
$$

Using back substitution to solve successively for x_4, x_3, x_2, x_1, we obtain the solution

$$
x_1 = 0, \qquad x_2 = 1, \qquad x_3 = -1, \qquad x_4 = 0
$$

Partial Pivoting

In the elimination, it was assumed that $a_{kk}^{(k)} \neq 0$ at each of the steps $k = 1$, $2, \ldots, n-1$. To remove this assumption when $a_{kk}^{(k)}$ equals zero, examine the equations following $E(k)$ in (8.23). From the assumption that the original system (8.22) is nonsingular, it can be shown that one of the equations following $E(k)$ must contain a term involving x_k with a nonzero coefficient (see problem 4). This equation can be interchanged with $E(k)$ to obtain a new system in which $a_{kk}^{(k)} \neq 0$ is satisfied. But having $a_{kk}^{(k)}$ be nonzero is not sufficient, as the following example demonstrates.

Example

Solve

$$6x_1 + 2x_2 + 2x_3 = -2$$

$$2x_1 + \frac{2}{3}x_2 + \frac{1}{3}x_3 = 1 \qquad (8.29)$$

$$x_1 + 2x_2 - x_3 = 0$$

This has the solution

$$x_1 = 2.6, \qquad x_2 = -3.8, \qquad x_3 = -5.0 \qquad (8.30)$$

We use a decimal computer to solve this system. The computer uses a floating-point representation with four digits in the mantissa, and all operations will be rounded. The augmented matrix will be used to display the steps of the elimination, beginning with the machine representation of (8.29).

$$\begin{bmatrix} 6.000 & 2.000 & 2.000 & \bigm| & -2.000 \\ 2.000 & 0.6667 & 0.3333 & \bigm| & 1.000 \\ 1.000 & 2.000 & -1.000 & \bigm| & 0.0 \end{bmatrix}$$

$$\downarrow \quad m_{21} = 0.3333, \qquad m_{31} = 0.1667$$

$$\begin{bmatrix} 6.000 & 2.000 & 2.000 & \bigm| & -2.000 \\ 0.0 & 0.0001000 & -0.3333 & \bigm| & 1.667 \\ 0.0 & 1.667 & -1.333 & \bigm| & 0.3334 \end{bmatrix} \qquad (8.31)$$

$$\downarrow \quad m_{32} = 16670$$

$$\begin{bmatrix} 6.000 & 2.000 & 2.000 & \bigm| & -2.000 \\ 0.0 & 0.0001000 & -0.3333 & \bigm| & 1.667 \\ 0.0 & 0.0 & 5555 & \bigm| & -27790 \end{bmatrix}$$

Using back substitution, we obtain the following approximate solution:

$$x_1 = 1.335, \qquad x_2 = 0.0, \qquad x_3 = -5.003$$

Compare it with the true solution in (8.30).

The difficulty with this elimination process is that in (8.31), the element in row 2, column 2 should have been zero, but rounding error prevented it. This means the coefficient in this (2,2) position had essentially infinite relative error, and this was carried through into computations involving this coefficient. To avoid this, interchange rows 2 and 3 in (8.31) and then continue the elimination. The final matrix row equals

$$\begin{bmatrix} 6.000 & 2.000 & 2.000 & | & -2.000 \\ 0.0 & 1.667 & -1.333 & | & 0.3334 \\ 0.0 & 0.0 & -0.3332 & | & 1.667 \end{bmatrix}$$

and the final multiplier was $m_{32} = 0.00005999$. With back substitutions we then obtain the approximate solution

$$x_1 = 2.602, \qquad x_2 = -3.801, \qquad x_3 = -5.003$$

This compares well with the true solution in (8.30).

To avoid the problem presented by this example, we use the following strategy. At step k [see (8.23)], calculate

$$c = \max_{k \leq i \leq n} |a_{ik}^{(k)}|$$

This is the maximum size of the elements in column k of the coefficient matrix of step k, beginning at row k and going downward. If the element $|a_{kk}^{(k)}| < c$, then interchange $E(k)$ with one of the following equations, to obtain a new equation $E(k)$ in which $|a_{kk}^{(k)}| = c$. This strategy makes $a_{kk}^{(k)}$ as far away from zero as possible. The element $a_{kk}^{(k)}$ is called the *pivot element* for step k of the elimination, and the process described in this paragraph is called *partial pivoting*, or more simply, pivoting. There are other forms of pivoting, but partial pivoting is quite satisfactory in practice.

Partial pivoting has been introduced here in order to avoid using coefficients that are nearly zero as pivot elements. But there is another equally important reason to use pivoting: in most cases, it decreases the propagated effects of rounding errors. With partial pivoting, the multipliers m_{ik} in (8.24) will satisfy

$$|m_{ik}| \leq 1, \qquad 1 \leq k < i \leq n$$

This will help reduce loss-of-significance errors, because multiplications by m_{ik} will not lead to much larger numbers.

Example

Using the four decimal place computer described in the last example, solve

$$0.729x_1 + 0.81x_2 + 0.9x_3 = 0.6867$$
$$x_1 + x_2 + x_3 = 0.8338$$
$$1.331x_1 + 1.21x_2 + 1.1x_3 = 1.000$$

Its exact solution, rounded to four places, is

$$x_1 = 0.2245, \qquad x_2 = 0.2814, \qquad x_3 = 0.3279 \qquad (8.32)$$

Solution Without Pivoting

$$\begin{bmatrix} 0.7290 & 0.8100 & 0.9000 & 0.6867 \\ 1.000 & 1.000 & 1.000 & 0.8338 \\ 1.331 & 1.210 & 1.100 & 1.000 \end{bmatrix}$$

$$\downarrow \quad \begin{matrix} m_{21} = 1.372 \\ m_{31} = 1.826 \end{matrix}$$

$$\begin{bmatrix} 0.7290 & 0.8100 & 0.9000 & 0.6867 \\ 0.0 & -0.1110 & -0.2350 & -0.1084 \\ 0.0 & -0.2690 & -0.5430 & -0.2540 \end{bmatrix}$$

$$\downarrow \quad m_{32} = 2.423$$

$$\begin{bmatrix} 0.7290 & 0.8100 & 0.9000 & 0.6867 \\ 0.0 & -0.1110 & -0.2350 & -0.1084 \\ 0.0 & 0.0 & 0.02640 & 0.008700 \end{bmatrix}$$

The solution is

$$x_1 = 0.2251, \qquad x_2 = 0.2790, \qquad x_3 = 0.3295 \qquad (8.33)$$

Solution with Pivoting   To indicate the interchange of rows i and j, we will use the notation $r_i \leftrightarrow r_j$.

$$\begin{bmatrix} 0.7290 & 0.8100 & 0.9000 & \bigg| & 0.6867 \\ 1.000 & 1.000 & 1.000 & \bigg| & 0.8338 \\ 1.331 & 1.210 & 1.100 & \bigg| & 1.000 \end{bmatrix}$$

$$r_1 \leftrightarrow r_3 \quad \Bigg\downarrow \quad \begin{array}{l} m_{21} = 0.7513 \\ m_{31} = 0.5477 \end{array}$$

$$\begin{bmatrix} 1.331 & 1.210 & 1.100 & \bigg| & 1.000 \\ 0.0 & 0.09090 & 0.1736 & \bigg| & 0.08250 \\ 0.0 & 0.1473 & 0.2975 & \bigg| & 0.1390 \end{bmatrix}$$

$$r_2 \leftrightarrow r_3 \quad \Bigg\downarrow \quad m_{32} = 0.6171$$

$$\begin{bmatrix} 1.331 & 1.210 & 1.100 & \bigg| & 1.000 \\ 0.0 & 0.1473 & 0.2975 & \bigg| & 0.1390 \\ 0.0 & 0.0 & -0.01000 & \bigg| & -0.003280 \end{bmatrix}$$

The solution is

$$x_1 = 0.2246, \qquad x_2 = 0.2812, \qquad x_3 = 0.3280 \qquad (8.34)$$

Comparing this to (8.32), we see that this is a much more accurate solution than (8.33).

Operations Count

It is important to know how long a computation will last and, for that reason, we count the number of arithmetic operations involved in Gaussian elimination. For reasons that will be apparent in later sections, the count will be divided into three parts. Also, as a notational convenience, let U and g denote the coefficient matrices in (8.26).

1. *The elimination step.* We count the additions/subtractions, multiplications, and divisions in going from the system (8.22) to the triangular system (8.26). We consider only the operations for the coefficients of A and not for the right-hand side b in (8.22). Table 8.1 contains the number of these operations for each of the steps in the elimination. The totals in the last row of the table are obtained using the formulas

$$\sum_{j=1}^{p} j = \frac{p(p+1)}{2}, \qquad \sum_{j=1}^{p} j^2 = \frac{p(p+1)(2p+1)}{6}, \qquad p \geq 1 \quad (8.35)$$

Table 8.1 Operations Count for $A \rightarrow U$

Step	Additions	Multiplications	Divisions
1	$(n-1)^2$	$(n-1)^2$	$n-1$
2	$(n-2)^2$	$(n-2)^2$	$n-2$
$\vdots$	$\vdots$	$\vdots$	$\vdots$
$n-2$	2^2	2^2	2
$n-1$	1	1	1
Total	$\dfrac{n(n-1)(2n-1)}{6}$	$\dfrac{n(n-1)(2n-1)}{6}$	$\dfrac{n(n-1)}{2}$

Generally, the divisions and multiplications are counted together since they are about the same in operation time. Doing this gives us

$$AS(A \rightarrow U) = \frac{n(n-1)(2n-1)}{6}$$

$$MD(A \rightarrow U) = \frac{n(n-1)(2n-1)}{6} + \frac{n(n-1)}{2} = \frac{n(n^2-1)}{3} \quad (8.36)$$

$AS(\cdot)$ denotes additions and subtractions for the enclosed operation, and $MD(\cdot)$ denotes multiplications and divisions. The $A \rightarrow U$ denotes the elimination process that converts A in (8.22) into U in (8.26).

2. *Modification of the right side b to g.* Proceeding as before, we get

$$AS(b \rightarrow g) = (n-1) + (n-2) + \ldots + 1 = \frac{n(n-1)}{2}$$
$$\quad (8.37)$$
$$MD(b \rightarrow g) = (n-1) + (n-2) + \ldots + 1 = \frac{n(n-1)}{2}$$

3. *The back substitution step, find* x *from (8.26).* As before,

$$AS(g \rightarrow x) = 0 + 1 + \ldots + (n-1) = \frac{n(n-1)}{2}$$
$$\quad (8.38)$$
$$MD(g \rightarrow x) = 1 + 2 + \ldots + n = \frac{n(n+1)}{2}$$

Combining these results, we see that the total number of operations to obtain x is

$$
\begin{aligned}
AS(x) &= AS(A \rightarrow U) + AS(b \rightarrow g) + AS(g \rightarrow x) \\
&= \frac{n(n-1)(2n-1)}{6} + \frac{n(n-1)}{2} + \frac{n(n-1)}{2} \\
&= \frac{n(n-1)(2n+5)}{6}
\end{aligned}
\tag{8.39}
$$

$$
MD(x) = \frac{n(n^2 + 3n - 1)}{3}
$$

Since AS and MD are almost the same in all of these counts, only MD is discussed; these are also the more expensive operations in running time. For larger values of n, the operation count for Gaussian elimination is about $\frac{1}{3}n^3$. This means that as n is doubled, the cost of solving the linear system goes up by a factor of 8. In addition, most of the cost of Gaussian elimination is in the elimination step, $A \rightarrow U$, since for the remaining steps

$$
MD(b \rightarrow g) + MD(g \rightarrow x) = \frac{n(n-1)}{2} + \frac{n(n+1)}{2} = n^2 \tag{8.40}
$$

Thus, once the $A \rightarrow U$ step has been completed, it is much less expensive to solve the linear system. We return to this later in section 8.5, where we give a simple and inexpensive way to estimate the error in the answer obtained by Gaussian elimination.

Computer Programs

A great deal of mathematical research and program development has been carried out on producing efficient and accurate programs for Gaussian elimination, and your computer center should possess such a code. Some of the best such codes are contained in the book *Linpack: User's Guide* by Dongarra, Bunch, Moler, and Stewart (1979). Good programs will carry out the elimination in the most efficient manner and will arrange operations to minimize rounding errors. They will also offer error estimates, and this is needed because many linear systems are not as well-behaved as our examples have been.

We give a subroutine LINSYS implementing Gaussian elimination as described in this section. It is intended as a pedagogical tool; actual production computing should use a code of the type described in the preceding paragraph. The comment statements in LINSYS describe the values of the input and output variables. The use of matrix notation is based on ideas introduced in the next section.

```
C        TITLE: THIS IS A DEMO PROGRAM FOR SUBROUTINE LINSYS.
C
C        IT WILL SOLVE A LINEAR SYSTEM A*X=B, GIVEN BY THE USER.
C
         DIMENSION A(10,10),B(10)
C
C        INPUT ORDER OF LINEAR SYSTEM.
10       PRINT *, ' GIVE THE ORDER OF THE LINEAR SYSTEM.'
         READ *, N
         IF(N .EQ. 0) STOP
C
C        INPUT LINEAR SYSTEM.
         PRINT *, ' GIVE THE LINEAR SYSTEM, ONE EQUATION AT A TIME.'
         PRINT *, ' CONCLUDE EACH EQUATION WITH ITS RIGHT-HAND CONSTANT.'
         DO 20 I=1,N
            PRINT *, ' GIVE COEFFICIENTS OF EQUATION',I
20          READ *, (A(I,J),J=1,N),B(I)
C
C        SOLVE THE LINEAR SYSTEM.
         CALL LINSYS(A,B,N,10,IER)
C
C        PRINT THE RESULTS.
         PRINT 1000, N,IER
1000     FORMAT(///,' N=',I2,5X,'IER=',I3,//,'   I          SOLUTION',/)
         PRINT 1001, (I,B(I),I=1,N)
1001     FORMAT(I3,1PE20.10)
         GO TO 10
         END

         SUBROUTINE LINSYS(MAT,B,N,MD,IER)
C
C        THIS ROUTINE SOLVES A SYSTEM OF LINEAR EQUATIONS
C                A*X = B
C        THE METHOD USED IS GAUSSIAN ELIMINATION WITH
C        PARTIAL PIVOTING.
C
C        INPUT:
C        THE COEFFICIENT MATRIX A IS STORED IN THE ARRAY MAT.
C        THE RIGHT SIDE CONSTANTS ARE IN THE ARRAY B.
C        THE ORDER OF THE LINEAR SYSTEM IS N.
C        THE VARIABLE MD IS THE NUMBER OF ROWS THAT MAT
C        IS DIMENSIONED AS HAVING IN THE CALLING PROGRAM.
C
C        OUTPUT:
C        THE ARRAY B CONTAINS THE SOLUTION X.
C        MAT CONTAINS THE UPPER TRIANGULAR MATRIX U
C        OBTAINED BY ELIMINATION. THE ROW MULTIPLIERS
C        USED IN THE ELIMINATION ARE STORED IN THE
C        LOWER TRIANGULAR PART OF MAT.
C        IER=0 MEANS THE MATRIX A WAS COMPUTATIONALLY
C        NONSINGULAR, AND THE GAUSSIAN ELIMINATION
C        WAS COMPLETED SATISFACTORILY.
C        IER=1 MEANS THAT THE MATRIX A WAS
C        COMPUTATIONALLY SINGULAR.
C
```

```
      INTEGER PIVOT
      REAL MULT,MAT
      DIMENSION MAT(MD,*),B(*),PIVOT(100)

C
C     BEGIN ELIMINATION STEPS.
      DO 40 K=1,N-1
C       CHOOSE PIVOT ROW.
        PIVOT(K)=K
        AMAX=ABS(MAT(K,K))
        DO 10 I=K+1,N
          ABSA=ABS(MAT(I,K))
          IF(ABSA .GT. AMAX) THEN
              PIVOT(K)=I
              AMAX=ABSA
            END IF
10        CONTINUE
C
        IF(AMAX .EQ. 0.0) THEN
C           COEFFICIENT MATRIX IS SINGULAR.
            IER=1
            RETURN
          END IF
C
        IF(PIVOT(K) .NE. K) THEN
C           SWITCH ROWS K AND PIVOT(K).
            I=PIVOT(K)
            TEMP=B(K)
            B(K)=B(I)
            B(I)=TEMP
            DO 20 J=K,N
              TEMP=MAT(K,J)
              MAT(K,J)=MAT(I,J)
20            MAT(I,J)=TEMP
          END IF
C
C       PERFORM STEP #K OF ELIMINATION.
        DO 30 I=K+1,N
          MULT=MAT(I,K)/MAT(K,K)
          MAT(I,K)=MULT
          B(I)=B(I)-MULT*B(K)
          DO 30 J=K+1,N
30          MAT(I,J)=MAT(I,J)-MULT*MAT(K,J)
40      CONTINUE
C
      IF(MAT(N,N) .EQ. 0.0) THEN
C         COEFFICIENT MATRIX IS SINGULAR.
          IER=1
          RETURN
        END IF
C
C     SOLVE FOR SOLUTION X USING BACK SUBSTITUTION.
      DO 60 I=N,1,-1
        SUM=0.0
        DO 50 J=I+1,N
50        SUM=SUM+MAT(I,J)*B(J)
```

60   B(I)=(B(I)-SUM)/MAT(I,I)
            IER=0
            RETURN
            END

PROBLEMS

1. Solve the following linear systems using Gaussian elimination without pivoting. The systems are specified by A and b, with the system then written as in (8.22).

   (a)
   $$A = \begin{bmatrix} 2 & 1 & -1 \\ 4 & 0 & -1 \\ -8 & 2 & 2 \end{bmatrix}, \quad b = \begin{bmatrix} 6 \\ 6 \\ -8 \end{bmatrix}$$

   (b)
   $$A = \begin{bmatrix} 2 & 1 & -1 & -2 \\ 4 & 4 & 1 & 3 \\ -6 & -1 & 10 & 10 \\ -2 & 1 & 8 & 4 \end{bmatrix}, \quad b = \begin{bmatrix} 2 \\ 4 \\ -5 \\ 1 \end{bmatrix}$$

   (c)
   $$A = \begin{bmatrix} 1 & -1 & 2 \\ -1 & 5 & 4 \\ 2 & 4 & 29 \end{bmatrix}, \quad b = \begin{bmatrix} 1 \\ -3 \\ 15 \end{bmatrix}$$

2. Implement the subroutine LINSYS given in this section. Using it, solve the linear systems of order n whose coefficients are given in (8.3) to (8.4) of section 8.1. Do so for various values of n, say $n = 2$, 5, 10, 20.

3. (a) Modify LINSYS to print out the final coefficients stored in MAT, obtained from the elimination performed on the original coefficient matrix A. Apply it to the systems given in problem 1.

   (b) Which statements would have to be deleted from LINSYS if partial pivoting is to be eliminated?

4. In step 2 for the $n = 3$ case (8.15), suppose that both $a_{22}^{(2)}$ and $a_{32}^{(2)}$ are zero. Then the system (8.17) becomes

   $$a_{11}x_1 + a_{12}x_2 + a_{13}x_3 = b_1$$
   $$a_{23}^{(2)}x_3 = b_2^{(2)}$$
   $$a_{33}^{(2)}x_3 = b_3^{(2)}$$

   Show that this system is not uniquely solvable, but rather has either no solution or an infinity of solutions. The same type of argument can be used with larger systems to show there must be a nonzero pivot element for a nonsingular system.

5. Investigate the linear equation solvers available from your computer center. Using one of these, repeat problem 2.

6. Repeat problem 2 or 5 for the systems with the following coefficients.

(a)
$$A = \begin{bmatrix} 5 & 7 & 6 & 5 \\ 7 & 10 & 8 & 7 \\ 6 & 8 & 10 & 9 \\ 5 & 7 & 9 & 10 \end{bmatrix}, \quad b = \begin{bmatrix} 1 \\ -1 \\ -1 \\ 1 \end{bmatrix}, \quad x = \begin{bmatrix} 136 \\ -82 \\ -35 \\ 21 \end{bmatrix}$$

(b)
$$A = \begin{bmatrix} 1 & \frac{1}{2} & \frac{1}{3} & \frac{1}{4} \\ \frac{1}{2} & \frac{1}{3} & \frac{1}{4} & \frac{1}{5} \\ \frac{1}{3} & \frac{1}{4} & \frac{1}{5} & \frac{1}{6} \\ \frac{1}{4} & \frac{1}{5} & \frac{1}{6} & \frac{1}{7} \end{bmatrix}, \quad b = \begin{bmatrix} 1 \\ -1 \\ 1 \\ -1 \end{bmatrix}, \quad x = \begin{bmatrix} 516 \\ -5700 \\ 13620 \\ -8820 \end{bmatrix}$$

The true answers are given for comparison with your calculated answers. Note any anomalies in the computed values.

7. In the system (8.22), let $a_{ij} = 0$ whenever $i - j \geq 2$. Write out the general form of this system. Use Gaussian elimination to solve it, taking advantage of the elements that are known to be zero. Do an operations count for $A \to U$ in this case.

8.3 MATRIX ARITHMETIC

In working with linear systems, it is very useful to introduce arithmetic operations for the matrices that were introduced and used in the first two sections. Such matrix operations can simplify many derivations, and they lead to a conceptually clearer notation for writing linear systems of equations. In this section, the arithmetic of matrices will be studied and related to the solution of systems of linear equations.

A matrix is a rectangular array of numbers,

$$B = \begin{bmatrix} b_{11} & b_{12} & \cdots & b_{1n} \\ \vdots & & & \vdots \\ b_{m1} & b_{m2} & \cdots & b_{m,n} \end{bmatrix} \tag{8.41}$$

It is said to have *order* $m \times n$ if m is the number of rows and n is the number of columns. Square matrices with n rows and n columns are said to have order n. Matrices consisting of a single row or column are called row and column matrices, respectively. They are also called row or column vectors, or more simply, vectors. This is because they can be identified with geometric vectors drawn from the origin of a space to the point with coordinates given by the row or column matrix. This adds an important geometric perspective to the study of matrices, but we will omit it here.

As notation, capital letters will be used to denote matrices, except for row and column matrices; and lowercase letters will usually be used to denote matrix coefficients. Two matrices will be said to be equal if (i) they have the same order, and (ii) corresponding coefficients are the same. Finally, the *transpose* of the general matrix in (8.41) is

$$
B^T = \begin{bmatrix} b_{11} & \cdots & b_{m1} \\ \vdots & & \vdots \\ b_{1n} & \cdots & b_{mn} \end{bmatrix}
\tag{8.42}
$$

It is obtained by interchanging the rows and columns of B; the order of B^T is $n \times m$.

Example

$$
A \equiv \begin{bmatrix} a & b & 1 \\ 2 & c & d \end{bmatrix} = \begin{bmatrix} 5 & 6 & 1 \\ 2 & 2 & a + 1 \end{bmatrix}
$$

implies

$$
a = 5, \qquad b = 6, \qquad c = 2, \qquad d = a + 1 = 6
$$

Also

$$
A^T = \begin{bmatrix} a & 2 \\ b & c \\ 1 & d \end{bmatrix}
$$

A has order 2×3, and A^T has order 3×2.

Arithmetic Operations

Let B be the $m \times n$ matrix in (8.41) and let α be an arbitrary real number. Then αB is a matrix of order $m \times n$, defined by

$$\alpha B = \begin{bmatrix} \alpha b_{11} & \alpha b_{12} & \cdots & \alpha b_{1n} \\ \vdots & & & \vdots \\ \alpha b_{m1} & \alpha b_{m2} & \cdots & \alpha b_{m,n} \end{bmatrix} \tag{8.43}$$

Each element of B is multiplied by α to get the corresponding element in αB.

Let A and B be matrices of order $m \times n$. Then $A + B$ is a new matrix of order $m \times n$, defined by

$$A + B = \begin{bmatrix} a_{11} + b_{11} & a_{12} + b_{12} & \cdots & a_{1n} + b_{1n} \\ \vdots & & & \vdots \\ a_{m1} + b_{m1} & a_{m2} + b_{m2} & \cdots & a_{mn} + b_{mn} \end{bmatrix} \tag{8.44}$$

The (i, j) element of $A + B$ is $a_{ij} + b_{ij}$. For matrix addition, there is a zero matrix. Define the zero matrix of order $m \times n$ as having all zero entries. It is denoted by $O_{m \times n}$, or more commonly, O. It has the property that

$$A + O = A \tag{8.45}$$

for any matrix A.

Examples

$$-5 \begin{bmatrix} 2 & 3 \\ 6 & -1 \end{bmatrix} = \begin{bmatrix} -10 & -15 \\ -30 & 5 \end{bmatrix}$$

$$\begin{bmatrix} 3 & 2 & 1 \\ 1 & 2 & 3 \end{bmatrix} + \begin{bmatrix} 1 & 2 & 3 \\ 3 & 2 & 1 \end{bmatrix} = \begin{bmatrix} 4 & 4 & 4 \\ 4 & 4 & 4 \end{bmatrix} = 4 \begin{bmatrix} 1 & 1 & 1 \\ 1 & 1 & 1 \end{bmatrix}$$

$$\begin{bmatrix} 2 & 3 \\ -1 & 5 \end{bmatrix} + \begin{bmatrix} -2 & -3 \\ 1 & -5 \end{bmatrix} = \begin{bmatrix} 0 & 0 \\ 0 & 0 \end{bmatrix} = O$$

Matrix multiplication is a more complicated operation to define and to calculate. Let A have order $m \times n$ and B have order $n \times p$. Then

$C = AB$, the product of A and B, is a matrix of order $m \times p$, and its general element c_{ij} is defined by

$$c_{ij} = \sum_{k=1}^{n} a_{ik}b_{kj}, \qquad 1 \le i \le m, \qquad 1 \le j \le p \qquad (8.46)$$

This is the sum of the products of corresponding elements from row i of A and column j of B. If A and B are square matrices, then AB is a square matrix of the same order. And if A is $m \times n$ and B is a column matrix of order $n \times 1$, then AB is a column matrix of order $m \times 1$.

Examples

(a)

$$\begin{bmatrix} 2 & 3 & 2 \\ -1 & 2 & -1 \end{bmatrix} \begin{bmatrix} 1 & 0 \\ 2 & -1 \\ 3 & 1 \end{bmatrix} = \begin{bmatrix} 14 & -1 \\ 0 & -3 \end{bmatrix}$$

$$\begin{bmatrix} 1 & 0 \\ 2 & -1 \\ 3 & 1 \end{bmatrix} \begin{bmatrix} 2 & 3 & 2 \\ -1 & 2 & -1 \end{bmatrix} = \begin{bmatrix} 2 & 3 & 2 \\ 5 & 4 & 5 \\ 5 & 11 & 5 \end{bmatrix}$$

(b)

$$\begin{bmatrix} 3 & 2 & 1 \\ 1 & 2 & 3 \\ 1 & -1 & 1 \end{bmatrix} \begin{bmatrix} 1 \\ 1 \\ 1 \end{bmatrix} = \begin{bmatrix} 6 \\ 6 \\ 1 \end{bmatrix}$$

$$\begin{bmatrix} 3 & 2 & 1 \\ 1 & 2 & 3 \\ 1 & -1 & 1 \end{bmatrix} \begin{bmatrix} x_1 \\ x_2 \\ x_3 \end{bmatrix} = \begin{bmatrix} 3x_1 + 2x_2 + x_3 \\ x_1 + 2x_2 + 3x_3 \\ x_1 - x_2 + x_3 \end{bmatrix}$$

As suggested by the last example, a system of linear equations can be written as a matrix equation by using matrix multiplication. For example, the system (8.12),

$$x_1 + 2x_2 + x_3 = 0$$
$$2x_1 + 2x_2 + 3x_3 = 3$$
$$-x_1 - 3x_2 = 2$$

can be written as

$$
\begin{bmatrix} 1 & 2 & 1 \\ 2 & 2 & 3 \\ -1 & -3 & 0 \end{bmatrix} \begin{bmatrix} x_1 \\ x_2 \\ x_3 \end{bmatrix} = \begin{bmatrix} 0 \\ 3 \\ 2 \end{bmatrix}
$$

The reader should check that (8.12) will follow after performing the matrix multiplication on the left side and matching the corresponding elements on the two sides of the equation.

The general linear system

$$
\begin{aligned}
a_{11}x_1 + \ldots + a_{1n}x_n &= b_1 \\
&\vdots \\
a_{n1}x_1 + \ldots + a_{nn}x_n &= b_n
\end{aligned} \tag{8.47}
$$

can be written as the matrix equation

$$
Ax = b \tag{8.48}
$$

with A, x, and b the usual coefficient matrices introduced in section 8.1. When written in this way, many results become conceptually clearer. Also, results from the theory of matrices and linear algebra can be applied to (8.48), to obtain both theoretical results and computational methods for the linear system (8.47).

The Matrix Inverse

The number 1 satisfies

$$
1 \cdot x = x \cdot 1 = x
$$

for all real numbers x. With matrix multiplication, the analog of 1 is the identity matrix I. Let $n \geq 1$, and define I_n to be the square matrix of order n whose (i, j) element is

$$
\delta_{ij} = \begin{cases} 1, & i = j \\ 0, & i \neq j \end{cases}
$$

With $n = 3$,

$$I_3 = \begin{bmatrix} 1 & 0 & 0 \\ 0 & 1 & 0 \\ 0 & 0 & 1 \end{bmatrix}$$

The most important property of the identity matrix is

$$AI_n = A, \qquad I_m A = A \tag{8.49}$$

where A has order $m \times n$. This is the analog of the above identity with the real number 1.

There is an identity matrix of each order $n \geq 1$. But instead of writing I_n, the subscript n is usually omitted and I denotes any possible identity, with the order chosen suitably for the given situation. For example, (8.49) would be written as

$$AI = IA = A$$

Again in analogy with the real numbers, we can generalize the concept of the reciprocal of a number. Let A be a square matrix. If there is a square matrix B such that

$$AB = I, \qquad BA = I \tag{8.50}$$

then B is called the *inverse* of A. It is shown later that if A has an inverse, then it has exactly one inverse. The inverse of A is denoted by A^{-1}, in analogy with the reciprocal of a real number. Moreover, only one of the above statements about B need be true in order for the other to be also true.

Example

(a)

$$A = \begin{bmatrix} 2 & 1 & 0 \\ 1 & 2 & 1 \\ 0 & 1 & 2 \end{bmatrix}, \qquad A^{-1} = \begin{bmatrix} \dfrac{3}{4} & -\dfrac{1}{2} & \dfrac{1}{4} \\[2mm] -\dfrac{1}{2} & 1 & -\dfrac{1}{2} \\[2mm] \dfrac{1}{4} & -\dfrac{1}{2} & \dfrac{3}{4} \end{bmatrix}$$

(b)   Let

$$A = \begin{bmatrix} a & b \\ c & d \end{bmatrix}$$

Then it can be shown that A has an inverse if and only if $ad - bc \neq 0$, and in that case

$$A^{-1} = \frac{1}{ad - bc} \begin{bmatrix} d & -b \\ -c & a \end{bmatrix} \qquad (8.51)$$

In both (a) and (b), check the formula for A^{-1} by multiplying with A and seeing whether the product is the identity matrix.

With real numbers, the number b has a reciprocal (or inverse) if and only if $b \neq 0$. For matrices, this generalizes as follows.

Theorem 8.2

Let A be a square matrix. Then A has an inverse if and only if $\det (A) \neq 0$.

Referring back to Theorem 8.1, this result says that the existence of A^{-1} is equivalent to any of the statements (S1) to (S4) of that earlier theorem. Thus A has an inverse if and only if the linear system $Ax = b$ has a unique solution x for all right-hand column vectors b.

To calculate A^{-1}, we can use the Gaussian elimination method of section 8.2. To explain what is involved, consider matrices of order $n = 3$. Let X denote the unknown inverse matrix, $X = A^{-1}$, and let the three columns of X be denoted by x_{*1}, x_{*2}, x_{*3}. Also, let the columns of I_3 be denoted by $e_1, e_2,$ and e_3,

$$e_1 = \begin{bmatrix} 1 \\ 0 \\ 0 \end{bmatrix}, \quad e_2 = \begin{bmatrix} 0 \\ 1 \\ 0 \end{bmatrix}, \quad e_3 = \begin{bmatrix} 0 \\ 0 \\ 1 \end{bmatrix} \qquad (8.52)$$

The statement

$$AX = I$$

becomes

$$\begin{bmatrix} a_{11} & a_{12} & a_{13} \\ a_{21} & a_{22} & a_{23} \\ a_{31} & a_{32} & a_{33} \end{bmatrix} \begin{bmatrix} x_{11} & x_{12} & x_{13} \\ x_{21} & x_{22} & x_{23} \\ x_{31} & x_{32} & x_{33} \end{bmatrix} = \begin{bmatrix} 1 & 0 & 0 \\ 0 & 1 & 0 \\ 0 & 0 & 1 \end{bmatrix} \qquad (8.53)$$

Consider only the first column of the product. It is

$$\begin{bmatrix} a_{11}x_{11} + a_{12}x_{21} + a_{13}x_{31} \\ a_{21}x_{11} + a_{22}x_{21} + a_{23}x_{31} \\ a_{31}x_{11} + a_{32}x_{21} + a_{33}x_{31} \end{bmatrix} = Ax_{*1}$$

Similarly, the second and third columns of AX are Ax_{*2} and Ax_{*3}. Thus (8.53) can be written as

$$AX = [Ax_{*1}, Ax_{*2}, Ax_{*3}] = [e_1, e_2, e_3] \qquad (8.54)$$

Matching corresponding columns, we get

$$Ax_{*1} = e_1, \qquad Ax_{*2} = e_2, \qquad Ax_{*3} = e_3 \qquad (8.55)$$

Thus the columns of X are the solutions of three simultaneous linear systems, all with the same matrix of coefficients A. Gaussian elimination can be used to solve these systems.

Example

Find the inverse of

$$A = \begin{bmatrix} 1 & 1 & -1 \\ 1 & 2 & -2 \\ -2 & 1 & 1 \end{bmatrix}$$

Perform Gaussian elimination on the augmented matrix

$$\left[\begin{array}{ccc|ccc} 1 & 1 & -1 & 1 & 0 & 0 \\ 1 & 2 & -2 & 0 & 1 & 0 \\ -2 & 1 & 1 & 0 & 0 & 1 \end{array} \right]$$

This simultaneously takes account of the three systems in (8.55). The steps of the elimination are

$$
\left[\begin{array}{ccc|ccc}
1 & 1 & -1 & 1 & 0 & 0 \\
0 & 1 & -1 & -1 & 1 & 0 \\
0 & 3 & -1 & 2 & 0 & 1
\end{array}\right]
\begin{array}{l}
m_{21} = 1 \\
m_{31} = -2
\end{array}
$$

$$\downarrow$$

$$
\left[\begin{array}{ccc|ccc}
1 & 1 & -1 & 1 & 0 & 0 \\
0 & 1 & -1 & -1 & 1 & 0 \\
0 & 0 & 2 & 5 & -3 & 1
\end{array}\right]
\quad m_{32} = 3
$$

Using back substitution, we solve the system with x_{*1} as its solution. This is

$$x_{11} + x_{21} - x_{31} = 1$$
$$x_{21} - x_{31} = -1$$
$$2x_{31} = 5$$

The solution is

$$x_{11} = 2, \qquad x_{21} = \tfrac{3}{2}, \qquad x_{31} = \tfrac{5}{2}$$

The second and third columns can be found similarly. And then

$$
X = A^{-1} =
\begin{bmatrix}
2 & -1 & 0 \\
\dfrac{3}{2} & -\dfrac{1}{2} & \dfrac{1}{2} \\
\dfrac{5}{2} & -\dfrac{3}{2} & \dfrac{1}{2}
\end{bmatrix}
$$

This can be checked by multiplication by A, to obtain the identity.

The above procedure generalizes to matrices A of arbitrary order $n \geq 0$. Augment A by the identity of the same order and then apply Gaussian elimination in the manner illustrated above.

Matrix Algebra Rules

Many of the rules of arithmetic with real numbers are still valid for arithmetic with matrices, but there are some important differences. We simply list these rules, illustrating and showing a few of them. The matrices in these statements are not restricted to be square, except when referring to inverses and determinants.

$$(A + B) + C = A + (B + C)$$ Associative laws   (8.56)
$$(AB)C = A(BC)$$ (8.57)

$$A + B = B + A$$ Commutative law   (8.58)

$$A(B + C) = AB + AC$$ Distributive law   (8.59)
$$(A + B)C = AC + BC$$

$$(AB)^T = B^T A^T$$ (8.60)
$$(A + B)^T = A^T + B^T$$

$$(cA)^{-1} = \frac{1}{c} A^{-1},$$ $c = \text{constant}$ (8.61)
$$(AB)^{-1} = B^{-1} A^{-1}$$

$$\det (AB) = \det (A) \det (B)$$ (8.62)

$$\det (A^T) = \det (A)$$ (8.63)
$$\det (cA) = c^n \det (A),$$ $\text{order } (A) = n, c = \text{constant}$

The result (8.57) is useful and important and, for that reason, we look at its proof. Let A, B, and C have orders $m \times n$, $n \times p$, and $p \times q$, respectively. Then AB has order $m \times p$, and BC has order $n \times q$. In the following, D_{ij} will denote the (i, j) element in the matrix D. Then

$$[(AB)C]_{ij} = \sum_{l=1}^{p} (AB)_{il} C_{lj} = \sum_{l=1}^{p} \left[\sum_{k=1}^{n} A_{ik} B_{kl} \right] C_{lj}$$

$$[A(BC)]_{ij} = \sum_{k=1}^{n} A_{ik} (BC)_{kj} = \sum_{k=1}^{n} A_{ik} \left[\sum_{l=1}^{p} B_{kl} C_{lj} \right]$$

These two right sides are the same, being rearrangements of the sum of all products $A_{ik} B_{kl} C_{lj}$ for $1 \leq k \leq n$, $1 \leq l \leq p$. This proves (6.57).

Example

Show that if A has an inverse, then it is unique. To show this, suppose there are two such inverses B_1 and B_2, with

$$B_1A = AB_1 = I, \qquad B_2A = AB_2 = I$$

Examine B_1AB_2. First,

$$(B_1A)B_2 = B_1(AB_2)$$

Then,

$$(B_1A)B_2 = IB_2 = B_2$$
$$B_1(AB_2) = B_1I = B_1$$

Putting these together proves $B_1 = B_2$ and, thus, A can have only one inverse.

Example

Let

$$A = \begin{bmatrix} 1 & 2 \\ 3 & 4 \end{bmatrix}, \qquad B = \begin{bmatrix} 1 & -1 \\ 1 & 1 \end{bmatrix}$$

Then

$$\det(A) = -2, \qquad \det(B) = 2$$
$$AB = \begin{bmatrix} 3 & 1 \\ 7 & 1 \end{bmatrix}$$
$$\det(AB) = -4 = \det(A)\det(B)$$

This illustrates (8.62).

Missing from the above list of rules is the commutative law for multiplication,

$$AB = BA \tag{8.64}$$

because this is not true for most square matrices A and B. Those matrices that do satisfy (8.64) are said to *commute* with one another.

Example

Use the matrices A and B from the last example. Then

$$AB = \begin{bmatrix} 3 & 1 \\ 7 & 1 \end{bmatrix}, \quad BA = \begin{bmatrix} -2 & -2 \\ 4 & 6 \end{bmatrix}$$

Solving Linear Systems

Consider solving the linear system of equations

$$Ax = b$$

where A has order n and is nonsingular. Then A^{-1} exists and

$$A^{-1}(Ax) = A^{-1}b \qquad (8.65)$$
$$x = A^{-1}b$$

Thus, if A^{-1} is known, then x can be found by a matrix multiplication.

From this, it might at first seem reasonable to find A^{-1} and to solve for x using (8.65). But this is not an efficient procedure, because of the greater cost needed to find A^{-1}. The operations cost for finding A^{-1} can be shown to be

$$MD(A \rightarrow A^{-1}) = n^3 \qquad (8.66)$$

This is about three times the cost of finding x by the Gaussian elimination method of section 8.2. Even if we wish to solve $Ax = b$ with several different right sides b, it is no more efficient to use (8.65) than to use Gaussian elimination, doing the elimination step, $A \rightarrow U$ in (8.22) → (8.26), only once. From (8.40), the cost of elimination (when $A \rightarrow U$ has already been completed) is only n^2, and this is exactly the same as the number of multiplications to calculate $x = A^{-1}b$. Thus there is no savings in using A^{-1}.

The chief value of A^{-1} is as a theoretical tool for examining the solution of nonsingular systems of linear equations. With a few exceptions, one seldom needs to calculate A^{-1} explicitly.

PROBLEMS

1. Simplify the following matrix expressions to obtain a single matrix for each case.

   (a) $2 \begin{bmatrix} 1 & 0 \\ -1 & 3 \end{bmatrix} + \begin{bmatrix} 1 & 1 \\ 1 & -1 \end{bmatrix} \begin{bmatrix} 2 & 3 \\ 1 & 3 \end{bmatrix}$

   (b) $\begin{bmatrix} 1 & 2 \\ 0 & 3 \end{bmatrix} \begin{bmatrix} a & b \\ 0 & c \end{bmatrix}$

   (c) $A = I_3 - 2ww^T, \qquad w^T = [\frac{1}{3}, \frac{2}{3}, \frac{2}{3}]$

   $B = A^2$

   (d) $A = \begin{bmatrix} \cos(\theta) & \sin(\theta) \\ -\sin(\theta) & \cos(\theta) \end{bmatrix}, \qquad B = AA^T$

   (e)
   $$A = \begin{bmatrix} 2 & 1 & 0 & 0............0 \\ 1 & 2 & 1 & 0............0 \\ 0 & 1 & 2 & 1 & 0.........0 \\ \vdots & & & \ddots & & \vdots \\ \vdots & & & & & 0 \\ \vdots & & & & 1 & 2 & 1 \\ 0............& & & 0 & 1 & 2 \end{bmatrix}, \qquad B = A^2$$

2. Let A be an arbitrary square matrix of order 3, and let

   $$D = \begin{bmatrix} \lambda_1 & 0 & 0 \\ 0 & \lambda_2 & 0 \\ 0 & 0 & \lambda_3 \end{bmatrix}$$

   calculate AD and DA. The matrix D is called a *diagonal matrix*. Give a simple rule for multiplication of a general matrix by a diagonal matrix.

3. Let A be a square matrix of order n, and let it satisfy

   $$a_{ij} = 0 \qquad \text{for } i > j$$

   Such a matrix is called *upper triangular*. Show that the sum and products of such matrices are also upper triangular. Is the inverse upper triangular?

4. Let w be a column matrix for which $w^T w = 1$. The product $A = ww^T$ is a square matrix. Show that $A^2 = A$.

5. A matrix B is called *symmetric* if $B^T = B$. Write a general formula for all symmetric matrices of order 2×2.

6. Combining ideas from problems 4 and 5, define a new matrix

$$B = I - 2ww^T$$

where $w^Tw = 1$. Show that B is symmetric and that $B^2 = I$.

7. Let A be $m \times n$ and let B be $n \times p$. Do an operations count for calculating AB. Consider, in particular, the cases $m = n = p$ and $m = n, p = 1$.

8. Let A, B, and C have orders $m \times n, n \times p$, and $p \times q$, respectively. Do operation counts for the multiplications involved in calculating $(AB)C$ and $A(BC)$. Show with particular values of m, n, p, and q that these can be very different.

9. Produce two square matrices A and B of order 2 for which $AB = 0$, but A and B have all nonzero elements.

10. In order that

$$A = \begin{bmatrix} a & b \\ c & d \end{bmatrix}$$

will commute with

$$B = \begin{bmatrix} 1 & 2 \\ 2 & 1 \end{bmatrix}$$

what conditions must be satisfied by a, b, c, and d.

11. Derive the formula (8.51) for the inverse of an order 2 matrix.

12. Calculate the inverses of the following matrices.

(a) $\begin{bmatrix} 3 & 1 & 1 \\ 1 & 3 & 1 \\ 1 & 1 & 3 \end{bmatrix}$
(b) $\begin{bmatrix} 0 & 1 & 2 \\ 1 & 0 & 1 \\ 2 & 1 & 0 \end{bmatrix}$

(c) $\begin{bmatrix} -3 & 1 & 0 & 0 \\ 1 & -2 & 1 & 0 \\ 0 & 1 & -2 & 1 \\ 0 & 0 & 1 & -1 \end{bmatrix}$
(d) $\begin{bmatrix} 1 & 1 & 1 & 1 \\ 1 & 2 & 3 & 4 \\ 1 & 3 & 6 & 10 \\ 1 & 4 & 10 & 20 \end{bmatrix}$

13. Check that the inverse of the $n \times n$ matrix

$$A = \begin{bmatrix} 1 & -1 & 0 & \dots & & & & 0 \\ -1 & 2 & -1 & & & & & \vdots \\ 0 & -1 & 2 & -1 & & & & \vdots \\ \vdots & & -1 & \ddots & & & & \vdots \\ \vdots & & & & \ddots & & & 0 \\ \vdots & & & & & -1 & 2 & -1 \\ 0 & \dots & & & & 0 & -1 & 2 \end{bmatrix}$$

is given by

$$A^{-1} = \begin{bmatrix} n & n-1 & n-2 \ldots\ldots\ldots 1 \\ n-1 & n-1 & n-2 & n-3\ldots 1 \\ n-2 & n-2 & n-2 & n-3\ldots 1 \\ \vdots & & & \vdots \\ 1 & 1 & 1\ldots\ldots\ldots\ldots 1 \end{bmatrix}$$

14. Let A be $m \times n$ and B be $n \times p$. Show $(AB)^T = B^T A^T$.
   Hint: Follow the type of proof used for showing $(AB)C = A(BC)$, following (8.63).

15. Let A and B be nonsingular matrices of the same order. Show that AB is nonsingular and $(AB)^{-1} = B^{-1}A^{-1}$.
   Hint: Multiply by AB.

16. Using Theorem 8.1, if det $(A) = 0$, there is a vector $x \neq 0$ for which $Ax = 0$. For the following singular matrix, find such an x.

$$\begin{bmatrix} 0 & 4 & 1 & 1 \\ 4 & 0 & 1 & 1 \\ 1 & 1 & -1 & 2 \\ 1 & 1 & 2 & -1 \end{bmatrix}$$

17. (a) Find the values of λ for which

$$\det [\lambda I - B] = 0, \qquad B = \begin{bmatrix} 2 & 1 \\ 1 & 2 \end{bmatrix}$$

   (b) Using the λ values found in (a), and using the first sentence of problem 16 with $A = \lambda I - B$, find an x for which $(\lambda I - B)x = 0$ or, equivalently, $Bx = \lambda x$, with $x \neq 0$.

18. Consider the following matrices, called elementary matrices. Let I be the identity matrix of order n, and modify its column #k below the diagonal, for some $1 \leq k \leq n - 1$: Define

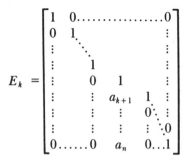

   (a) Show that the products of elementary matrices are lower triangular matrices.

(b)   If A is a matrix of order $n \times m$, show that $E_k A$ is formed from A as follows: Add a_i times row #k of A to row #i of A, forming the new row #i of $E_k A$, for $k + 1 \leq i \leq n$.

(c)   Calculate the inverse of E_k.
      Hint: Use the result of part (b), reversing the operations used in forming $E_k A$.

19.   Write a section of Fortran code to multiply two $n \times n$ matrices A and B. To minimize the effect of rounding errors, use the ideas embodied in the function SUMPRD given at the end of Chapter 3.

8.4   THE LU FACTORIZATION

Using matrix multiplication, another meaning can be given to the Gaussian elimination method of section 8.2. The matrix of coefficients of the linear system being solved can be factored into the product of two triangular matrices. We begin this section with that result, and then we discuss implications and applications of it. Let $Ax = b$ denote the system to be solved, as before, with A the $n \times n$ coefficient matrix. In the elimination steps (8.22) to (8.25) of section 8.2, the linear system was reduced to the upper triangular system $Ux = g$ of (8.26), with

$$
U = \begin{bmatrix} u_{11}, \dots . u_{1n} \\ 0 \ddots \vdots \\ \vdots \ddots \vdots \\ 0 \dots . . u_{nn} \end{bmatrix}
\tag{8.67}
$$

and $u_{ij} = a_{ij}^{(i)}$. Introduce an auxiliary lower triangular matrix L based on the multipliers m_{ij} of (8.24),

$$
L = \begin{bmatrix} 1, 0 \dots \dots . 0 \\ m_{21} \ddots \vdots \\ \vdots \ddots \vdots \\ m_{n1} \dots \dots m_{n,n-1} 1 \end{bmatrix}
\tag{8.68}
$$

The relationship of the matrices L and U to the original matrix A is given by the following theorem.

Theorem 8.3

Let A be a nonsingular matrix, and let L and U be defined by (8.67) and (8.68). Then, if U is produced without pivoting, as in (8.22) to (8.25), then

$$LU = A \tag{8.69}$$

This is called the *LU factorization* of A.

We omit the proof, but the $n = 3$ case is considered in problems 3 and 4. Also, there is a result analogous to (8.69) when pivoting is used, but it will not be needed here.

Example

Recall the system (8.28). Then

$$
L = \begin{bmatrix} 1 & 0 & 0 & 0 \\ \dfrac{3}{4} & 1 & 0 & 0 \\ \dfrac{1}{2} & \dfrac{6}{7} & 1 & 0 \\ \dfrac{1}{4} & \dfrac{5}{7} & \dfrac{5}{6} & 1 \end{bmatrix}, \quad
U = \begin{bmatrix} 4 & 3 & 2 & 1 \\ 0 & \dfrac{7}{4} & \dfrac{3}{2} & \dfrac{5}{4} \\ 0 & 0 & \dfrac{12}{7} & \dfrac{10}{7} \\ 0 & 0 & 0 & \dfrac{5}{3} \end{bmatrix} \tag{8.70}
$$

and

$$
LU = \begin{bmatrix} 4 & 3 & 2 & 1 \\ 3 & 4 & 3 & 2 \\ 2 & 3 & 4 & 3 \\ 1 & 2 & 3 & 4 \end{bmatrix} = A
$$

The *LU* factorization leads to another perspective on Gaussian elimination. From (8.69), solving $Ax = b$ is equivalent to solving $LUx = b$. And this is equivalent to solving the two systems

$$Lg = b, \quad Ux = g \tag{8.71}$$

$Ux = g$ is the upper triangular system whose solution was given in (8.27), based on back substitution. $Lg = b$ is the lower triangular system

$$
\begin{array}{rl}
g_1 & = b_1 \\
m_{21}g_1 + g_2 & = b_2 \\
\vdots & \vdots \\
m_{n1}g_1 + m_{n2}g_2 + \cdots + m_{n,n-1}g_{n-1} + g_n & = b_n
\end{array} \tag{8.72}
$$

It is solved by forward substitution, yielding the column matrix g that appears in $Ux = g$. Thus, once the factorization $A = LU$ is known, the solution of $Ax = b$ is reduced to solving two triangular systems, a relatively simple task.

Example

Again recall the system (8.28), with

$$
b = [1, 1, -1, -1]^T
$$

Use the factorization $A = LU$ given in (8.70). Then $Lg = b$ has the solution

$$
g = \left[1, \frac{1}{4}, -\frac{12}{7}, 0 \right]^T
$$

and $Ux = g$ has the solution

$$
x = [0, 1, -1, 0]^T
$$

The reader should check these results.

Compact Variants of Gaussian Elimination

Rather than constructing L and U using the elimination steps (8.22) to (8.25) of section 8.2, it is possible to solve directly for these matrices. The number of operations will be the same as with Gaussian elimination, but there are some advantages to such *compact variants* of Gaussian elimination.

To illustrate the direct computation of L and U, consider the $n = 3$ case. Write $A = LU$ as

$$\begin{bmatrix} a_{11} & a_{12} & a_{13} \\ a_{21} & a_{22} & a_{23} \\ a_{31} & a_{32} & a_{33} \end{bmatrix} = \begin{bmatrix} 1 & 0 & 0 \\ m_{21} & 1 & 0 \\ m_{31} & m_{32} & 1 \end{bmatrix} \begin{bmatrix} u_{11} & u_{12} & u_{13} \\ 0 & u_{22} & u_{23} \\ 0 & 0 & u_{33} \end{bmatrix} \quad (8.73)$$

Multiply L and U, and match the elements of the product with the corresponding elements in A. Doing this for the first row and column of the product yields

$$a_{11} = u_{11} \qquad a_{12} = u_{12} \qquad a_{13} = u_{13} \qquad (8.74)$$
$$a_{21} = m_{21}u_{11} \qquad a_{31} = m_{31}u_{11}$$

This gives the first column of L and the first row of U. Next, multiply row 2 of L to obtain

$$a_{22} = m_{21}u_{12} + u_{22} \qquad a_{23} = m_{21}u_{13} + u_{23} \qquad (8.75)$$

These can be solved for u_{22} and u_{23}. Next multiply row 3 of L to obtain

$$m_{31}u_{12} + m_{32}u_{22} = a_{32} \qquad (8.76)$$
$$m_{31}u_{13} + m_{32}u_{23} + u_{33} = a_{33}$$

These yield values for m_{32} and u_{33}, completing the construction of L and U. In this process, we must have $u_{11} \neq 0$, $u_{22} \neq 0$ in order to solve for L. There are modifications of the method to avoid this assumption, using pivoting, but we will simply consider those cases where the necessary elements will be nonzero.

Example

Let

$$A = \begin{bmatrix} 1 & 1 & -1 \\ 1 & 2 & -2 \\ -2 & 1 & 1 \end{bmatrix}$$

Then using (8.74) we get

$$u_{11} = a_{11} = 1 \qquad u_{12} = a_{12} = 1 \qquad u_{13} = a_{13} = -1$$
$$m_{21} = a_{21}/u_{11} = 1 \qquad m_{31} = a_{31}/u_{11} = -2$$

From (8.75) and (8.76),

$$u_{22} = a_{22} - m_{21}u_{12} = 2 - 1 \cdot 1 = 1$$
$$u_{23} = a_{23} - m_{21}u_{13} = -2 - 1 \cdot (-1) = -1$$
$$m_{32} = (a_{32} - m_{31}u_{12})/u_{22} = (1 - (-2) \cdot 1)/1 = 3$$
$$u_{33} = a_{33} - m_{31}u_{13} - m_{32}u_{23} = 1 - (-2)(-1) - 3(-1) = 2$$

Thus

$$A = \begin{bmatrix} 1 & 0 & 0 \\ 1 & 1 & 0 \\ -2 & 3 & 1 \end{bmatrix} \cdot \begin{bmatrix} 1 & 1 & -1 \\ 0 & 1 & -1 \\ 0 & 0 & 2 \end{bmatrix}$$

The above procedure (8.74) to (8.76) can be extended to any $n \geq 2$; it is called *Doolittle's method*. It can also be modified to allow for the pivoting procedure discussed in section 8.2. The detailed definition of Doolittle's method will not be given here, but it does have an important computational advantage over Gaussian elimination.

The two methods for computing L and U, Gaussian elimination and Doolittle's method, are mathematically equivalent and have equal operation counts. But Doolittle's method can be used to greatly reduce the number of rounding errors, with only a minimal increase in cost. The formulas for L and U, with larger values of n, will involve many sums of products, as suggested in (8.76). If we assume that the numbers are all stored in single precision, then a limited use of double precision can be used to decrease greatly the number of single-precision rounding errors. Multiply all of the products in double precision, sum them in double precision, and then round back to single precision. Thus the computation of a sum of products of single-precision numbers will involve only one single-precision rounding error, rather than many. The number of such rounding errors can be reduced from about $\frac{2}{3}n^3$ for regular Gaussian elimination to a number proportional to n^2 for Doolittle's method, a significant reduction for larger values of n. This limited use of double-precision arithmetic was discussed earlier at the end of Chapter 3, including giving a Fortran function SUMPRD. Well-written programs for solving linear systems use Doolittle's method or a close relation to it, and they use the limited double-precision arithmetic described above.

One topic not yet discussed is the storage of the matrix L. Returning to the elimination steps (8.22) to (8.25) in section 8.2, we see that the multipliers m_{ik} can be stored in those memory positions of the machine array A that have been zeroed in the elimination step. Thus m_{ik} is stored in $A(i, k)$ in the machine, for $i > k$. And as discussed previously, u_{ij} is stored in $A(i, j)$ for $j \geq i$. The diagonal elements of L are all 1 and, thus, they do not need to be stored. Thus, after Gaussian elimination or Doolittle's method is applied to $Ax = b$, the elements of L and U will have replaced the original elements of A in the computer, as in subroutine LINSYS in section 8.2.

Tridiagonal Systems

When the coefficient matrix A has a special form, it is often possible to simplify Gaussian elimination. We will do this for tridiagonal systems of linear equations.

A system $Ax = f$ is called tridiagonal if its matrix has the form

$$A = \begin{bmatrix} b_1 & c_1 & 0 & 0 & \ldots & \ldots & 0 \\ a_2 & b_2 & c_2 & 0 & & & 0 \\ 0 & a_3 & b_3 & c_3 & & & 0 \\ \vdots & & & \ddots & & & \vdots \\ 0 & & & & a_{n-1} & b_{n-1} & c_{n-1} \\ \vdots & \ldots & \ldots & \ldots & 0 & a_n & b_n \end{bmatrix} \tag{8.77}$$

This is called a *tridiagonal matrix*. Tridiagonal systems occur commonly in solving problems in many different areas of numerical analysis. In Chapter 5, spline function interpolation led to the tridiagonal system in (5.69).

When Gaussian elimination is applied to $Ax = f$, most of the multipliers $m_{ik} = 0$ and most of the elements of U are also zero. With this in mind, it can be shown that the LU factorization will have the following general form, provided that pivoting is not used.

$$LU = \begin{bmatrix} 1 & 0 & 0 & \ldots & 0 \\ \alpha_2 & 1 & 0 & & 0 \\ 0 & \alpha_3 & 1 & & \vdots \\ \vdots & & & \ddots & \vdots \\ 0 & \ldots & \ldots & \alpha_n & 1 \end{bmatrix} \cdot \begin{bmatrix} \beta_1 & c_1 & 0 & \ldots & \ldots & 0 \\ 0 & \beta_2 & c_2 & 0 & & 0 \\ \vdots & & \ddots & & & \vdots \\ \vdots & & & & \beta_{n-1} & c_{n-1} \\ 0 & \ldots & \ldots & \ldots & 0 & \beta_n \end{bmatrix} \tag{8.78}$$

Multiply in succession each row of L times the various columns of U, and then equate the results to the corresponding elements of A. This yields

$$\beta_1 = b_1 \qquad\qquad\qquad\qquad\qquad : \text{ row 1 of } LU$$
$$\alpha_2\beta_1 = a_2, \qquad \alpha_2 c_1 + \beta_2 = b_2 \ : \text{ row 2 of } LU \qquad (8.79)$$
$$\alpha_j\beta_{j-1} = a_j, \qquad \alpha_j c_{j-1} + \beta_j = b_j : \text{ row } j \text{ of } LU$$

for $j = 3, \ldots, n$. The reader should check these equations for the first few rows of A.

The equations (8.79) are easily solved for the elements $\{\alpha_i\}$ and $\{\beta_i\}$.

$$\beta_1 = b_1 \qquad\qquad\qquad\qquad\qquad\qquad\qquad (8.80)$$
$$\alpha_j = a_j/\beta_{j-1}, \qquad \beta_j = b_j - \alpha_j c_{j-1}, \qquad j = 2, \ldots, n$$

To solve the system $Ax = f$, solve the pair of triangular systems

$$Lg = f \qquad Ux = g$$

Forward substitution in $Lg = f$ gives

$$g_1 = f_1 \qquad\qquad\qquad\qquad\qquad\qquad\qquad (8.81)$$
$$g_j = f_j - \alpha_j g_{j-1}, \qquad j = 2, 3, \ldots, n$$

Back substitution in $Ux = g$ yields

$$x_n = g_n/\beta_n \qquad\qquad\qquad\qquad\qquad\qquad (8.82)$$
$$x_j = (g_j - c_j x_{j+1})/\beta_j, \qquad j = n - 1, \ldots, 1$$

The entire procedure (8.80) to (8.82) is extremely rapid, with an operations count of only $5n - 4$ multiplications and divisions. Showing this is left as problem 11. Also, to store the matrix A, only three one-dimensional arrays for the three diagonals of A are needed. This means that very large systems can be solved, and systems of order $n = 10,000$ are not unusual in some applications.

Example

Let

$$A = \begin{bmatrix} 2 & 1 & 0 & 0 & 0 \\ 1 & 2 & 1 & 0 & 0 \\ 0 & 1 & 2 & 1 & 0 \\ 0 & 0 & 1 & 2 & 1 \\ 0 & 0 & 0 & 1 & 2 \end{bmatrix} \qquad (8.83)$$

In the notation of (8.77),

$$a_j = 1, \qquad b_j = 2, \qquad c_j = 1, \qquad \text{all } j$$

From (8.80), $\beta_1 = 2$ and

$$\alpha_j = 1/\beta_{j-1}, \qquad \beta_j = 2 - \alpha_j, \qquad j = 2, 3, 4, 5$$

This leads to

$$A = LU = \begin{bmatrix} 1 & 0 & 0 & 0 & 0 \\ \frac{1}{2} & 1 & 0 & 0 & 0 \\ 0 & \frac{2}{3} & 1 & 0 & 0 \\ 0 & 0 & \frac{3}{4} & 1 & 0 \\ 0 & 0 & 0 & \frac{4}{5} & 1 \end{bmatrix} \begin{bmatrix} 2 & 1 & 0 & 0 & 0 \\ 0 & \frac{3}{2} & 1 & 0 & 0 \\ 0 & 0 & \frac{4}{3} & 1 & 0 \\ 0 & 0 & 0 & \frac{5}{4} & 1 \\ 0 & 0 & 0 & 0 & \frac{6}{5} \end{bmatrix} \qquad (8.84)$$

The procedure (8.78) to (8.82) assumes that pivoting is not necessary. This can be shown to be true if the following conditions are satisfied.

$$|b_1| > |c_1|$$
$$|b_j| \geq |a_j| + |c_j|, \qquad j = 2, 3, \ldots, n - 1 \qquad (8.85)$$
$$|b_n| > |a_n|$$

These conditions are satisfied in the most important cases that occur in applications. The case of interpolating cubic splines is considered in problem 12.

A Program for Tridiagonal Systems

The following Fortran subroutine solves tridiagonal systems by means of the method (8.78) to (8.82). The coefficients $\{a_j, b_j, c_j\}$ are given by the user in the one-dimensional arrays A, B, and C, and the right-hand constants are supplied in F. On exit, the arrays A and B contain the constants $\{\alpha_j, \beta_j\}$ of (8.80), and F contains the solution x. In many applications, a tridiagonal system is solved with many right sides f. To allow for this option, the variable $IFLAG$ indicates whether or not the LU factorization has already been computed and stored in the arrays, A, B, and C. See the comment statements in the program for additional information.

```
      SUBROUTINE TRIDGL(A,B,C,F,N,IFLAG,IER)
C
C     THIS SOLVES A TRIDIAGONAL SYSTEM OF LINEAR EQUATIONS M*X=F.
C
C     INPUT:
C     THE ORDER OF THE LINEAR SYSTEM IS GIVEN BY N.
C     THE SUBDIAGONAL,DIAGONAL, AND SUPERDIAGONAL OF M ARE GIVEN
C     BY THE ARRAYS A,B, AND C, RESPECTIVELY. MORE PRECISELY,
C         M(I,I-1)=A(I),   I=2,...,N
C         M(I,I)  =B(I),   I=1,...,N
C         M(I,I+1)=C(I),   I=1,...,N-1
C     IFLAG=0 MEANS THAT THE ORIGINAL MATRIX M IS GIVEN AS
C     SPECIFIED ABOVE.
C     IFLAG=1 MEANS THAT THE LU FACTORIZATION OF M IS ALREADY KNOWN
C     AND IS STORED IN A,B, AND C. THIS WILL HAVE BEEN ACCOMPLISHED
C     BY A PREVIOUS CALL TO THIS SUBROUTINE.
C
C     OUTPUT:
C     UPON EXIT, THE LU FACTORIZATION OF M WILL BE STORED
C     IN A,B, AND C.
C     THE SOLUTION VECTOR X WILL BE RETURNED IN F.
C     IER=0 MEANS THE PROGRAM WAS COMPLETED SATISFACTORILY.
C     IER=1 MEANS THAT A ZERO PIVOT ELEMENT WAS ENCOUNTERED, AND
C     NO SOLUTION WAS ATTEMPTED.
C
      DIMENSION A(*),B(*),C(*),F(*)
C
      IF(IFLAG .EQ. 0) THEN
C         COMPUTE LU FACTORIZATION OF MATRIX M.
          DO 10 J=2,N
            IF(B(J-1) .EQ. 0.0) THEN
                IER=1
                RETURN
            END IF
            A(J)=A(J)/B(J-1)
10          B(J)=B(J)-A(J)*C(J-1)
          IF(B(N) .EQ. 0.0) THEN
                IER=1
                RETURN
            END IF
C     END IF
```

```
C       COMPUTE SOLUTION X TO M*X=F.
        DO 20 J=2,N
20        F(J)=F(J)-A(J)*F(J-1)
        F(N)=F(N)/B(N)
        DO 30 J=N-1,1,-1
30        F(J)=(F(J)-C(J)*F(J+1))/B(J)
        IER=0
        RETURN
        END
```

PROBLEMS

1. Recall the solution of the system (8.12) in section 8.2. Using that computation, give the LU factorization of the matrix of coefficients of that system.

2. Calculate the LU factorization of the matrices in problem 1 of section 8.2.

3. Consider the proof of Theorem 8.3 when $n = 3$. To do so, recall the general derivation (8.15) to (8.21) for Gaussian elimination with 3 equations. Form the product

$$LU = \begin{bmatrix} 1 & 0 & 0 \\ m_{21} & 1 & 0 \\ m_{31} & m_{32} & 1 \end{bmatrix} \begin{bmatrix} a_{11} & a_{12} & a_{13} \\ 0 & a_{22}^{(2)} & a_{23}^{(2)} \\ 0 & 0 & a_{33}^{(3)} \end{bmatrix}$$

Use the definitions of m_{ij} and $a_{ij}^{(i)}$ to simplify the results.

4. Recall the definition of elementary matrices from problem 18 of section 8.3. Define

$$E_1 = \begin{bmatrix} 1 & 0 & 0 \\ -m_{21} & 1 & 0 \\ -m_{31} & 0 & 1 \end{bmatrix}, \quad E_2 = \begin{bmatrix} 1 & 0 & 0 \\ 0 & 1 & 0 \\ 0 & -m_{32} & 1 \end{bmatrix}$$

Define $A_2 = E_1A$, $A_3 = E_2A_2$. Show $A_3 = U$, the upper triangular matrix obtained by Gaussian elimination. Thus $E_2E_1A = U$. Use this to show $A = LU$, with L derived from E_2 and E_1. How would this proof generalize to A a matrix of order n?

5. The LU factorization of A is not unique if one only requires that L be lower triangular and U be upper triangular.

(a) Given an LU factorization of A, define

$$L_1 = LD \qquad U_1 = D^{-1}U$$

for some nonsingular diagonal matrix D. Show that L_1U_1 is another such factorization of A.

(b) Let $L_1U_1 = L_2U_2 = A$, with L_1, L_2 lower triangular, U_1, U_2 upper triangular. Also let A be nonsingular. Show that $L_1 = L_2D$, $U_1 = D^{-1}U_2$ for some diagonal matrix D.
Hint: Derive $L_2^{-1}L_1 = U_2U_1^{-1}$. Show that these matrices must be diagonal. To make matters simpler, initially let $n = 2$ be the order of A. Also recall problem 3 of section 8.3.

6. When A is a symmetric matrix, it is often possible to factor A in the form

$$A = LL^T$$

for some lower triangular matrix.
(a) For

$$A = \begin{bmatrix} 1 & -1 \\ -1 & 5 \end{bmatrix}$$

let

$$L = \begin{bmatrix} l_{11} & 0 \\ l_{21} & l_{22} \end{bmatrix}$$

Find L such that $LL^T = A$.
(b) Repeat this process for

$$A = \begin{bmatrix} 2.25 & -3 & 4.5 \\ -3 & 5 & -10 \\ 4.5 & -10 & 34 \end{bmatrix}$$

7. Write a section of Fortran code to solve the linear systems (8.71). Assume the storage conventions for L and U described in the paragraph preceding the discussion of tridiagonal systems.

8. For the A in (8.83), solve the linear system with

$$f = [1, 0, 0, 0, 0]^T$$

9. Solve the tridiagonal system

$$
\begin{aligned}
2x_1 + x_2 \qquad\qquad\ &= 3 \\
-x_1 + 2x_2 + x_3 \qquad &= 2 \\
-x_2 + 2x_3 + x_4 \quad\ &= 2 \\
-x_3 + 2x_4 + x_5 &= 2 \\
-x_4 + 2x_5 &= 1
\end{aligned}
$$

10. Solve the tridiagonal system $Ax = f$ with

$$A_{ii} = 4, \qquad A_{i,i-1} = A_{i,i+1} = 1$$

for all i. Let the order of the system be $n = 100$, and let

$$f = [1, 1, \ldots, 1]^T.$$

11. Let $Ax = f$ be a tridiagonal system of order n.
(a) Give an operation count for forming $A = LU$, using (8.80).
(b) Assuming LU is known, give an operations count for solving $Ax = f$ using (8.81) to (8.82).

12. Recall the tridiagonal system (5.69) to (5.70) for cubic spline interpolation. Show that the conditions (8.85) are satisified by this system.

8.5 ERROR IN SOLVING LINEAR SYSTEMS

On a computer or calculator, Gaussian elimination involves rounding errors through its arithmetic operations, and these errors lead to error in the computed solution of the linear system being solved. In this section, we look at a method for predicting and correcting the error in the computed solution. Then, we examine the sensitivity of the solution to small changes such as rounding errors.

Let $Ax = b$ denote the system being solved, and let $\hat{x}$ denote the computed solution. Define

$$r = b - A\hat{x} \qquad\qquad (8.86)$$

This is called the *residual* in the approximation of b by $A\hat{x}$. If $\hat{x}$ is the exact solution, then r equals zero. To relate r to the error in x, substitute $b = Ax$ in (8.86), yielding

$$r = Ax - A\hat{x} = A(x - \hat{x})$$

Let $e = x - \hat{x}$, the error in $\hat{x}$. Then

$$Ae = r \tag{8.87}$$

The error e satisfies a linear system with the same coefficient matrix A as in the original system $Ax = b$.

To find e, solve (8.87). In the original solution of $Ax = b$, the matrices L and U should have been saved, along with a knowledge of any possible row interchanges. This information can be used to solve $Ae = r$ at a relatively small cost. Recalling (8.40), the operations count for finding e from (8.87) is

$$MD(r \to e) = n^2 \tag{8.88}$$

In addition, the residual r must be computed from (8.86), and for it,

$$MD(r) = n^2 \tag{8.89}$$

Thus the operation count for calculating e from $\hat{x}$ is only $2n^2$, and this is relatively small compared to the original operation count of about $\frac{1}{3}n^3$ for solving $Ax = b$.

In the practical implementation of this procedure for computing the error e, there are two possible sources of difficulty. First, the computation of the residual r in (8.86) will involve loss-of-significance errors. Generally each component

$$r_i = b_i - \sum_{j=1}^{n} a_{ij}\hat{x}_j \tag{8.90}$$

will be very close to zero, and this occurs by the subtraction of the nearly equal quantities b_i and $\sum a_{ij}\hat{x}_j$. To avoid obtaining a very inaccurate value for r_i, the calculation of (8.90) should be carried out in a higher-precision arithmetic. If the solution of $Ax = b$ was done in single-precision arithmetic, then r should be computed using double-precision arithmetic and then rounded back to single precision.

The second source of difficulty in finding e is that solving $Ae = r$ will involve the same rounding errors that led to the error in the calculated value $\hat{x}$. Thus we will not obtain e, but only an approximation $\hat{e}$ of e. How closely $\hat{e}$ approximates e will depend on the sensitivity or conditioning of the matrix, a topic taken up later in this section. As a general rule, the approximation $\hat{e}$ will have at least one digit of accuracy relative to e, except for the most ill-behaved of matrices A.

Example

Recall the system preceding (8.32) in section 8.2,

$$0.729x_1 + 0.81x_2 + 0.9x_3 = 0.6867$$
$$x_1 + x_2 + x_3 = 0.8338 \qquad (8.91)$$
$$1.331x_1 + 1.21x_2 + 1.1x_3 = 1.000$$

As before, we use a four digit decimal machine with rounding. The true solution of this system is

$$x_1 = 0.2245 \qquad x_2 = 0.2814 \qquad x_3 = 0.3279 \qquad (8.92)$$

correctly rounded to four digits.

To illustrate the estimation of the error e, we consider the solution of the system by Gaussian elimination without pivoting. This leads to the answers given earlier in (8.33),

$$\hat{x}_1 = 0.2252 \qquad \hat{x}_2 = 0.2790 \qquad \hat{x}_3 = 0.3295 \qquad (8.93)$$

Calculating the residual r in the manner described above, using eight digit floating-point decimal arithmetic, and rounding the resultant components of the residual to four significant digits,

$$r = [0.00006210, \quad 0.0002000, \quad 0.0003519]^T$$

Solving the linear system in (8.91) with this new right-hand side, we obtain the approximation

$$\hat{e} = [-0.0004471, \quad 0.002150, \quad -0.001504]^T \qquad (8.94)$$

Compare this to the true error of

$$e = x - \hat{x} = [-0.0007, \quad 0.0024, \quad -0.0016]^T$$

obtained from (8.92), (8.93). This $\hat{e}$ gives a fairly good idea of the size of the error e in the computed solution $\hat{x}$.

The Residual Correction Method

A further use of this error estimation procedure is to define an iterative method for improving the computed value x. Let $x^{(0)} = \hat{x}$, the initial computed value for x, generally obtained using Gaussian elimination. Define

$$r^{(0)} = b - Ax^{(0)} \tag{8.95}$$

Then as before in (8.87),

$$Ae^{(0)} = r^{(0)}, \qquad e^{(0)} = x - x^{(0)}$$

Solving by Gaussian elimination, we obtain an approximate value $\hat{e}^{(0)} \doteq e^{(0)}$. Using it, define an improved approximation

$$x^{(1)} = x^{(0)} + \hat{e}^{(0)} \tag{8.96}$$

Now repeat the entire process, calculating

$$r^{(1)} = b - Ax^{(1)}$$
$$x^{(2)} = x^{(1)} + \hat{e}^{(1)}$$

where $\hat{e}^{(1)}$ is the approximate solution of

$$Ae^{(1)} = r^{(1)}, \qquad e^{(1)} = x - x^{(1)} \tag{8.97}$$

Continue this process until there is no further decrease in the size of $\hat{e}^{(k)}$, $k \geq 0$.

This method usually gives a way to obtain answers that are as accurate as possible relative to the number of digits in the arithmetic being used. The procedure is called the *residual correction method*.

Example

Use a computer with four digit floating-point decimal arithmetic with rounding, and use Gaussian elimination with pivoting. The system to be solved is

$$
\begin{aligned}
x_1 + \quad 0.5x_2 + 0.3333x_3 &= 1 \\
0.5x_1 + 0.3333x_2 + \quad 0.25x_3 &= 0 \\
0.3333x_1 + \quad 0.25x_2 + \quad 0.2x_3 &= 0
\end{aligned} \tag{8.98}
$$

Then

$$x^{(0)} = [8.968, -35.77, 29.77]^T$$
$$r^{(0)} = [-0.005341, -0.004359, -0.0005344]^T$$
$$\hat{e}^{(0)} = [0.09216, -0.5442, 0.5239]^T$$
$$x^{(1)} = [9.060, -36.31, 30.29]$$
$$r^{(1)} = [-0.0006570, -0.0003770, -0.0001980]^T$$
$$\hat{e}^{(2)} = [0.001707, -0.01300, 0.01241]^T$$
$$x^{(2)} = [9.062, -36.32, 30.30]^T$$

The iterate $x^{(2)}$ is the correctly rounded solution of the system (8.98). This illustrates the usefulness of the residual correction method.

Stability in Solving Linear Systems

Rounding errors in Gaussian elimination will lead to errors in the computed solution of $Ax = b$, as has been illustrated above. We would now like to examine the degree of sensitivity of the solution x to these errors. The complexity of Gaussian elimination with its many arithmetic operations makes it difficult to do a direct analysis of the effects of these rounding errors. Instead, we look for the effect on the solution x of making a small change in the right side b of the linear system.

Let $\hat{b}$ be an approximation to b, and consider the solution $\hat{x}$ of

$$A\hat{x} = \hat{b} \tag{8.99}$$

The residual equation (8.86) fits into this framework, with $\hat{b} = b - r$. To examine the sensitivity of x to small changes, we ask what is the size of $x - \hat{x}$ in comparison to $b - \hat{b}$.

Example

The linear system

$$5x_1 + 7x_2 = 0.7$$
$$7x_1 + 10x_2 = 1 \tag{8.100}$$

has the solution

$$x_1 = 0, \qquad x_2 = 0.1$$

The perturbed system

$$5\hat{x}_1 + 7\hat{x}_2 = 0.69$$
$$7\hat{x}_1 + 10\hat{x}_2 = 1.01$$

has the solution

$$\hat{x}_1 = -0.17, \qquad \hat{x}_2 = 0.22$$

A relatively small change in the right side of (8.100) has led to a relatively large change in the solution.

To more easily express the relationship of $x - \hat{x}$ to $b - \hat{b}$, we introduce the following measures for the size of a matrix. Define

$$\|z\| = \max_{1 \le i \le n} |z_i|, \qquad \text{order } (z) = n \times 1 \tag{8.101}$$

$$\|A\| = \max_{1 \le i \le n} \sum_{j=1}^{n} |a_{ij}|, \qquad \text{order } (A) = n \times n$$

These quantities are called vector and matrix norms, respectively, and they can be shown to satisfy the following inequalities.

$$
\begin{aligned}
&\text{(P1)} \quad \|y + z\| \le \|y\| + \|z\| \\
&\text{(P2)} \quad \|A + B\| \le \|A\| + \|B\| \\
&\text{(P3)} \quad \|AB\| \le \|A\| \, \|B\| \\
&\text{(P4)} \quad \|Az\| \le \|A\| \, \|z\|
\end{aligned}
\tag{8.102}
$$

To examine the error $x - \hat{x}$, we look at

$$\|x - \hat{x}\| = \max_{1 \le i \le n} |x_i - \hat{x}_i|$$

the maximum of the errors in the components of $\hat{x}$.

Theorem 8.4

Let A be nonsingular. Then, the solutions of $Ax = b$ and $A\hat{x} = \hat{b}$ satisfy

$$\frac{\|x - \hat{x}\|}{\|x\|} \le \|A\| \, \|A^{-1}\| \frac{\|b - \hat{b}\|}{\|b\|} \tag{8.103}$$

And for suitable choices of b and $\hat{b}$, the inequality becomes an equality.

Proof Subtracting $A\hat{x} = \hat{b}$ from $Ax = b$, we get

$$A(x - \hat{x}) = b - \hat{b}$$
$$x - \hat{x} = A^{-1}(b - \hat{b})$$

Using (P4) of (8.102) yields

$$\|x - \hat{x}\| = \|A^{-1}(b - \hat{b})\| \leq \|A^{-1}\|\,\|b - \hat{b}\|$$

Divide by $\|x\|$ to obtain

$$\frac{\|x - \hat{x}\|}{\|x\|} \leq \frac{\|A^{-1}\|\,\|b - \hat{b}\|}{\|x\|} = \|A\|\,\|A^{-1}\|\frac{\|b - \hat{b}\|}{\|A\|\,\|x\|} \tag{8.104}$$

Use (P4) again to obtain

$$\|b\| = \|Ax\| \leq \|A\|\,\|x\|$$

and substitute this on the right side of (8.104) to obtain (8.103). We omit the proof of the final statement in the theorem.

The result (8.103) compares the relative error in $\hat{x}$ to the relative error in $\hat{b}$. It says that if $\|A\|\,\|A^{-1}\|$ is very large, then the relative error in $\hat{x}$ may be much larger than that for $\hat{b}$; this is guaranteed to happen for certain choices of b and $\hat{b}$, although the actual choices are unknown in practice. The number

$$\text{cond}\,(A) = \|A\|\,\|A^{-1}\| \tag{8.105}$$

is called a *condition number*. When it is very large, the solution of $Ax = b$ will be very sensitive to relatively small changes in b. Or in the earlier notation of the residual in (8.86), a relatively small residual will quite possibly lead to a relatively large error in $\hat{x}$ as compared to x. Such linear systems are called *ill-conditioned*, while systems with a small condition number are called *well-conditioned*.

These comments are also valid when the changes are made to A rather than to b. In particular, if we compare the solutions of $Ax = b$ and $A\hat{x} = b$, then it can be shown that

$$\frac{\|x - \hat{x}\|}{\|x\|} \le \frac{\text{cond } (A)}{1 - \text{cond } (A) \dfrac{\|A - \hat{A}\|}{\|A\|}} \cdot \frac{\|A - \hat{A}\|}{\|A\|} \qquad (8.106)$$

provided that

$$\|A - \hat{A}\| < 1/\|A^{-1}\|$$

Again, small changes in A will lead to possibly large changes in x, if cond (A) is a large number. Much of the current computer software for solving linear systems contain techniques for estimating cond (A), without calculating A^{-1}; then the condition number is used in estimating the error in the computed solution.

Example

A well-known example of an ill-conditioned matrix is the *Hilbert matrix:*

$$H_n = \begin{bmatrix} 1 & \dfrac{1}{2} & \dfrac{1}{3} & \cdots\cdots & \dfrac{1}{n} \\ \dfrac{1}{2} & \dfrac{1}{3} & \dfrac{1}{4} & \cdots & \dfrac{1}{n+1} \\ \vdots & & & & \vdots \\ \dfrac{1}{n} & \dfrac{1}{n+1} & \cdots\cdots & & \dfrac{1}{2n-1} \end{bmatrix} \qquad (8.107)$$

As n increases, H_n becomes increasingly ill-conditioned. To illustrate the ill-conditioning, let $n = 4$ and define $\hat{H}_4$ by rounding the elements of H_4 to five significant digits. Then

$$\hat{H}_4 = \begin{bmatrix} 1.0000 & 0.50000 & 0.33333 & 0.25000 \\ 0.50000 & 0.33333 & 0.25000 & 0.20000 \\ 0.33333 & 0.25000 & 0.20000 & 0.16667 \\ 0.25000 & 0.20000 & 0.16667 & 0.14286 \end{bmatrix}$$

To see that H_4 is ill-conditioned, compare H_4^{-1} and $\hat{H}_4^{-1}$. These are the solutions of $H_4 X = I$ and $\hat{H}_4 \hat{X} = I$. Then

$$
H_4^{-1} = \begin{bmatrix}
16 & -120 & 240 & -140 \\
-120 & 1200 & -2700 & 1680 \\
240 & -2700 & 6480 & -4200 \\
-140 & 1680 & -4200 & 2800
\end{bmatrix}
$$

$$
\hat{H}_4^{-1} = \begin{bmatrix}
16.248 & -122.72 & 246.49 & -144.20 \\
-122.72 & 1229.9 & -2771.3 & 1726.1 \\
246.49 & -2771.3 & 6650.1 & -4310.0 \\
-144.20 & 1726.1 & -4310.0 & 2871.1
\end{bmatrix}
$$

The errors in $\hat{H}_4$ are in the sixth decimal place, but the errors in $\hat{H}_4^{-1}$ are in the second decimal place. This is a significant change in $\hat{H}_4^{-1}$ as compared to that in $\hat{H}_4$. Also, by direct computation,

$$
\text{cond } (H_4) = \|H_4\| \, \|H_4^{-1}\| = \left(\frac{25}{12}\right)(13620) \doteq 28000
$$

and this is consistent with the loss of four significant digits in $\hat{H}_4^{-1}$ when compared to H_4^{-1}.

Ill-conditioned matrices do not occur in most applications, but they do occur. The Hilbert matrix arises naturally out of the application taken up in the next section, and there are other ill-conditioned matrices that arise as naturally. For that reason, it is best to use linear equation solvers that have some way to detect ill-conditioning, or else, the residual correction technique described earlier should be used to estimate the accuracy in the computed solution $\hat{x}$ to your system $Ax = b$.

PROBLEMS

1. Write a program to solve a linear system $Ax = b$, using the subroutine LINSYS given in section 8.2. Have the program contain the following features. (i) Create or read A and b. (ii) Solve $Ax = b$ using LINSYS, calling the computed solution $\hat{x}$. (iii) Compute the residual $r = b - A\hat{x}$, using the procedure described following (8.90). (iv) Solve for the error in $Ae = r$, calling the solution $\hat{e}$. (v) Print $\hat{e}$ and $\hat{x} + \hat{e}$.

Also print $\|\hat{e}\|/\|\hat{x}\|$ to approximate $\|e\|/\|x\|$. Apply this program to solving the following linear systems:

(a) $4x_1 - 6x_2 + 4x_3 - x_4 = 0.43$
$\quad -6x_1 + 14x_2 - 11x_3 + 3x_4 = -1$
$\quad 4x_1 - 11x_2 + 10x_3 - 3x_4 = 0.82$
$\quad -x_1 + 3x_2 - 3x_3 + x_4 = -0.23$

(b) $5x_1 + 7x_2 + 6x_3 + 5x_4 = 23$
$\quad 7x_1 + 10x_2 + 8x_3 + 7x_4 = 32$
$\quad 6x_1 + 8x_2 + 10x_3 + 9x_4 = 33$
$\quad 5x_1 + 7x_2 + 9x_3 + 10x_4 = 31$
$\quad$ Its solution is $x_1 = x_2 = x_3 = x_4 = 1$

(c) $H_5x = [1, 0.6, 0.4, 0.3, 0.3]^T$, where H_5 is defined in (8.107).

2.   Verify the norm properties (P1) to (P4) of (8.102).

3.   (a)   Calculate cond (A) for the system (8.100).
     (b)   Graph the equations in (8.100). What is the effect geometrically of changing the right-hand constants in the equations? Interpret the moderate ill-conditioning of this system geometrically.

4.   Calculate cond (A) for

$$A = \begin{bmatrix} 1 & c \\ c & 1 \end{bmatrix} \quad , c \neq 1$$

When does A become ill-conditioned? What does this say about the linear system $Ax = b$? How is cond (A) related to det (A)?

5.   Prove that cond $(A) \geq 1$ for any A.
     Hint: First show $\|I\| = 1$.

6.   Use a package program from your computer center to solve the systems in problem 1. Use a program that will predict the error in the solution, if possible.

7.   Define the matrix A_n of order n by

$$\begin{bmatrix} 1 & -1 & -1 & -1 & \cdots & -1 \\ 0 & 1 & -1 & -1 & \cdots & -1 \\ \vdots & & & \ddots & & \vdots \\ \vdots & & & & 1 & -1 \\ 0 & \cdots & & & 0 & 1 \end{bmatrix}$$

(a)   Find the inverse of A_n, explicitly.
     Hint: Find the inverse of A_6, by solving $BA_6 = I$, with B an unknown upper triangular matrix. Then use this result to "guess" the inverse of A_n in general.

(b)   Calculate cond (A_n).

(c)   With $b = [-n + 2, -n + 3, \ldots, -1, 0, 1]^T$, the solution of
$A_n x = b$ is $x = [1, 1, \ldots, 1]^T$. Perturb b to $\hat{b} = b + [0, \ldots 0, \epsilon]^T$.
Solve for $\hat{x}$ in $A_n \hat{x} = \hat{b}$.
Hint: Use $\hat{x} = x + A_n^{-1}(\hat{b} - b)$.
Show that these values of b, $\hat{b}$, x, $\hat{x}$ satisfy Theorem 8.4.

8.6   LEAST SQUARES DATA FITTING

In scientific and engineering experiments, the measurement of physical
quantities will generally be somewhat inaccurate. This may be due to
human error, but more often, it is due to limitations inherent in the
equipment being used to make the measurements. In this section, we
present a method for obtaining relationships between empirically obtained variables, with a minimization of the effects of the measurement
errors present in these variables.

Suppose an experiment involves measuring two related variables x and
y in order to obtain an unknown relationship

$$y = f(x) \tag{8.107}$$

In the experiment, one chooses various values of x, say $x_1, x_2, \ldots, x_n$,
and then one measures a corresponding set of values for y. Let the actual
measurements be denoted by $y_1, \ldots, y_n$, and let

$$\epsilon_i = f(x_i) - y_i \tag{8.108}$$

denote the measurement errors. The experimenter wants to use the
points $(x_1, y_1), \ldots, (x_n, y_n)$ to determine the analytic relationship (8.107)
as accurately as possible.

An analytic relation (8.107) can always be obtained by using interpolation with polynomials, spline functions, or some other form of interpolating function. But this ignores the presence of the errors $\{\epsilon_i\}$, and
it often leads to a poor estimate of the true function $f(x)$. In many cases,
experimenters will know or will suspect that the unknown function $f(x)$
lies within some known class of functions, for example, polynomials.
Then they will want to choose the member of that class of functions that
will best approximate their unknown function $f(x)$, taking into account
the experimental errors $\{\epsilon_i\}$. As an example of such a situation, consider
the data in Table 8.2 and the plot of it in Figure 8.1. From this plot, it
is reasonable to expect $f(x)$ to be a linear polynomial,

$$f(x) = mx + b \tag{8.109}$$

Table 8.2 Empirical Data

x_i	y_i	x_i	y_i
1.0	− 1.945	3.2	0.764
1.2	− 1.253	3.4	0.532
1.4	− 1.140	3.6	1.073
1.6	− 1.087	3.8	1.286
1.8	− 0.760	4.0	1.502
2.0	− 0.682	4.2	1.582
2.2	− 0.424	4.4	1.993
2.4	− 0.012	4.6	2.473
2.6	− 0.190	4.8	2.503
2.8	0.452	5.0	2.322
3.0	0.337		

Assuming this to be the true form, the problem of determining $f(x)$ is now reduced to that of determining the constants m and b.

To determine a reasonable approximation to $f(x)$, we must assume something more is known about the errors $\{\epsilon_i\}$. A standard assumption is that each of these errors is a random variable chosen from a *normal probability distribution*. This is a topic developed in probability theory and statistics courses, and it is too involved to consider here. But errors satisfying this assumption have the following general characteristics. (1) If the experiment is repeated many times for the same $x = x_i$, then the associated unknown errors ϵ_i in the empirical values y_i will be likely to

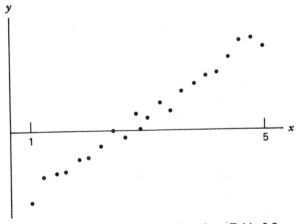

Figure 8.1 Plot of empirical data from Table 8.2

have an average of zero; and (2) for this same experimental case with $x = x_i$, as the size of ϵ_i increases, the likelihood of its occurring will decrease rapidly. There are very good reasons for accepting this assumption on the behavior of the errors $\{\epsilon_i\}$, and it will be denoted as the *normal error assumption* in what follows below. Also, to further simplify the following presentation, we will assume that the individual errors ϵ_i, $1 \leq i \leq n$, are all random variables from the same probability distribution function, meaning that the size of ϵ_i is unrelated to the size of x_i or y_i.

Let the true function $f(x)$ be assumed to lie in a known class of functions, call it C. An example is the assumption that $f(x)$ is linear for the data in Table 8.2. Then among all functions $\hat{f}(x)$ in C, it can be shown that the function $\hat{f}*$ that is most likely to equal f will also minimize the expression

$$E = \sqrt{\frac{1}{n} \sum_{i=1}^{n} [\hat{f}(x_i) - y_i]^2} \qquad (8.110)$$

among all functions $\hat{f}$ in C. This is called the *root-mean-square error* in the approximation of the data $\{y_i\}$ by $\hat{f}(x)$. The function $\hat{f}*(x)$ that minimizes E relative to all $\hat{f}$ in C is called the *least squares approximation* to the data $\{(x_i, y_i)\}$.

The Linear Least Squares Approximation

We consider the use of $\hat{f}(x) = mx + b$ in E. Before attempting to minimize (8.110) relative to such functions $\hat{f}(x)$, note that minimizing E is equivalent to minimizing the sum in (8.110), although the minimum values will be different. For this reason, we seek to minimize

$$G(b, m) = \sum_{i=1}^{n} [mx_i + b - y_i]^2 \qquad (8.111)$$

as b and m are allowed to vary arbitrarily.

From multivariable calculus, the values of (b, m) that minimize $G(b, m)$ will satisfy

$$\frac{\partial G(b, m)}{\partial b} = 0, \qquad \frac{\partial G(b, m)}{\partial m} = 0 \qquad (8.112)$$

using the partial derivatives of $G(b, m)$. From (8.111),

$$\frac{\partial G}{\partial b} = \sum_{i=1}^{n} 2[mx_i + b - y_i]$$

$$\frac{\partial G}{\partial m} = \sum_{i=1}^{n} 2[mx_i + b - y_i]x_i = 2\sum_{i=1}^{n} [mx_i^2 + bx_i - x_i y_i]$$

Combining these with (8.112), we obtain the linear system of order 2,

$$nb + \left(\sum_{i=1}^{n} x_i\right) m = \sum_{i=1}^{n} y_i \tag{8.113}$$

$$\left(\sum_{i=1}^{n} x_i\right) b + \left(\sum_{i=1}^{n} x_i^2\right) m = \sum_{i=1}^{n} x_i y_i$$

This can be solved for a unique solution (b, m), provided that the determinant is nonzero,

$$n \sum_{i=1}^{n} x_i^2 - \left(\sum_{i=1}^{n} x_i\right)^2 \neq 0 \tag{8.114}$$

This condition will be satisfied except in the one trivial case

$$x_1 = x_2 = \ldots = x_n = \text{constant}$$

and it cannot happen here since the numbers $\{x_j\}$ are assumed to be distinct.

Example

Determine the least squares fit for the data in Table 8.2. Then $n = 21$,

$$\sum_{1}^{n} x_i = 63.0 \qquad \sum_{1}^{n} x_i^2 = 219.8$$

$$\sum_{1}^{n} y_i = 9.326 \qquad \sum_{1}^{n} x_i y_i = 60.7302$$

Using this in (8.113) and then solving for b and m gives the solution

$$b = -2.74605 \qquad m = 1.06338$$

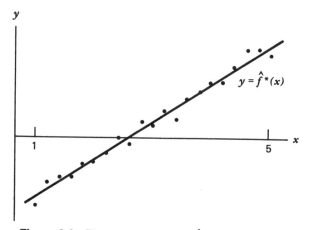

Figure 8.2 The least squares fit $\hat{f}^*(x)$ of (8.115).

The linear least squares fit to the data in Table 8.2 is

$$\hat{f}^*(x) = 1.06338x - 2.74605 \qquad (8.115)$$

This is optimum in the sense of minimizing the root-mean-square error E in (8.110), relative to all linear functions $\hat{f}(x) = mx + b$. The value of E for (8.115) is

$$E = 0.171$$

It can be considered as an average error in the approximation of the data in Table 8.2 by the least squares fit (8.115). A graph of (8.115) is given in Figure 8.2, along with the data points from Table 8.2.

Polynomial Least Squares Approximation

Most data will not correspond to a linear function, but to something more complicated. Let the functions $\hat{f}(x)$ that are being used to represent the data $\{(x_i, y_i)\}$ be given by

$$\hat{f}(x) = a_1 \varphi_1(x) + a_2 \varphi_2(x) + \ldots + a_m \varphi_m(x) \qquad (8.116)$$

where $a_1, a_2, \ldots, a_m$ are arbitrary numbers and $\varphi_1(x), \ldots, \varphi_m(x)$ are given functions. For example, if $\hat{f}(x)$ was to be a quadratic polynomial, we could write

$$\hat{f}(x) = a_1 + a_2 x + a_3 x^2 \qquad (8.117)$$

where

$$\varphi_1(x) \equiv 1, \qquad \varphi_2(x) = x, \qquad \varphi_3(x) = x^2$$

Under the normal error assumption, the function $\hat{f}(x)$ is to be chosen to minimize the root-mean-square error of (8.110). To show what is involved, we will consider the specific case $m = 3$. Then

$$\hat{f}(x) = a_1 \varphi_1(x) + a_2 \varphi_2(x) + a_3 \varphi_3(x) \qquad (8.118)$$

and the constants a_1, a_2, a_3 are to be chosen so as to minimize the expression

$$G(a_1, a_2, a_3) = \sum_{j=1}^{n} [a_1 \varphi_1(x_j) + a_2 \varphi_2(x_j) + a_3 \varphi_3(x_j) - y_j]^2 \qquad (8.119)$$

The points (a_1, a_2, a_3) that minimize $G(a_1, a_2, a_3)$ will satisfy

$$\frac{\partial G}{\partial a_1} = 0, \qquad \frac{\partial G}{\partial a_2} = 0, \qquad \frac{\partial G}{\partial a_3} = 0$$

This leads to the three equations

$$0 = \frac{\partial G}{\partial a_i} = \sum_{j=1}^{n} 2[a_1 \varphi_1(x_j) + a_2 \varphi_2(x_j) + a_3 \varphi_3(x_j) - y_j] \varphi_i(x_j)$$

for $i = 1, 2, 3$. Simplifying, we get

$$\left[\sum_{j=1}^{n} \varphi_1(x_j) \varphi_i(x_j) \right] a_1 + \left[\sum_{j=1}^{n} \varphi_2(x_j) \varphi_i(x_j) \right] a_2$$

$$+ \left[\sum_{j=1}^{n} \varphi_3(x_j) \varphi_i(x_j) \right] a_3 = \sum_{j=1}^{n} y_j \varphi_i(x_j), \qquad i = 1, 2, 3 \quad (8.120)$$

These are three linear equations for the three unknowns a_1, a_2, a_3.

Example

Apply (8.120) to the quadratic formula for $f(x)$ in (8.117). Then the three equations are

$$na_1 + \left[\sum_{j=1}^{n} x_j\right] a_2 + \left[\sum_{j=1}^{n} x_j^2\right] a_3 = \sum_{j=1}^{n} y_j$$

$$\left[\sum_{j=1}^{n} x_j\right] a_1 + \left[\sum_{j=1}^{n} x_j^2\right] a_2 + \left[\sum_{j=1}^{n} x_j^3\right] a_3 = \sum_{j=1}^{n} y_j x_j \qquad (8.121)$$

$$\left[\sum_{j=1}^{n} x_j^2\right] a_1 + \left[\sum_{j=1}^{n} x_j^3\right] a_2 + \left[\sum_{j=1}^{n} x_j^4\right] a_3 = \sum_{j=1}^{n} y_j x_j^2$$

This will be a nonsingular system due to the assumption that the points $\{x_j\}$ are distinct.

When $\hat{f}(x)$ takes the more general form given in (8.116), the choice for which the root-mean-square error E in (8.110) is minimized has coefficients $a_1, \ldots, a_m$ determined by the linear system

$$\sum_{k=1}^{m} a_k \left[\sum_{j=1}^{n} \varphi_k(x_j)\varphi_i(x_j)\right] = \sum_{j=1}^{n} y_j \varphi_i(x_j), \qquad i = 1, \ldots, m \quad (8.122)$$

The derivation of this is left as problem 6. The special case in which $\hat{f}(x)$ is a polynomial of degree $m - 1$ is very important, and we consider it in greater detail.

Write $\hat{f}(x)$ in the standard form for a polynomial of degree $m - 1$,

$$\hat{f}(x) = a_1 + a_2 x + a_3 x^2 + \ldots + a_m x^{m-1} \qquad (8.123)$$

Then this corresponds to (8.116) with

$$\varphi_1(x) = 1, \qquad \varphi_2(x) = x, \qquad \varphi_3(x) = x^2, \ldots, \varphi_m(x) = x^{m-1} \quad (8.124)$$

The system (8.122) for determining $\{a_j\}$ becomes

$$\sum_{k=1}^{m} a_k \left[\sum_{j=1}^{n} x_j^{i+k-2}\right] = \sum_{j=1}^{n} y_j x_j^{i-1}, \qquad i = 1, 2, \ldots, m \quad (8.125)$$

When $m = 3$, this yields the system (8.121) obtained earlier.

The system (8.125) is nonsingular (for $m < n$), but unfortunately it becomes increasingly ill-conditioned as the degree of $\hat{f}(x)$ increases; furthermore the size of the condition number for the matrix of coefficients can be very large for fairly small values of m, say $m = 4$. For this reason, it is seldom advisable to use (8.125) to do a least squares polynomial fit, except for degree ≤ 2.

To do a least squares fit to data $\{(x_i, y_i)\}$ with a higher degree polynomial $\hat{f}(x)$, it is necessary to write it in a form

$$\hat{f}(x) = a_1\varphi_1(x) + \ldots + a_m\varphi_m(x) \qquad (8.126)$$

where $\varphi_1(x), \ldots, \varphi_m(x)$ are so chosen that the matrix of coefficients in (8.122) is not ill-conditioned. There are optimal choices of these functions $\varphi_j(x)$, with $\text{degree}(\varphi_j) = j - 1$ and with the coefficient matrix for (8.122) becoming diagonal. A nonoptimal but still satisfactory choice in general can be based on the Chebyshev polynomials $\{T_k(x)\}$ of Chapter 6.

Suppose that the nodes $\{x_i\}$ are chosen from an interval $[\alpha, \beta]$. Then introduce modified Chebyshev polynomials

$$\varphi_k(x) = T_{k-1}\left(\frac{2x - \alpha - \beta}{\beta - \alpha}\right), \qquad \alpha \leq x \leq \beta, \qquad k \geq 1 \quad (8.127)$$

Then degree $(\varphi_k) = k - 1$; and any polynomial $\hat{f}(x)$ of degree $m - 1$ can be written as a combination of $\varphi_1(x), \ldots, \varphi_m(x)$.

Example

Consider the data in Table 8.3 and the plot of it in Figure 8.3. We first find the least squares cubic polynomial fit $\hat{f}(x)$ using the normal representation

$$\hat{f}(x) = a_1 + a_2x + a_3x^2 + a_4x^3 \qquad (8.128)$$

The resulting linear system (8.125), denoted here by $La = b$, is given by

$$L = \begin{bmatrix} 21 & 10.5 & 7.175 & 5.5125 \\ 10.5 & 7.175 & 5.5125 & 4.51666 \\ 7.175 & 5.5125 & 4.51666 & 3.85416 \\ 5.5125 & 4.51666 & 3.85416 & 3.38212 \end{bmatrix}$$

$$a = [a_1, a_2, a_3, a_4]^T$$

$$b = [24.1180, 13.2345, 9.46836, 7.55944]^T$$

Table 8.3 Data for a Cubic Least Squares Fit

x_i	y_i	x_i	y_i
0.00	0.486	0.55	1.102
0.05	0.866	0.60	1.099
0.10	0.944	0.65	1.017
0.15	1.144	0.70	1.111
0.20	1.103	0.75	1.117
0.25	1.202	0.80	1.152
0.30	1.166	0.85	1.265
0.35	1.191	0.90	1.380
0.40	1.124	0.95	1.575
0.45	1.095	1.00	1.857
0.50	1.122		

The solution is

$$a = [0.5747, 4.7259, -11.1282, 7.6687]^T$$

The condition number (8.105) can be calculated to be

$$\mathrm{cond}\,(L) = \|L\|\,\|L^{-1}\| \doteq 22000 \qquad (8.129)$$

This is very large, and it indicates possible difficulty in finding an accurate answer for $La = b$. To verify this, perturb b above by adding to

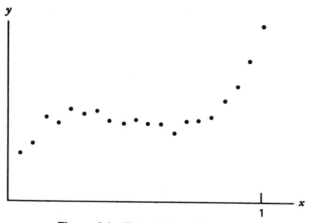

Figure 8.3 Plot of data of Table 8.3.

it the perturbation

$$[0.01, -0.01, 0.01, -0.01]^T$$

This will change b in its second place to the right of the decimal point within the range of possible perturbations due to errors in the data. The solution of the new perturbed system is

$$a = [0.7408, 2.6825, -6.1538, 4.4550]^T$$

which is dramatically different from the earlier result for a. The main point here is that use of (8.128) leads to a very ill-conditioned system of linear equations.

To find a cubic least squares approximation for the data in Table 8.3, with a better behaved linear system, use the functions in (8.127) on $[0, 1]$ and write

$$f(x) = a_1\varphi_1(x) + a_2\varphi_2(x) + a_3\varphi_3(x) + a_4\varphi_4(x)$$
$$\varphi_1(x) = T_0(2x - 1) \equiv 1, \ \varphi_2(x) = T_1(2x - 1) = 2x - 1$$
$$\varphi_3(x) = T_2(2x - 1) = 2(2x - 1)^2 - 1 = 8x^2 - 8x + 1 \quad (8.130)$$
$$\varphi_4(x) = T_3(2x - 1) = 4(2x - 1)^3 - 3(2x - 1)$$
$$= 32x^3 - 48x^2 + 18x - 1$$

Because of the different functions used in the representation of $\hat{f}(x)$ of (8.130), the values a_1, a_2, a_3, a_4 will be completely different than in the representation (8.128). The linear system (8.122) will again be denoted by $La = b$, now

$$L = \begin{bmatrix} 21 & 0 & -5.6 & 0 \\ 0 & 7.7 & 0 & -2.8336 \\ -5.6 & 0 & 10.4664 & 0 \\ 0 & -2.8336 & 0 & 11.01056 \end{bmatrix}$$

$$b = [24.118, 2.351, -6.01108, 1.523576]^T$$

The solution is

$$a = [1.160969, 0.393514, 0.046850, 0.239646]^T$$

The linear system is very stable with respect to the type of perturbation made in b with the earlier approach to the cubic least squares fit, using (8.128). This also is implied by the small condition number of L.

$$\text{cond } (L) = \|L\| \, \|L^{-1}\| = (26.6)(0.1804) \doteq 4.8 \qquad (8.131)$$

From the result (8.103) in section 8.5, relatively small perturbations in b will lead to correspondingly small changes in the solution a. Also, compare (8.131) to the condition number (8.129) for the earlier system. To give some idea of the accuracy of $\hat{f}(x)$ in approximating the data in Table 8.3, we easily compute the root-mean-square error from (8.110) to be

$$E \doteq 0.0421$$

a fairly small value when compared to the function values of $\hat{f}(x)$.

PROBLEMS

1. Calculate the linear least squares fit for the following data. Graph the data and the least squares fit. Also, find the root-mean-square error in the least squares fit.

x_i	y_i	x_i	y_i	x_i	y_i
0	−1.466	1.2	1.068	2.4	4.148
0.3	−0.062	1.5	1.944	2.7	4.464
0.6	0.492	1.8	2.583	3.0	5.185
0.9	0.822	2.1	3.239		

2. Repeat problem 1 with the following data:

x_i	y_i	x_i	y_i	x_i	y_i
−1.0	1.032	−0.3	1.139	0.4	−0.415
−0.9	1.563	−0.2	0.646	0.5	−0.112
−0.8	1.614	−0.1	0.474	0.6	−0.817
−0.7	1.377	0	0.418	0.7	−0.234
−0.6	1.179	0.1	0.067	0.8	−0.623
−0.5	1.189	0.2	0.371	0.9	−0.536
−0.4	0.910	0.3	0.183	1.0	−1.173

3. Do a quadratic least squares polynomial fit to the following data. Use the standard form

$$\hat{f}(x) = a_1 + a_2 x + a_3 x^2$$

with a_1, a_2, a_3 the solution of the system (8.121). Also, calculate the root-mean-square error, plot the data, and plot the least squares fit.

x_i	y_i	x_i	y_i	x_i	y_i
-1.0	7.904	-0.3	0.335	0.4	-0.711
-0.9	7.452	-0.2	-0.271	0.5	0.224
-0.8	5.827	-0.1	-0.963	0.6	0.689
-0.7	4.400	0	-0.847	0.7	0.861
-0.6	2.908	0.1	-1.278	0.8	1.358
-0.5	2.144	0.2	-1.335	0.9	2.613
-0.4	.581	0.3	-0.656	1.0	4.599

4. Plot the solution $\hat{f}(x)$ given following (8.130) for the data given in Table 8.3. Also plot the data from that table.

5. The process described in this section is called the least squares approximation method for discrete data. There is a corresponding method for approximating data given as a continuous function, although the reason for doing this is different than in the discrete case. To keep matters somewhat simple, consider approximating $f(x)$ by a quadratic polynomial

$$\hat{f}(x) = a_1 + a_2 x + a_3 x^2$$

on the interval $0 \le x \le 1$. Do so by choosing a_1, a_2, and a_3 to minimize the root-mean-square error

$$E = \left[\int_0^1 [a_1 + a_2 x + a_3 x^2 - f(x)]^2 \, dx \right]^{1/2}$$

Derive the linear system satisfied by the optimum choices of a_1, a_2, a_3. What is its relation to the Hilbert matrix of section 8.5?

6. Derive the linear system (8.122) by imitating the derivation of the system (8.120).

7. Replace the assumption of $\hat{f}(x) = mx + b$ with

$$\hat{f}(x) = a + be^{-x}$$

Find a and b by minimizing the root-mean-square error

$$E = \left[\frac{1}{n} \sum_{j=1}^{n} [a + be^{-x_j} - y_j]^2 \right]^{1/2}$$

Find a linear system satisfied by the optimum choice of a and b.

Hint: To simplify the work, minimize the sum in E, and then follow the derivation of (8.111) to (8.113).

8.7 THE EIGENVALUE PROBLEM

For a square matrix A, we define the concept of an eigenvalue of A as follows. The number λ is called an *eigenvalue* of A and the column matrix v is a corresponding *eigenvector* if

$$Av = \lambda v \tag{8.132}$$

and $v \neq 0$, the zero column matrix. The eigenvalue-eigenvector pairs $\{\lambda, v\}$ of a matrix A can be used in understanding what happens when a general column matrix x is multiplied by A and when solving a linear system $Ax = b$. The eigenvalues and eigenvectors also arise naturally in many physical problems, often leading to certain "basic solutions" that are of intrinsic physical interest, as in the natural frequencies of vibration of a taut string or a musical instrument. In this section, we give a brief introduction to the theory of eigenvalues and eigenvectors, and then we present a numerical method for calculating the largest eigenvalue of a matrix.

Example

For the matrix

$$A = \begin{bmatrix} 1.25 & 0.75 \\ 0.75 & 1.25 \end{bmatrix} \tag{8.133}$$

The eigenvalue–eigenvector pairs are

$$\begin{aligned} \lambda_1 = 2 \qquad & v^{(1)} = \begin{bmatrix} 1 \\ 1 \end{bmatrix} \\ \lambda_2 = 0.5 \qquad & v^{(2)} = \begin{bmatrix} -1 \\ +1 \end{bmatrix} \end{aligned} \tag{8.134}$$

The reader should verify that (8.132) is satisfied with both of these pairs.

We now illustrate the use of the eigenvalues and eigenvectors in further understanding the multiplication of a general column matrix x by A. For $x = [x_1, x_2]^T$, it can be verified directly that

$$x = c_1 v^{(1)} + c_2 v^{(2)} \tag{8.135}$$

with

$$c_1 = \frac{x_1 + x_2}{2}, \qquad c_2 = \frac{x_2 - x_1}{2}$$

Consider x as a vector from the origin to the point with coordinates (x_1, x_2) in the x_1x_2-plane. Then (8.135) says that x can be written as the sum of the two vectors $c_1v^{(1)}$ and $c_2v^{(2)}$, which is illustrated in Figure 8.4a.

Now compute Ax:

$$\begin{aligned} Ax &= A[c_1v^{(1)} + c_2v^{(2)}] \\ &= c_1Av^{(1)} + c_2Av^{(2)} \\ &= c_1\lambda_1v^{(1)} + c_2\lambda_2v^{(2)} \\ Ax &= 2c_1v^{(1)} + 0.5c_2v^{(2)} \end{aligned} \qquad (8.136)$$

Thus Ax is based on doubling the length of one part of x, the *component* $c_1v^{(1)}$, and halving the other part, the component $c_2v^{(2)}$. The resulting value of Ax is shown in Figure 8.4b. The figure suggests that the problem that led to calculating Ax might have been better formulated in the coordinate system determined by axes containing the eigenvectors $v^{(1)}$ and $v^{(2)}$. These same ideas extend to most other matrices of order 2 and to most other square matrices of higher order.

The Characteristic Polynomial

Rewrite the definition (8.132) as

$$(\lambda I - A)v = 0, \qquad v \neq 0 \qquad (8.137)$$

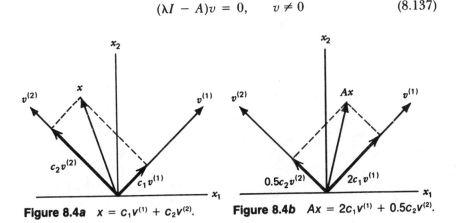

Figure 8.4a $x = c_1v^{(1)} + c_2v^{(2)}$.   **Figure 8.4b** $Ax = 2c_1v^{(1)} + 0.5c_2v^{(2)}$.

This is a homogeneous system of linear equations with the coefficient matrix $\lambda I - A$ and the nonzero solution v. From Theorem 8.1 of section 8.1, this can be true if and only if

$$\det (\lambda I - A) = 0 \qquad (8.138)$$

The expression

$$f(\lambda) = \det (\lambda I - A)$$

is called the *characteristic polynomial* of A, and its roots are the eigenvalues of A. Using the properties of determinants, and assuming A has order n, it can be shown that

$$f(\lambda) = \lambda^n + \alpha_{n-1}\lambda^{n-1} + \ldots + \alpha_1\lambda + \alpha_0 \qquad (8.139)$$

a polynomial of degree n. For the case $n = 2$,

$$f(\lambda) = \begin{bmatrix} \lambda - a_{11} & -a_{12} \\ -a_{21} & \lambda - a_{22} \end{bmatrix}$$

$$= (\lambda - a_{11})(\lambda - a_{22}) - a_{21}a_{12}$$

$$= \lambda^2 - (a_{11} + a_{22})\lambda + a_{11}a_{22} - a_{21}a_{12}$$

The formula (8.139) shows that a matrix A of order n can have at most n distinct eigenvalues.

Example

For the earlier example A of (8.133),

$$f(\lambda) = \det \begin{bmatrix} \lambda - 1.25 & -0.75 \\ -0.75 & \lambda - 1.25 \end{bmatrix}$$

$$= \lambda^2 - 2.5\lambda + 1$$

This has the roots $\lambda = 0.5, 2$, as given earlier in (8.134).

Once the eigenvalues are known, then the eigenvectors can be found by solving the linear system (8.137), with λ replaced by the actual eigenvalue. Note that the solution v is not unique, because any constant multiple is also a solution. The solution process for v is nontrivial, since there are places where rounding errors can propagate and expand rapidly in size. To illustrate the general solution process with a simple case, where rounding error is not a problem, we include the following example.

Example

Let

$$A = \begin{bmatrix} -7 & 13 & -16 \\ 13 & -10 & 13 \\ -16 & 13 & -7 \end{bmatrix} \tag{8.140}$$

Then

$$f(\lambda) = \det \begin{bmatrix} \lambda + 7 & -13 & 16 \\ -13 & \lambda + 10 & -13 \\ 16 & -13 & \lambda + 7 \end{bmatrix}$$
$$= \lambda^3 + 24\lambda^2 - 405\lambda + 972$$

The roots are

$$\lambda_1 = -36, \qquad \lambda_2 = 9, \qquad \lambda_3 = 3 \tag{8.141}$$

For $\lambda = \lambda_1 = -36$, we find the associated eigenvector $v = v^{(1)}$. If we substitute into (8.137), v satisfies

$$(-36I - A)v = 0$$
$$\begin{bmatrix} -29 & -13 & 16 \\ -13 & -26 & -13 \\ 16 & -13 & -29 \end{bmatrix} \begin{bmatrix} v_1 \\ v_2 \\ v_3 \end{bmatrix} = \begin{bmatrix} 0 \\ 0 \\ 0 \end{bmatrix}$$

If $v_1 = 0$ in this system, then it is straightforward to show that $v_2 = v_3 = 0$, as well. Thus, $v_1 \neq 0$ if we want $v \neq 0$. Since any multiple of v is also a solution, choose $v_1 = 1$. This leads to

$$-13v_2 + 16v_3 = 29$$
$$-26v_2 - 13v_3 = 13$$
$$-13v_2 - 29v_3 = -16$$

These have the solution $v_2 = -1$, $v_3 = 1$. Thus

$$v^{(2)} = [1, -1, 1]^T \qquad (8.142)$$

We leave the construction of $v^{(2)}$ and $v^{(3)}$ to problem 2.

In general, eigenvalues are not computed by finding the roots of the characteristic polynomial. Other techniques are used to transform A into an equivalent matrix with the same eigenvalues, but a matrix with which it is easier to work. We will not consider such methods, but excellent computer packages are available that implement such procedures in an easy to use manner.

Eigenvalues for Symmetric Matrices

The matrix $A = [a_{ij}]$ is symmetric if

$$A^T = A$$

or equivalently,

$$a_{ij} = a_{ji}, \qquad 1 \leq i, j \leq n$$

This means that A is visually symmetric about the diagonal of A, the line between the elements a_{11} and a_{nn}. The matrix

$$A = \begin{bmatrix} a & b & c \\ b & d & e \\ c & e & f \end{bmatrix}$$

is a general symmetric matrix of order 3. Symmetric matrices occur commonly in applications, and both the theory and the numerical analysis of the eigenvalue problem is simpler in this case. For symmetric matrices we have the following important result.

Theorem 8.5

Let A be symmetric and have order n. Then, there is a set of n eigenvalue-eigenvector pairs $\{\lambda_i, v^{(i)}\}$, $1 \leq i \leq n$, which satisfy the following properties.

i. The numbers $\lambda_1, \lambda_2, \ldots, \lambda_n$ are all of the roots of the characteristic polynomial $f(\lambda)$ of A, repeated according to their multiplicity. Moreover, all the λ_i are real numbers.

ii. If the column matrices $v^{(i)}$, $1 \leq i \leq n$ are regarded as vectors in n-dimensional space, then they are mutually perpendicular and of length 1.

iii. For each column matrix $x = [x_1, x_2, \ldots, x_n]^T$, there is a unique choice of constants $c_1, \ldots, c_n$ for which

$$x = c_1 v^{(1)} + \ldots + c_n v^{(n)} \tag{8.143}$$

The constants are given by

$$c_i = \sum_{j=1}^{n} x_j v_j^{(i)}, \qquad 1 \leq i \leq n \tag{8.144}$$

where $v^{(i)} = [v_1^{(i)}, \ldots, v_n^{(i)}]^T$.

iv. Define the matrix U of order n by

$$U = [v^{(1)}, v^{(2)}, \ldots, v^{(n)}] \tag{8.145}$$

Then

$$U^T A U = D \equiv \begin{bmatrix} \lambda_1 & 0 \ldots \ldots \ldots 0 \\ 0 & \lambda_2 \;\; 0 \ldots \ldots 0 \\ \vdots & \ddots \\ 0 \ldots \ldots \ldots 0 & \lambda_n \end{bmatrix} \tag{8.146}$$

and

$$U^T U = I \tag{8.147}$$

We omit the proof of this theorem, as it is quite lengthy and it requires other concepts of linear algebra that have not been included in this chapter. Matrices satisfying (8.147) are called *orthogonal matrices*, and as a class, they are important in the numerical analysis of matrix algebra problems. For examples of such orthogonal matrices, see problem 9, and also see problem 6 of section 8.3.

Example

Recall the matrix A in (8.133) and its eigenvalue pairs in (8.134). Figure 8.4 illustrates that they are perpendicular. To have them be of length 1, replace (8.134) by

$$\lambda_1 = 2, \qquad v^{(1)} = \begin{bmatrix} 1/\sqrt{2} \\ 1/\sqrt{2} \end{bmatrix}$$

$$\lambda_2 = 0.5, \qquad v^{(2)} = \begin{bmatrix} -1/\sqrt{2} \\ +1/\sqrt{2} \end{bmatrix}$$

The matrix U of (8.145) is given by

$$U = \begin{bmatrix} 1/\sqrt{2} & -1/\sqrt{2} \\ 1/\sqrt{2} & +1/\sqrt{2} \end{bmatrix}$$

Easily, $U^T U = I$. Also,

$$U^T A U = \begin{bmatrix} 1/\sqrt{2} & 1/\sqrt{2} \\ -1/\sqrt{2} & +1/\sqrt{2} \end{bmatrix} \begin{bmatrix} 1.25 & 0.75 \\ 0.75 & 1.25 \end{bmatrix} \begin{bmatrix} 1/\sqrt{2} & -1/\sqrt{2} \\ 1/\sqrt{2} & +1/\sqrt{2} \end{bmatrix}$$

$$= \begin{bmatrix} 2 & 0 \\ 0 & 0.5 \end{bmatrix}$$

as specified in (8.146).

The result (8.146) is often rewritten as

$$A = UDU^T \tag{8.148}$$

This decomposition of A is useful in many applications, both inside and outside of mathematics.

The Nonsymmetric Eigenvalue Problem

With nonsymmetric matrices, there are many more possibilities for the eigenvalues and eigenvectors. The roots of the characteristic polynomial $f(\lambda)$ may be complex numbers, and these lead to eigenvectors with complex numbers as components. Also, for multiple roots of $f(\lambda)$, it may not be possible to write an arbitrary column matrix x as a combination of eigenvectors, as was done in (8.143) for symmetric matrices; and there is no longer a simple formula such as (8.144) for the coefficients. The nonsymmetric eigenvalue problem is important, but it is too sophisticated in its general theory to be discussed here.

Example

We illustrate the existence of complex eigenvalues. Let

$$A = \begin{bmatrix} 0 & 1 \\ -1 & 0 \end{bmatrix}$$

The characteristic polynomial is

$$f(\lambda) = \det \begin{bmatrix} \lambda & -1 \\ 1 & \lambda \end{bmatrix} = \lambda^2 + 1$$

The roots of $f(\lambda)$ are complex,

$$\lambda_1 = i = \sqrt{-1}, \qquad \lambda_2 = -i$$

and the corresponding eigenvectors are

$$v^{(1)} = \begin{bmatrix} i \\ -1 \end{bmatrix}, \qquad v^{(2)} = \begin{bmatrix} i \\ 1 \end{bmatrix}$$

The Power Method

We now describe a numerical method that calculates the eigenvalue of A that is largest in size. It has some limitations as a general procedure, but it is simple to program and it is a satisfactory method for many matrices of interest. The matrix A need not be symmetric; but we will need to assume that the eigenvalues λ_i of A satisfy

$$|\lambda_1| > |\lambda_2| \geq \ldots \geq |\lambda_n| \tag{8.149}$$

where these eigenvalues are repeated according to their multiplicity as a root of the characteristic polynomial.

We define an iteration process for computing λ_1 and its associated eigenvector $v^{(1)}$. Choose an initial guess to $v^{(1)}$, calling it $z^{(0)}$. Generally it is chosen randomly, choosing each component using a random number generator; this is usually sufficient. Define

$$w^{(1)} = Az^{(0)}$$

and let α_1 be the maximum component of $w^{(1)}$, in size. If there is more than one such component, choose the first such component as α_1. Then define

$$z^{(1)} = \frac{1}{\alpha_1} w^{(1)}$$

Repeat the process iteratively. Define

$$w^{(m)} = Az^{(m-1)} \tag{8.150}$$

and let α_m be the maximum component of $w^{(m)}$, in size. Define

$$z^{(m)} = \frac{1}{\alpha_m} w^{(m)} \tag{8.151}$$

Then roughly speaking the vectors $z^{(m)}$ will converge to some multiple of $v^{(1)}$. To find λ_1 by this process, also pick some nonzero component of the vectors $z^{(m)}$ and $w^{(m)}$, say component k, and fix k. Often this is picked as the maximal component of $z^{(l)}$, for some sufficiently large l. Define

$$\lambda_1^{(m)} = \frac{w_k^{(m)}}{z_k^{(m-1)}} \tag{8.152}$$

where $z_k^{(m-1)}$ denotes component k of $z^{(m-1)}$. It can be shown that $\lambda_1^{(m)}$ converges to λ_1 as $m \to \infty$.

Example

We apply the power method to the earlier example (8.133), where

$$A = \begin{bmatrix} 1.25 & 0.75 \\ 0.75 & 1.25 \end{bmatrix} \tag{8.153}$$

The results are shown in Table 8.4. The column of successive differences $\lambda_1^{(m)} - \lambda_1^{(m-1)}$ and their successive ratios are included to show that there is a regular pattern to the convergence. The reason for this will be explained below. Note that in this example, $\lambda_1 = 2$ and $v^{(1)} = [1, 1]^T$, and the numerical results are converging to these.

Table 8.4 Power Method Example for (8.153)

m	$z_1^{(m)}$	$z_2^{(m)}$	$\lambda_1^{(m)}$	$\lambda_1^{(m)} - \lambda_1^{(m-1)}$	Ratio
0	1.0	0.5			
1	1.0	0.8461538	1.6250000		
2	1.0	0.9591837	1.8846154	2.596E − 1	
3	1.0	0.9896373	1.9693878	8.477E − 2	.327
4	1.0	0.9973992	1.9922280	2.284E − 2	.269
5	1.0	0.9993492	1.9980494	5.821E − 3	.255
6	1.0	0.9998373	1.9995119	1.462E − 3	.251
7	1.0	0.9999593	1.9998779	3.660E − 4	.250

Convergence of the Power Method

To analyze the convergence of the power method, we consider only the case when A is a symmetric matrix. Also assume that λ_1 is dominant, as in (8.149). We begin the analysis by showing that

$$z^{(m)} = \sigma_m \cdot \frac{A^m z^{(0)}}{\|A^m z^{(0)}\|} \tag{8.154}$$

where $\sigma_m = \pm 1$. For $m = 1$,

$$\alpha_1 = \pm\|w^{(1)}\| = \pm\|Az^{(0)}\|$$

with α_1 the maximal component of $w^{(1)}$. Then

$$z^{(1)} = \frac{1}{\alpha_1} w^{(1)} = \pm \frac{Az^{(0)}}{\|Az^{(0)}\|}$$

The maximum component of $z^{(1)}$, or the first such component if there are several maximal components in size, is forced to equal 1 in this definition. Thus the sign on the right side is to be chosen in accordance with that requirement on the maximal component. The proof of the general case of (8.154) can be done inductively, using the same ideas as for the $m = 1$ case. The sign σ_m is to be so chosen that the maximum component on the right side of (8.154) is 1, since that is true for $z^{(m)}$ by its definition (8.151).

Using (8.143), write

$$z^{(0)} = c_1 v^{(1)} + \ldots + c_n v^{(n)}, \tag{8.155}$$

where the c_j's are given in (8.144), with x replaced by $z^{(0)}$. We assume $c_1 \neq 0$. If $z^{(0)}$ is chosen randomly, then it is extremely likely that $c_1 \neq 0$; even if $c_1 = 0$, the power method will usually introduce a modified $c_1 \neq 0$, due to rounding errors. Apply A to $z^{(0)}$ in (8.155), to get

$$Az^{(0)} = c_1 Av^{(1)} + \ldots + c_n Av^{(n)}$$
$$= \lambda_1 c_1 v^{(1)} + \ldots + \lambda_n c_n v^{(n)}$$

Apply A repeatedly to get

$$A^m z^{(0)} = \lambda_1^m c_1 v^{(1)} + \ldots + \lambda_n^m c_n v^{(n)}$$
$$= \lambda_1^m \left[c_1 v^{(1)} + \left(\frac{\lambda_2}{\lambda_1}\right)^m c_2 v^{(2)} + \ldots + \left(\frac{\lambda_n}{\lambda_1}\right)^m c_n v^{(n)} \right]$$

From (8.154),

$$z^{(m)} = \sigma_m \left(\frac{\lambda_1}{|\lambda_1|}\right)^m \frac{c_1 v^{(1)} + \left(\frac{\lambda_2}{\lambda_1}\right)^m c_2 v^{(2)} + \ldots + \left(\frac{\lambda_n}{\lambda_1}\right)^m c_n v^{(n)}}{\left\| c_1 v^{(1)} + \left(\frac{\lambda_2}{\lambda_1}\right)^m c_2 v^{(2)} + \ldots + \left(\frac{\lambda_n}{\lambda_1}\right)^m c_n v^{(n)} \right\|} \tag{8.156}$$

As $m \to \infty$, the terms $(\lambda_j/\lambda_1)^m \to 0$, for $2 \leq j \leq n$, with $(\lambda_2/\lambda_1)^m$ the largest. Also,

$$\sigma_m \left(\frac{\lambda_1}{|\lambda_1|}\right)^m = \pm 1$$

Thus as $m \to \infty$, most terms in (8.156) are converging to zero, and we can cancel c_1 from numerator and denominator to see that $z^{(m)}$ looks like

$$z^{(m)} \doteq \hat{\sigma}_m \frac{v^{(1)}}{\|v^{(1)}\|}$$

where $\hat{\sigma}_m = \pm 1$. If the normalization of $z^{(m)}$ is modified, to always have some particular component be positive, then

$$z^{(m)} \to \pm \frac{v^{(1)}}{\|v^{(1)}\|} \equiv \hat{v}^{(1)} \tag{8.157}$$

with a fixed sign independent of m. Our earlier normalization of sign, dividing by α_m, will usually accomplish this, but not always. See problem 12. The error in $z^{(m)}$ will satisfy

$$\|z^{(m)} - \hat{v}^{(1)}\| \le c \left| \frac{\lambda_2}{\lambda_1} \right|^m, \qquad m \ge 0 \tag{8.158}$$

for some constant $c > 0$.

An analysis similar to the above can be given for the convergence of $\lambda_1^{(m)}$ to λ_1, with the same kind of error bound. Moreover, if we also assume

$$|\lambda_1| > |\lambda_2| > |\lambda_3| \ge |\lambda_4| \ge \ldots \ge |\lambda_n| \ge 0 \tag{8.159}$$

then we can show

$$\lambda_1 - \lambda_1^{(m)} \doteq c \left(\frac{\lambda_2}{\lambda_1} \right)^m \tag{8.160}$$

for some constant c, as $m \to \infty$. For example, recall the calculations in Table 8.3 for the matrix A in (8.153). There $\lambda_1 = 2$, $\lambda_2 = 0.5$, and $\lambda_2/\lambda_1 = 0.25$. Note that the ratios in the table are approaching 0.25, agreeing with (8.160).

Example

Apply the power method to the symmetric matrix

$$A = \begin{bmatrix} -7 & 13 & -16 \\ 13 & -10 & 13 \\ -16 & 13 & -7 \end{bmatrix} \tag{8.161}$$

From earlier, in (8.141), the eigenvalues are

$$\lambda_1 = -36, \qquad \lambda_2 = 9, \qquad \lambda_3 = 3$$

and the eigenvector $v^{(1)}$ associated with λ_1 is

$$v^{(1)} = [1, -1, 1]^T$$

The results of using the power method are shown in Table 8.5. Note that the ratios of the successive differences of $\lambda_1^{(m)}$ are approaching

$$\frac{\lambda_2}{\lambda_1} = -0.25$$

Also note that the maximal component of $z^{(m)}$ changes from one iteration to the next. The initial guess $z^{(0)}$ was chosen closer to $v^{(1)}$ than would be usual in actual practice. It was so chosen for purposes of illustration.

The error result (8.160) for the iterates $\{\lambda_1^{(m)}\}$ can be used to produce an acceleration method. From (8.160),

$$\lambda_1 - \lambda_1^{(m+1)} \doteq r(\lambda_1 - \lambda_1^{(m)}), \qquad r = \lambda_2/\lambda_1 \qquad (8.162)$$

for all sufficiently large values of m. Choose r using

$$r \doteq \frac{\lambda_1^{(m+1)} - \lambda_1^{(m)}}{\lambda_1^{(m)} - \lambda_1^{(m-1)}} \qquad (8.163)$$

The justification of this is left as problem 13(d). Using this r, solve for λ_1 in (8.162),

$$\lambda_1 \doteq \frac{1}{1 - r} [\lambda_1^{(m+1)} - r\lambda_1^{(m)}]$$

$$= \lambda_1^{(m+1)} + \frac{r}{1 - r} [\lambda_1^{(m+1)} - \lambda_1^{(m)}]$$

Table 8.5 The Power Method for (8.161)

m	$z_1^{(m)}$	$z_2^{(m)}$	$z_3^{(m)}$	$\lambda_1^{(m)}$	$\lambda_1^{(m)} - \lambda_1^{(m-1)}$	Ratio
0	1.000000	−0.800000	0.900000			
1	−0.972477	1.000000	−1.000000	−31.80000		
2	1.000000	−0.995388	0.993082	−36.82075	−5.03E0	
3	0.998265	−0.999229	1.000000	−35.82936	9.91E−1	−0.197
4	1.000000	−0.999775	0.999566	−36.04035	−2.11E−1	−0.213
5	0.999891	−0.999946	1.000000	−35.99013	5.02E−2	−0.238
6	1.000000	−0.999986	0.999973	−36.00245	−1.23E−2	−0.245
7	0.999993	−0.999997	1.000000	−35.99939	3.06E−3	−0.249

This is called *Aitken's extrapolation formula*. It also gives us the error estimate

$$\lambda_1 - \lambda_1^{(m+1)} \doteq \frac{r}{1-r} [\lambda_1^{(m+1)} - \lambda_1^{(m)}] \tag{8.165}$$

which is called Aitken's error estimation formula. Under the same assumption (8.159), similar results can also be shown for the iterates $\{z^{(m)}\}$.

Example

In Table 8.5, take $m + 1 = 7$. Then (8.165) yields

$$\lambda_1 - \lambda_1^{(7)} \doteq \frac{r}{1-r} [\lambda_1^{(7)} - \lambda_1^{(6)}]$$

$$= \frac{-0.249}{1 + 0.249} [0.00306] = -0.00061$$

which is the actual error. Also the Aitken formula (8.164) will give the exact answer for λ_1, to seven significant digits.

PROBLEMS

1. For the following matrices, find the characteristic polynomial, the eigenvalues, and the eigenvectors. Do it by hand computation.

    (a) $\begin{bmatrix} 1 & 2 \\ 5 & 4 \end{bmatrix}$ (b) $\begin{bmatrix} 2 & 36 \\ 36 & 23 \end{bmatrix}$ (c) $\begin{bmatrix} -9 & 6 \\ 3 & 8 \end{bmatrix}$

2. Find the eigenvectors $v^{(2)}$ and $v^{(3)}$ for (8.140), corresponding to the eigenvalues λ_2 and λ_3.

3. Find the characteristic polynomial, eigenvalues, and eigenvectors of

$$A = \begin{bmatrix} 2 & 1 & 0 \\ 1 & 3 & 1 \\ 0 & 1 & 2 \end{bmatrix}$$

4. Let A be a square matrix of order n, and let $f(\lambda)$ be its characteristic polynomial, as in (8.139).

    (a) Show that the constant term α_0 of $f(\lambda)$ is

$$\alpha_0 = (-1)^n \det(A)$$

(b)   For readers knowledgeable about determinants, show that the coefficient α_{n-1} is given by

$$\alpha_{n-1} = -[a_{11} + a_{22} + \ldots + a_{nn}]$$

The term in brackets is called the *trace* of A, and it is the sum of the diagonal terms of A.

5.   Continuing problem 4, show that

$$\det (A) = \lambda_1 \lambda_2 \ldots \lambda_n$$

$$\text{trace } (A) = \lambda_1 + \lambda_2 + \ldots + \lambda_n$$

where $\lambda_1, \ldots, \lambda_n$ are the roots of $f(\lambda)$, each repeated as many times as its multiplicity.

Hint: Expand $f(\lambda) = (\lambda - \lambda_1) \ldots (\lambda - \lambda_n)$.

6.   Show that the eigenvectors of (8.140) satisfy Theorem 8.5, provided that they are properly normalized to represent vectors of length 1. For example, change $v^{(1)}$ in (8.142) to

$$v^{(1)} = \left[\frac{1}{\sqrt{3}}, \frac{-1}{\sqrt{3}}, \frac{1}{\sqrt{3}} \right]^T$$

(The length of a vector is the square root of the sum of the squares of its components. For a general vector

$$x = [x_1, \ldots, x_n]^T$$

the length of x is $\sqrt{x_1^2 + \ldots + x_n^2}$.) Produce the matrix U of (8.145), and verify that it satisfies $U^T U = I$.

7.   Repeat problem 6 for the matrix in problem 1(b).

8.   Calculate the characteristic polynomial, eigenvalues, and eigenvectors of

(a) $\begin{bmatrix} 1 & 0 \\ 0 & 1 \end{bmatrix}$ (b) $\begin{bmatrix} 1 & 0 \\ 1 & 1 \end{bmatrix}$

Do your results contradict Theorem 8.5?

9.   Show that for any angle θ,

$$U = \begin{bmatrix} \cos \theta & \sin \theta \\ -\sin \theta & \cos \theta \end{bmatrix}$$

is an orthogonal matrix. These are called rotation matrices, because of their relationship to changes of rectangular coordinates by rotation of the coordinate axes.

10.   Apply the power method to the matrices of problem 1. In each

case, also calculate the successive differences $\lambda_1^{(m)} - \lambda_1^{(m-1)}$ and their successive ratios. Using (8.160) and these ratios, try to estimate λ_2.

11. Repeat problem 10 for the following matrices.

(a) $\begin{bmatrix} 1 & \frac{1}{2} \\ \frac{1}{2} & \frac{1}{3} \end{bmatrix}$

(b) $\begin{bmatrix} 1 & 5 & -8 \\ 5 & -2 & 5 \\ -8 & 5 & 1 \end{bmatrix}$

(c) $\begin{bmatrix} 1 & 2 & 3 \\ 2 & 3 & 4 \\ 3 & 4 & 5 \end{bmatrix}$

(d) $\begin{bmatrix} 33 & 16 & 72 \\ -24 & -10 & -57 \\ -8 & -4 & -17 \end{bmatrix}$

(e) $\begin{bmatrix} 1 & \frac{1}{2} & \frac{1}{3} \\ \frac{1}{2} & \frac{1}{3} & \frac{1}{4} \\ \frac{1}{3} & \frac{1}{4} & \frac{1}{5} \end{bmatrix}$

(f) $\begin{bmatrix} 6 & 4 & 4 & 1 \\ 4 & 6 & 1 & 4 \\ 4 & 1 & 6 & 4 \\ 1 & 4 & 4 & 6 \end{bmatrix}$

12. Use the power method on the example of (8.161), and choose $z^{(0)} = [1, 0, 1]^T$. Describe the pattern on convergence of the approximate $z^{(m)}$.

13. (a) Using the definition (8.152) of $\lambda_1^{(m)}$ and the assumption (8.149), show

$$\lim_{m \to \infty} \lambda_1^{(m)} = \lambda_1$$

(b) Show

$$|\lambda_1 - \lambda_1^{(m)}| \le c \, |\lambda_2/\lambda_1|^m, \qquad m \ge 0$$

for some constant c.

(c) Under the assumption (8.159), prove formula

$$\lambda_1 - \lambda_1^{(m)} \doteq c(\lambda_2/\lambda_1)^m, \qquad c = \text{constant}$$

with the approximation becoming better as $m \to \infty$. More precisely,

$$\lim_{m \to \infty} \frac{\lambda_1 - \lambda_1^{(m)}}{(\lambda_2/\lambda_1)^m} = c$$

a constant that is usually nonzero.

(d) Use (8.160) to show

$$\lim_{m \to \infty} \frac{\lambda_1^{(m+1)} - \lambda_1^{(m)}}{\lambda_1^{(m)} - \lambda_1^{(m-1)}} = \frac{\lambda_2}{\lambda_1}$$

thus justifying (8.163)

14. For symmetric matrices A, define the following variant of the power method, for computing λ_1:

$$\lambda_1^{(m)} = \frac{(z^{(m)})^T A (z^{(m)})}{(z^{(m)})^T (z^{(m)})} = \frac{(z^{(m)})^T (w^{(m+1)})}{(z^{(m)})^T (z^{(m)})}$$

Using Theorem 8.5, verify that

$$|\lambda_1 - \lambda_1^{(m)}| \leq c \left(\frac{\lambda_2}{\lambda_1}\right)^{2m}, \qquad m \geq 0$$

This is a faster rate of linear convergence, with the error decreasing by $(\lambda_2/\lambda_1)^2$ rather than (λ_2/λ_1). Illustrate the method by applying it to the example (8.161).

15. Compare the eigenvalues of

$$A = \begin{bmatrix} 101 & -90 \\ 110 & -98 \end{bmatrix}, \qquad \hat{A} = \begin{bmatrix} 100.999 & -90.001 \\ 110 & -98 \end{bmatrix}$$

How would you describe the sensitivity of the eigenvalues of A?

NINE

THE NUMERICAL SOLUTION OF DIFFERENTIAL EQUATIONS

Differential equations are one of the most important mathematical tools used in producing models of physical and biological processes. In this chapter, we will consider numerical methods for solving one class of problems for some of these equations: the initial value problem for ordinary differential equations. Many processes in the physical and biological sciences, engineering, and other areas are modeled by such equations.

The differential equations we will consider are of the form

$$y'(x) = f(x, y(x)), \qquad x \geqq x_0$$

where $y(x)$ is an unknown function that is being sought. The given function $f(x, z)$ of two variables defines the differential equation; examples will be given in section 9.1. This equation is called a first order differential equation, because it contains a first order derivative of the unknown function, but no higher order derivative. A higher order differential equation can be reformulated as a system of first order equations, and the numerical methods of this chapter can be extended to this new system. The reformulation of higher order equations and the numerical solution of systems of first order differential equations is discussed in the final section of this chapter.

A discussion of the theory of the initial value problem for ordinary differential equation is given in section 9.1; the idea of stability of differential equations is also introduced. The simplest of the numerical methods for solving (9.1), Euler's method, is given in sections 9.2 and 9.3. It is not an efficient method, but it is an intuitive way to introduce many important ideas. Sections 9.4 and 9.5 cover more sophisticated and rapidly convergent methods; and section 9.6 discusses stability of numerical methods. Higher order equations and systems of first order equations are taken up in section 9.7.

9.1 THEORY OF DIFFERENTIAL EQUATIONS: AN INTRODUCTION

The simplest form of differential equation is

$$y'(x) = f(x) \tag{9.1}$$

with $f(x)$ a given function. The general solution of this equation is

$$y(x) = \int f(x) \, dx + c \tag{9.2}$$

with c an arbitrary constant. A particular solution is obtained by specifying the value of $y(x)$ at some given point,

$$y(x_0) = y_0$$

thus determining the value of c in (9.2).

Example

Solve

$$y'(x) = \sin(x), \qquad y(\pi/3) = 2$$

Then

$$y(x) = -\cos(x) + c$$

The given value of y yields

$$2 = -\cos (\pi/3) + c = -0.5 + c$$
$$c = 2.5$$

Thus the desired solution is

$$y(x) = 2.5 - \cos (x)$$

The more general equation

$$y'(x) = f(x, y(x))$$

is approached in a similar spirit, but first we consider some examples. The equation

$$y'(x) = a(x)y(x) + b(x) \tag{9.3}$$

is called a first order *linear equation*. It can be solved exactly. For this equation,

$$f(x, z) = a(x)z + b(x)$$

and $a(x)$, $b(x)$ are given functions. Because of their relative simplicity, linear equations are often used in analytically investigating numerical methods for differential equations. A particularly useful case is the linear equation

$$y'(x) = \lambda y(x) + b(x), \qquad x \geq x_0 \tag{9.4}$$

with λ a given constant. The general solution is

$$y(x) = ce^{\lambda x} + \int_{x_0}^{x} e^{\lambda(x-t)} b(t) \, dt = e^{\lambda x} \left[c + \int_{x_0}^{x} e^{-\lambda t} b(t) \, dt \right] \tag{9.5}$$

with c an arbitrary constant.

Another interesting and useful class of equations is given by

$$y'(x) = a(x)[y(x)]^2 + b(x)y(x) + c(x) \tag{9.6}$$

This is an example of a nonlinear equation. For it,

$$f(x, z) = a(x)z^2 + b(x)z + c(x)$$

By choosing various functions $a(x)$, $b(x)$, and $c(x)$, one can get a wide variety of behavior in the solutions $y(x)$. For example, the equation

$$y'(x) = -[y(x)]^2 \tag{9.7}$$

has the general solution

$$y(x) = 1/(x + c) \tag{9.8}$$

with c an arbitrary constant. For

$$y'(x) = -[y(x)]^2 + y(x) \tag{9.9}$$

the general solution is

$$y(x) = 1/(1 + ce^{-x})$$

with c arbitrary.

As can be guessed from these examples, the differential equation $y'(x) = f(x, y(x))$ will have a general solution dependent on an arbitrary constant c. As with the earlier case (9.1) to (9.2), this constant is calculated by specifying the value of $y(x)$ at a particular point x_0,

$$y(x_0) = y_0$$

The point x_0 is called the initial point, and y_0 is called the initial value. We call the problem of solving

$$y'(x) = f(x, y(x)) \tag{9.10}$$
$$y(x_0) = y_0$$

the *initial value problem*.

Example

1. Find the solution of equation (9.9) satisfying $y(0) = 4$. Then

$$4 = 1/(1 + c)$$
$$c = -0.75$$
$$y(x) = 1/(1 - 0.75e^{-x}), \qquad x \geq 0$$

2. The solution of

$$y'(x) = \lambda y(x) + e^{-x}, \qquad y(0) = 1, \qquad \lambda \neq -1 \qquad (9.11)$$

is obtained from (9.5) as

$$
\begin{aligned}
y(x) &= e^{\lambda x} + \int_0^x e^{\lambda(x-t)} e^{-t} \, dt \\
&= e^{\lambda x}\left[1 + \frac{1}{\lambda + 1}\{1 - e^{-(\lambda+1)x}\}\right]
\end{aligned}
\qquad (9.12)
$$

To make precise the preceding statements, we give the following theorem, which describes the existence and unique solvability of the initial value problem (9.10).

Theorem 9.1

Let $f(x, z)$ and $\partial f(x, z)/\partial z$ be continuous functions of x and z at all points (x, z) in some neighborhood of the initial point (x_0, y_0). Then there is a unique function $y(x)$ defined on some interval $[x_0 - \alpha, x_0 + \alpha]$ and satisfying

$$y'(x) = f(x, y(x)), \qquad x_0 - \alpha \leq x \leq x_0 + \alpha$$
$$y(x_0) = y_0$$

Example

Solve

$$y'(x) = 2x[y(x)]^2, \qquad y(0) = 1$$

Then

$$f(x, z) = 2xz^2, \qquad \partial f(x, z)/\partial z = 4xz$$

and both of these functions are continuous for all (x, z). Thus, there is a unique solution to this initial value problem. It is

$$y(x) = 1/(1 - x^2), \qquad -1 < x < 1$$

This also illustrates that the continuity of $f(x, z)$ and $\partial f(x, z)/\partial z$ for all (x, z) does not imply the existence of a $y(x)$ that is continuous for all x.

Stability of the Initial Value Problem

When numerically solving the initial value problem (9.10), we will generally assume that the solution $y(x)$ is being sought on a given finite interval $x_0 \leq x \leq b$. In that case, it is possible to obtain the following stability result. Make a small change in the initial value for the initial value problem, changing y_0 to $y_0 + \epsilon$. Call the resulting solution $y_\epsilon(x)$,

$$y_\epsilon'(x) = f(x, y_\epsilon(x)), \qquad x_0 \leq x \leq b, \qquad y_\epsilon(x_0) = y_0 + \epsilon \qquad (9.13)$$

Then under hypotheses similar to those of Theorem 9.1, it can be shown that for all small values of ϵ,

$$\max_{x_0 \leq x \leq b} |y_\epsilon(x) - y(x)| \leq c\epsilon, \qquad \text{some } c > 0 \qquad (9.14)$$

Thus small changes in the initial value y_0 will lead to small changes in the solution $y(x)$ of the initial value problem. This is a desirable property for a variety of very practical reasons.

Example

The problem

$$y'(x) = -y(x) + 1, \qquad 0 \leq x \leq b, \qquad y(0) = 1 \qquad (9.15)$$

has the solution $y(x) \equiv 1$. The perturbed problem

$$y_\epsilon'(x) = -y_\epsilon(x) + 1, \qquad 0 \leq x \leq b, \qquad y_\epsilon(0) = 1 + \epsilon$$

has the solution $y_\epsilon(x) = 1 + \epsilon e^{-x}$, $x \geq 0$. Thus

$$y(x) - y_\epsilon(x) = -\epsilon e^{-x}$$

$$|y(x) - y_\epsilon(x)| \leq |\epsilon|, \qquad x \geq 0$$

The problem (9.15) is said to be stable.

Virtually all initial value problems are stable in the sense specified in (9.14); but this is only a partial picture of the effect of small perturbations of the initial value y_0. If the maximum error in (9.14) is much larger than ϵ, then the initial value problem (9.10) is usually considered to be *ill-conditioned*. Attempting to numerically solve such a problem will usually lead to large errors in the computed solution.

Example

The problem

$$y'(x) = \lambda[y(x) - 1], \qquad 0 \leq x \leq b, \qquad y(0) = 1 \qquad (9.16)$$

has the solution

$$y(x) = 1, \qquad \text{all } x$$

The perturbed problem

$$y'_\epsilon(x) = \lambda[y_\epsilon(x) - 1], \qquad 0 \leq x \leq b, \qquad y_\epsilon(0) = 1 + \epsilon$$

has the solution

$$y_\epsilon(x) = 1 + \epsilon e^{\lambda x}, \qquad x \geq 0$$

For the error

$$y(x) - y_\epsilon(x) = -\epsilon e^{\lambda x} \qquad (9.17)$$

$$\max_{0 \leq x \leq b} |y(x) - y_\epsilon(x)| = \begin{cases} |\epsilon|, & \lambda \leq 0 \\ |\epsilon| e^{\lambda b}, & \lambda > 0 \end{cases} \qquad (9.18)$$

If $\lambda < 0$, the error $y(x) - y_\epsilon(x)$ decreases as x increases. We say that (9.16) is well-conditioned when $\lambda \leq 0$. In contrast, for $\lambda > 0$, the error $y(x) - y_\epsilon(x)$ increases as x increases. And for a λb that is moderately large, say $\lambda b \geq 10$, the change in $y(x)$ is quite significant at $x = b$. The problem (9.16) is increasingly ill-conditioned as λ increases.

For the more general initial value problem (9.10) and the perturbed problem (9.13), one can show that

$$y(x) - y_\epsilon(x) \doteq -\epsilon \, e^{\int_{x_0}^x g(t) \, dt} \qquad (9.19)$$

where

$$g(t) = \frac{\partial f(t, y(t))}{\partial z}$$

for x sufficiently close to x_0. Note that this formula correctly predicts (9.17), since in that case

$$f(x, z) = \lambda[z - 1], \qquad \partial f(x, z)/\partial z = \lambda$$
$$\int_0^x g(t) \, dt = \lambda x$$

Then (9.19) yields

$$y(x) - y_\epsilon(x) \doteq -\epsilon \, e^{\lambda x}$$

which agrees with the earlier formula (9.17).

Example

The problem

$$y'(x) = -[y(x)]^2, \qquad y(0) = 1 \qquad (9.20)$$

has the solution

$$y(x) = \frac{1}{x + 1}$$

For the perturbed problem,

$$y_\epsilon'(x) = -[y_\epsilon(x)]^2, \qquad y_\epsilon(0) = 1 + \epsilon \qquad (9.21)$$

We use (9.19) to estimate $y(x) - y_\epsilon(x)$. First,

$$f(x, z) = -z^2, \quad \partial f(x, z)/\partial z = -2z$$
$$g(t) = -2y(t) = -2/(t + 1)$$
$$\int_0^x g(t)\, dt = -2 \int_0^x \frac{dt}{t + 1} = -2 \ln (1 + x) = \ln (1 + x)^{-2}$$
$$\exp\left(\int_0^x g(t)\, dt\right) = e^{\log(x+1)^{-2}} = \frac{1}{(x + 1)^2}$$

Substituting into (9.19) gives us

$$y(x) - y_\epsilon(x) \doteq \frac{-\epsilon}{(1 + x)^2}, \quad x \geq 0 \qquad (9.22)$$

for x sufficiently small. This indicates that (9.20) is a well-conditioned problem.

In general, if

$$\frac{\partial f(x, y(x))}{\partial z} \leq 0, \quad x_0 \leq x \leq b \qquad (9.23)$$

then we say the initial value problem is well-conditioned. And although this test depends on $y(x)$ on the interval $[x_0, b]$, often one can show (9.23) without knowing $y(x)$ explicitly.

PROBLEMS

⑫ In each of the following cases, show that the given function $y(x)$ satisfies the associated differential equation. Then determine the value of c required by the initial condition. Finally, with reference to the general format in (9.10), identify $f(x, z)$ for each differential equation.

(a) $y'(x) = -y(x) + \sin (x) + \cos (x)$, $y(0) = 1$; $y(x) = \sin (x) + ce^{-x}$
(b) $y'(x) = (y - y^2)/x$, $y(1) = 2$; $y(x) = x/(x + c)$, $x > 0$
(c) $y'(x) = \cos^2 (y(x))$, $y(0) = \pi/4$; $y(x) = \tan^{-1} (x + c)$
(d) $y'(x) = y(x)[y(x) - 1]$, $y(0) = \frac{1}{2}$; $y(x) = 1/(1 + ce^x)$

2. Solve the following problem by using (9.5)
   (a) $y'(x) = \lambda y(x) + 1$, $y(0) = 1$
   (b) $y'(x) = \lambda y(x) + x$, $y(0) = 0$

3. Consider the differential equation

$$y'(x) = f_1(x)f_2(y(x)),$$

for some given functions $f_1(x)$ and $f_2(z)$. This is called a *separable* differential equation, and it can be solved by direct integration. Write the equation as

$$\frac{y'(x)}{f_2(y(x))} = f_1(x)$$

and find the antiderivative of each side:

$$\int \frac{y'(x)\, dx}{f_2(y(x))} = \int f_1(x)\, dx$$

On the left side, change the integration variable by letting $z = y(x)$. Then the equation becomes

$$\int \frac{dz}{f_2(z)} = \int f_1(x)\, dx$$

After integrating, replace z by $y(x)$; then solve for $y(x)$, if possible. If these integrals can be evaluated, then the differential equation can be solved. Do so for the following problems, finding the general solution and the solution satisfying the given initial condition.
   (a) $y'(x) = x/y(x)$, $y(0) = 2$
   (b) $y'(x) = xe^{-y(x)}$, $y(1) = 0$

4. To confirm the accuracy of (9.22), calculate the exact solution of the perturbed problem (9.21) by using (9.7) to (9.8). Then calculate the exact error $y(x) - y_\epsilon(x)$ and compare it to the estimate in (9.22).

5. Check the conditioning of the initial value problems in problem 1 above. Use the test (9.23).

6. Use (9.23) to discuss the conditioning of the problem

$$y'(x) = y^2 - 5 \sin (x) - 25 \cos^2 (x), \qquad y(0) = 5$$

You do not need to know the true solution.

9.2 EULER'S METHOD

The simplest method for solving the initial value problem is called Euler's method. We define it in this section, and then we analyze it in the next section. Euler's method is not an efficient numerical method, but many of the ideas involved in the numerical solution of differential equations are introduced most simply with it.

Before beginning, we establish some notation that will be used from hereon in this chapter. Let $Y(x)$ be the true solution of the initial value problem, with initial value Y_0:

$$Y'(x) = f(x, Y(x)), \qquad x_0 \leq x \leq b \qquad (9.24)$$
$$Y(x_0) = Y_0$$

The approximate solution will be denoted using $y(x)$, with some variations. Numerical methods for solving (9.24) will find an approximate solution $y(x)$ at only a discrete set of nodes,

$$x_0 < x_1 < x_2 < \ldots < x_N \leq b \qquad (9.25)$$

and for our work, we will take these nodes to be evenly spaced:

$$x_n = x_0 + nh, \qquad n = 0, 1, \ldots, N$$

The following notations are all used for the approximate solution at the node points:

$$y(x_n) = y_h(x_n) = y_n, \qquad n = 0, 1, \ldots, N$$

To obtain a solution $y(x)$ at points in $[x_0, b]$ other than those in (9.25), some form of interpolation must be used. We will not consider that problem here, although the techniques of Chapter 5 are easily applied.

To derive Euler's method, recall the derivative approximation (7.65) from Chapter 7,

$$Y'(x) \doteq \frac{1}{h} [Y(x + h) - Y(x)]$$

Applying this to the initial value problem (9.24) at $x = x_n$,

$$Y'(x_n) = f(x_n, Y(x_n))$$

we obtain

$$\frac{1}{h}[Y(x_{n+1}) - Y(x_n)] \doteq f(x_n, Y(x_n))$$

$$Y(x_{n+1}) \doteq Y(x_n) + hf(x_n, Y(x_n)) \tag{9.26}$$

Euler's method is defined by taking this to be exact:

$$y_{n+1} = y_n + hf(x_n, y_n), \qquad 0 \leq n \leq N - 1 \tag{9.27}$$

For the initial guess, use $y_0 = Y_0$ or some close approximation of Y_0. Sometimes Y_0 is obtained empirically and thus may be known only approximately. Formula (9.27) gives a rule for computing $y_1, y_2, \ldots, y_N$ in succession. This is typical of most numerical methods for solving ordinary differential equations.

Some geometric insight into Euler's method is given in Figure 9.1. The tangent line to the graph of $z = Y(x)$ at x_n has slope

$$Y'(x_n) = f(x_n, Y(x_n))$$

Using this tangent line to approximate the curve near the point $(x_n, Y(x_n))$, the value of the tangent line at $x = x_{n+1}$ is given by the right side of (9.26).

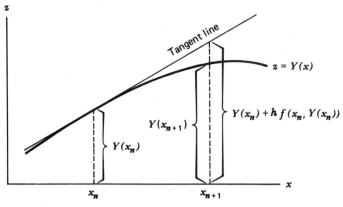

Figure 9.1   Illustration of Euler's method derivation.

Examples

(a) Solve

$$Y'(x) = -Y(x), \qquad Y(0) = 1 \qquad (9.28)$$

The true solution is $Y(x) = e^{-x}$. Then Euler's method is given by

$$y_{n+1} = y_n - hy_n, \qquad n \geq 0 \qquad (9.29)$$

with $y_0 = 1$ and $x_n = nh$. The solution $y(x)$ is given in Table 9.1, for three values of h and selected values of x. To illustrate the procedure, we compute y_1 and y_2 when $h = 0.1$. From (9.29),

$$y_1 = y_0 - hy_0 = 1 - (0.1)(1) = 0.9, \qquad x_1 = 0.1$$
$$y_2 = y_1 - hy_1 = 0.9 - (0.1)(0.9) = 0.81, \qquad x_2 = 0.2$$

For the error in these values,

$$Y(x_1) - y_1 = e^{-0.1} - y_1 = 0.004837$$
$$Y(x_2) - y_2 = e^{-0.2} - y_2 = 0.001873$$

Table 9.1 Euler's Method for (9.28)

h	x	$y_h(x)$	Error	Relative Error
0.2	1.0	3.2768E−1	4.02E−2	0.109
	2.0	1.0738E−1	2.80E−2	0.207
	3.0	3.5184E−2	1.46E−2	0.293
	4.0	1.1529E−2	6.79E−3	0.371
	5.0	3.7779E−3	2.96E−3	0.439
0.1	1.0	3.4867E−1	1.92E−2	0.0522
	2.0	1.2158E−1	1.38E−2	0.102
	3.0	4.2391E−2	7.40E−3	0.149
	4.0	1.4781E−2	3.53E−3	0.193
	5.0	5.1538E−3	1.58E−3	0.234
0.05	1.0	3.5849E−1	9.39E−3	0.0255
	2.0	1.2851E−1	6.82E−3	0.0504
	3.0	4.6070E−2	3.72E−3	0.0747
	4.0	1.6515E−2	1.80E−3	0.0983
	5.0	5.9205E−3	8.18E−4	0.121

(b) Solve

$$Y'(x) = [Y(x) + x^2 - 2]/(x + 1), \qquad Y(0) = 2 \qquad (9.30)$$

whose true solution is

$$Y(x) = x^2 + 2x + 2 - 2(x + 1) \log (x + 1)$$

Euler's method for this differential equation is

$$y_{n+1} = y_n + h(y_n + x_n^2 - 2)/(x_n + 1), \qquad n \geq 0$$

with $y_0 = 2$ and $x_n = nh$. The solution $y(x)$ is given in Table 9.2 for three values of h and selected values of x. A graph of the solution $y_h(x)$ for $h = 0.5$ is given in Figure 9.2. The node values $y_h(x_n)$ have been connected by straight line segments in the graph. Note that the horizontal and vertical scales are different.

In both examples, observe the behavior of the error as h decreases. For each fixed value of x, note that the errors decrease by a factor of about 2 when h is halved. As an illustration, take example (a) with $x = 5.0$. The errors for $h = 0.2$, 0.1, and 0.05, respectively, are

$$2.96E\text{-}3, \qquad 1.58E\text{-}3, \qquad 8.18E\text{-}4$$

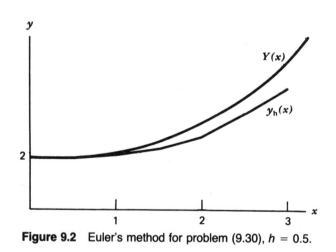

Figure 9.2 Euler's method for problem (9.30), $h = 0.5$.

Table 9.2 Euler's Method for (9.30)

h	x	$y_h(x)$	Error	Relative Error
0.2	1.0	2.1592	6.82E − 2	0.0306
	2.0	3.1697	2.39E − 1	0.0701
	3.0	5.4332	4.76E − 1	0.0805
	4.0	9.1411	7.65E − 1	0.129
	5.0	14.406	1.09E0	0.0703
	6.0	21.303	1.45E0	0.0637
0.1	1.0	2.1912	3.63E − 2	0.0163
	2.0	3.2841	1.24E − 1	0.0364
	3.0	5.6636	2.46E − 1	0.0416
	4.0	9.5125	3.93E − 1	0.0665
	5.0	14.939	5.60E − 1	0.0361
	6.0	22.013	7.44E − 1	0.0327
0.05	1.0	2.2087	1.87E − 2	0.00840
	2.0	3.3449	6.34E − 2	0.0186
	3.0	5.7845	1.25E − 1	0.0212
	4.0	9.7061	1.99E − 1	0.0337
	5.0	15.214	2.84E − 1	0.0183
	6.0	22.381	3.76E − 1	0.0165

and these decrease by successive factors of 1.87 and 1.93. The reader should do the same calculation for other values of x, in both examples (a) and (b). Also, note that the behavior of the error as x increases may be quite different from the behavior of the relative error. In the second example, the relative errors increase initially, and then they decrease with increasing x.

An Euler's Method Program

The following program implements Euler's method. The program allows for several differential equations to be solved in one run, and it is written for interative computing. The part of the program actually devoted to Euler's method is relatively small; but the program's structure allows it to be easily extended to the methods taken up in later sections.

```
C       TITLE: DEMONSTRATION OF EULER'S METHOD.
C
C       THIS SOLVES THE INITIAL VALUE PROBLEM
C           Y'(X) = F(X,Y(X)) ,   XO .LE. X .LE. B ,   Y(XO)=YO.
C       THE FUNCTION F(X,Z) IS DEFINED BELOW, ALONG WITH THE TRUE
C       SOLUTION Y(X).   THE NUMBER OF THE PROBLEM TO BE SOLVED
C       IS SPECIFIED BY THE INPUT VARIABLE 'NUMDE', WHICH IS USED
C       IN THE FUNCTIONS 'F' AND 'Y'.  THE PROGRAM WILL REQUEST
C       THE PROBLEM PARAMETERS, ALONG WITH THE VALUES OF 'H' AND
C       'IPRINT'.  'H' IS THE STEPSIZE, AND 'IPRINT' IS THE NUMBER
C       OF STEPS BETWEEN EACH PRINTING OF THE SOLUTION.
C       USE H=0 AND NUMDE=0 TO STOP THE PROGRAM.
C
        COMMON/BLOCKF/NUMDE
C
C       INPUT PROBLEM PARAMETERS.
10      PRINT *, ' NUMDE=?'
        READ *, NUMDE
        IF(NUMDE .EQ. 0) CALL EXIT
        PRINT *, ' GIVE XO,B, AND YO.'
        READ *, XZERO,B,YZERO
C
20      PRINT *, ' GIVE H AND IPRINT.'
        READ *, H,IPRINT
        IF(H .EQ. 0.0) GO TO 10
C
C       INITIALIZE.
        XO=XZERO
        YO=YZERO
        PRINT 1000, NUMDE,XZERO,B,YZERO,H,IPRINT
1000    FORMAT(//,' EQUATION',I2,5X,'XZERO=',1PE9.2,5X,'B=',E9.2,
     *         5X,'YZERO=',E12.5,/,' STEPSIZE=',E10.3,5X,
     *         'PRINT PARAMETER=',I3,/)
C
C       BEGIN THE MAIN LOOP FOR COMPUTING THE SOLUTION OF
C       THE DIFFERENTIAL EQUATION.
30      DO 40 K=1,IPRINT
          X1=XO+H
          IF(X1 .GT. B) GO TO 20
          Y1=YO+H*F(XO,YO)
          XO=X1
40        YO=Y1
C
C       CALCULATE ERROR AND PRINT RESULTS.
        TRUE=Y(X1)
        ERROR=TRUE-Y1
        PRINT 1001, X1,Y1,ERROR
1001    FORMAT(' X=',1PE10.3,5X,'Y(X)=',E17.10,5X,'ERROR=',E9.2)
        GO TO 30
        END

        FUNCTION F(X,Z)
C
C       THIS DEFINES THE RIGHT SIDE OF
C       THE DIFFERENTIAL EQUATION.
```

```
C
        COMMON/BLOCKF/NUMDE
C
        GO TO(10,20,30), NUMDE
10      F=-Z
        RETURN
20      F=(Z+X*X-2.0)/(X+1.0)
        RETURN
30      F=COS(Z)**2
        RETURN
        END

        FUNCTION Y(X)
C
C       THIS GIVES THE TRUE SOLUTION OF
C       THE INITIAL VALUE PROBLEM.
C
C       COMMON/BLOCKF/NUMDE
C
        GO TO(10,20,30), NUMDE
10      Y=EXP(-X)
        RETURN
20      X1=X+1.0
        Y=X*X-2.0*(X1*LOG(X1)-X1)
        RETURN
30      Y=ATAN(X)
        RETURN
        END
```

PROBLEMS

1. Solve the following problems using Euler's method with stepsizes of $h = 0.2, 0.1, 0.05$. Compute the error and relative error using the true answer $Y(x)$. For selected values of x, observe the ratio by which the error decreases when h is halved.

   (a) $Y'(x) = [\cos (Y(x))]^2, 0 \leq x \leq 10, Y(0) = 0.$

   $Y(x) = \tan^{-1} (x)$

   (b) $Y'(x) = \dfrac{1}{1 + x^2} - 2[Y(x)]^2, 0 \leq x \leq 10, Y(0) = 0.$

   $Y(x) = \dfrac{x}{1 + x^2}$

   (c) $Y'(x) = \dfrac{1}{4} Y(x)\left[1 - \dfrac{1}{20} Y(x)\right], 0 \leq x \leq 20, Y(0) = 1.$

   $Y(x) = 20/(1 + 19e^{-x/4})$

   (d) $Y'(x) = -[Y(x)]^2, 1 \leq x \leq 10, Y(1) = 1.$

   $Y(x) = 1/x$

2. Consider the linear equation

$$Y'(x) = \lambda Y(x) + (1 - \lambda) \cos (x) - (1 + \lambda) \sin (x), Y(0) = 1$$

from (9.4) of section 9.1. The true solution is $Y(x) = \sin (x) + \cos (x)$.
Solve this problem using Euler's method with several values of λ and
h, for $0 \leq x \leq 10$. Comment on the results.
  (a) $\lambda = -1; h = 0.5, 0.25, 0.125$
  (b) $\lambda = 1, h = 0.5, 0.25, 0.125$
  (c) $\lambda = -5; h = 0.5, 0.25, 0.125, 0.0625$
  (d) $\lambda = 5; h = 0.0625$

3. As a special case in which the error of Euler's method can be analyzed
directly, consider Euler's method applied to

$$Y'(x) = Y(x), \qquad Y(0) = 1$$

The true solution is e^x.
  (a) Show that the solution of Euler's method can be written as

$$y_h(x_n) = c(h)^{x_n}, \qquad n \geq 0$$

  where

$$c(h) = (1 + h)^{1/h}$$

  (b) Using L'hospital's rule from calculus, show

$$\lim_{h \to 0} c(h) = e$$

  This then proves that for fixed $x = x_n$,

$$\lim_{h \to 0} y_h(x) = e^x$$

  (c) Show that

$$\max_{0 \leq x \leq 1} |Y(x_n) - y_n(x_n)| = e - c(h) \doteq \frac{h}{2} e$$

  This shows that the error is proportional to h, just as with the
  earlier examples in Tables 9.1 and 9.2.

9.3 CONVERGENCE ANALYSIS OF EULER'S METHOD

The purpose of analyzing Euler's method is to understand how it works,
to be able to predict the error when using it, and to perhaps accelerate
its convergence. Being able to do this for Euler's method will also make

it easier to answer similar questions for other, more efficient numerical methods.

We begin by considering the error in approximation (9.26) that led directly to Euler's method. Using Taylor's theorem, write

$$Y(x_{n+1}) = Y(x_n) + hY'(x_n) + \frac{h^2}{2} Y''(\xi_n)$$

for some $x_n \leq \xi_n \leq x_{n+1}$. Using the fact that $Y(x)$ satisfies the differential equation, this becomes

$$Y(x_{n+1}) = Y(x_n) + hf(x_n, Y(x_n)) + \frac{h^2}{2} Y''(\xi_n) \qquad (9.31)$$

The term

$$T_{n+1} = \frac{h^2}{2} Y''(\xi_n) \qquad (9.32)$$

is called the *truncation error* for Euler's method, and it is the error in the earlier approximation (9.26),

$$Y(x_{n+1}) \doteq Y(x_n) + hf(x_n, Y(x_n))$$

To analyze the error in Euler's method, subtract

$$y_{n+1} = y_n + hf(x_n, y_n)$$

from (9.31), to obtain

$$Y(x_{n+1}) - y_{n+1} = Y(x_n) - y_n + h[f(x_n, Y(x_n)) - f(x_n, y_n)]$$
$$+ \frac{h^2}{2} Y''(\xi_n) \qquad (9.33)$$

The error in y_{n+1} consists of two parts: (1) the truncation error T_{n+1}, newly introduced at step x_{n+1}, and (2) the propagated error

$$Y(x_n) - y_n + h[f(x_n, Y(x_n)) - f(x_n, y_n)]$$

The propagated error can be simplified by applying the mean value theorem to $f(x, z)$, considering it as a function of z:

$$f(x_n, Y(x_n)) - f(x_n, y_n) \doteq \frac{\partial f(x_n, Y(x_n))}{\partial z} [Y(x_n) - y_n]$$

Let $e_k \equiv Y(x_k) - y_k$, $k \geq 0$, and then use the above to rewrite (9.33) as

$$e_{n+1} \doteq \left[1 + h \frac{\partial f(x_n, Y(x_n))}{\partial z} \right] e_n + \frac{h^2}{2} Y''(\xi_n) \qquad (9.34)$$

These results can be used to give a general error analysis of Euler's method for the initial value problem. But before doing so, we consider two special cases that will yield some intuitive understanding of the error in Euler's method.

Using Euler's method, solve the problem

$$Y'(x) = 2x, \qquad Y(0) = 0$$

whose true solution is $Y(x) = x^2$. The error formula (9.34) [or (9.33)] becomes

$$e_{n+1} = e_n + h^2, \qquad e_0 = 0$$

This leads, by induction, to

$$e_n = nh^2, \qquad n \geq 0$$

Since $nh = x_n$,

$$e_n = hx_n$$

For each fixed x_n, the error at x_n is proportional to h. The truncation error is h^2, but the cumulative effect of these errors is a total error proportional to h.

For the second case, assume that

$$\frac{\partial f(x, z)}{\partial z} \leq 0$$

for all z and for all $x_0 \leq x \leq b$, the interval on which the differential equation is being solved. Note the relation of this to the stability condition (9.23) in section 9.1. Also assume that h has been chosen so small that

$$1 + h \frac{\partial f(x, Y(x))}{\partial z} \geq -1$$

in (9.34). Then (9.34) implies

$$|e_{n+1}| \leq |e_n| + ch^2, \qquad x_0 \leq x_n \leq b \qquad (9.35)$$

where

$$c = \frac{1}{2} \max_{x_0 \leq x \leq b} |Y''(x)|$$

and $e_0 = 0$. Applying (9.35) inductively, we obtain

$$|e_n| \leq nch^2 = cx_n h \qquad (9.36)$$

The error is bounded by a quantity proportional to h, rather than to the h^2 of the truncation error. Again, the cumulative effect of the repeated truncation errors causes the total error to be proportional to h, not h^2.

Using (9.34) or (9.33), along with much algebraic manipulation, one can prove the following general result.

Theorem 9.2

Assume

$$K = \sup_{\substack{-\infty < z < \infty \\ x_0 \leq x \leq b}} \left| \frac{\partial f(x, z)}{\partial z} \right| < \infty \qquad (9.37)$$

Then the Euler method solution $y_h(x)$ satisfies the error bound

$$|Y(x_n) - y_h(x_n)| \leq e^{(b-x_0)K} |Y_0 - y_0|$$
$$+ h \left[\frac{e^{(b-x_0)K} - 1}{2K} \right] \max_{x_0 \leq x \leq b} |Y''(x)| \qquad (9.38)$$

for all $x_0 \leq x_n \leq b$.

The proof is omitted; it can be found in most higher level numerical analysis textbooks. The hypothesis (9.37) makes the proof easier, but it can be weakened to be much the same as in Theorem 9.1.

When $y_0 = Y_0$, as is commonly the case, (9.38) can be written as

$$|Y(x_n) - y_h(x_n)| \leqq ch, \qquad x_0 \leqq x_n \leqq b$$

with c a constant. This is consistent with the behavior observed in Tables 9.1 and 9.2 in the preceding section, and it agrees with the special cases preceding the statement of the theorem. When h is halved, the bound ch is also halved, and that is the behavior in the error that was observed earlier. Euler's method is said to converge with order 1, because that is the power of h that occurs in the error bound. In general, if we have

$$|Y(x_n) - y_h(x_n)| \leqq ch^p, \qquad x_0 \leqq x_n \leqq b \tag{9.39}$$

then we say that the numerical method is *convergent with order p*.

Example

The problem

$$Y'(x) = -Y(x), \qquad Y(0) = 1 \tag{9.40}$$

was solved in section 9.2, with the results given in Table 9.1. To apply (9.38), we have $\partial f(x, z)/\partial z = -1, K = 1$. The true solution is $Y(x) = e^{-x}$; thus

$$\max_{0 \leqq x \leqq b} |Y''(x)| = 1$$

With $y_0 = Y_0 = 1$, the bound (9.38) becomes

$$|e^{-x_n} - y_h(x_n)| \leqq \frac{h}{2}[e^b - 1], \qquad 0 \leqq x_n \leqq b \tag{9.41}$$

As $h \to 0$, this shows $y_h(x)$ converges to e^{-x}. However, this bound is excessively conservative. And as b increases, the bound increases exponentially. For $b = 5$, the above bound is

$$|e^{-x_n} - y_h(x_n)| \leq \frac{h}{2} [e^5 - 1] = 73.7h, \qquad 0 \leq x_n \leq 5$$

And this is far larger than the actual errors shown in Table 9.1, by several orders of magnitude. A more accurate estimate can be based on the following results.

Asymptotic Error Analysis

To obtain more accurate predictions of the error, we consider asymptotic error estimates, similar to what was done earlier in Chapter 7 with numerical integration. An asymptotic error formula is available for Euler's method, provided that certain conditions are satisfied. If the two functions

$$\frac{\partial f(x, z)}{\partial z}, \qquad \frac{\partial^2 f(x, z)}{\partial z^2} \tag{9.42}$$

are continuous for all values of (x, z) near to $(x, Y(x))$, $x_0 \leq x \leq b$, then one can prove that the error is Euler's method satisfies

$$Y(x_n) - y_h(x_n) = hD(x_n) + O(h^2), \qquad x_0 \leq x_n \leq b \tag{9.43}$$

The conditions on (9.42) are almost always satisfied in practice, leading to (9.43), for all sufficiently small values of h. The notation $O(h^2)$ means a quantity proportional to h^2 or something smaller; it goes to zero more rapidly than the term $hD(x_n)$, which involves h to the smaller power of 1. Assuming $y_0 = Y_0$, the usual case, the function $D(x)$ satisfies an initial value problem for a linear differential equation,

$$D'(x) = g(x) D(x) + \frac{1}{2} Y''(x), \qquad D(x_0) = 0 \tag{9.44}$$

where

$$g(x) = \frac{\partial f(x, Y(x))}{\partial z}$$

Obtaining $D(x)$ involves knowing the solution $Y(x)$, which we would not know in a practical situation. But the formula (9.43) is still useful in obtaining practical methods to predict the error, as will be done following an example.

Example

Consider again the example (9.40). Then $D(x)$ satisfies

$$D'(x) = -D(x) + \frac{1}{2}e^{-x}, \qquad D(0) = 0$$

The solution is

$$D(x) = \frac{1}{2}xe^{-x}$$

Using (9.43), the error satisfies

$$Y(x_n) - y_h(x_n) \doteq \frac{h}{2}x_n e^{-x_n} \tag{9.45}$$

We are neglecting the $O(h^2)$ term, since it should be smaller than the term $hD(x)$ in (9.43), for all sufficiently small values of h. To check the accuracy of (9.45), consider $x_n = 5.0$ with $h = 0.05$. Then

$$\frac{h}{2}x_n e^{-x_n} = 0.000842$$

From Table 9.1, the actual error is 0.000818, which is quite close to our estimate of it.

Richardson Extrapolation

The formula

$$Y(x) - y_h(x) \doteq hD(x) \tag{9.46}$$

is an asymptotic error estimate, and (9.45) is an example of it. Using it, we can obtain a practical way to estimate the error in Euler's method.

In (9.46), replace h by $2h$, to obtain

$$Y(x) - y_{2h}(x) \doteq 2hD(x)$$
$$\doteq 2[Y(x) - y_h(x)]$$

Solving for $Y(x)$, we get

$$Y(x) \doteq 2y_h(x) - y_{2h}(x) \equiv \bar{y}_n(x) \tag{9.47}$$

and for the error in $y_h(x)$,

$$Y(x) - y_h(x) \doteq [2y_h(x) - y_{2h}(x)] - y_h(x)$$
$$Y(x) - y_h(x) \doteq y_h(x) - y_{2h}(x) \tag{9.48}$$

This is Richardson's error estimate; the right side of (9.47), denoted by $\bar{y}_h(x)$, is *Richardson's extrapolation formula*. With these formulas, we can estimate the error in Euler's method and can obtain a more rapidly convergent solution, $\bar{y}_h(x)$.

Example

Consider (9.40) with stepsize $h = 0.05$, $2h = 0.1$. Then Table 9.3 contains the Richardson extrapolation results for selected values of h. Note that (9.48) is a fairly accurate estimator of the error, and that $\bar{y}_h(x)$ is much more accurate than $y_h(x)$.

Using (9.43) with the procedure that led to (9.47), we can show that

$$Y(x_n) - \bar{y}_h(x_n) = O(h^2) \tag{9.49}$$

an improvement on the speed of convergence of Euler's method. We will repeat this type of extrapolation for the methods introduced in later sections. However, the formulas will be different than (9.47) to (9.48), and they will depend on the order of the method.

Table 9.3 Euler's Method with Richardson Extrapolation

x	$Y(x) - y_h(x)$	$y_h(x) - y_{2h}(x)$	$\bar{y}_h(x)$	$Y(x) - \bar{y}_h(x)$
1.0	9.39E−3	9.80E−3	3.6829346E−1	−4.14E−4
2.0	6.82E−3	6.94E−3	1.3544764E−1	−1.12E−4
3.0	3.72E−3	3.68E−3	4.9748443E−2	3.86E−5
4.0	1.80E−3	1.73E−3	1.8249877E−2	6.58E−5
5.0	8.17E−4	7.67E−4	6.6872853E−3	5.07E−5

PROBLEMS

1. Check the accuracy of the error bound (9.41) for $b = 1, 2, 3, 4, 5$ and $h = 0.2, 0.1, 0.05$. Compute the error bound and compare it with Table 9.1.

2. Compute the error bound (9.38), assuming $y_0 = Y_0$, for the problem (9.30) of section 9.2. Compare the bound with the actual errors given in Table 9.2, for $b = 1, 2, 3, 4, 5$ and $h = 0.2, 0.1, 0.05$.

3. Repeat problem 2 for the equation in problem 1(a) of section 9.2.

4. For problems 1(b) to (d) of section 9.2, the constant K in (9.37) will be infinite. In order to use the error bound (9.38) in such cases, let

$$K = 2 \cdot \max_{x_0 \leq x \leq b} \left| \frac{\partial f(x, Y(x))}{\partial z} \right|$$

   This can be shown to be adequate for all sufficiently small values of h. Then repeat problem 2 above for problems 1(b) to (d) of section 9.2.

5. The solution of

$$Y'(x) = \lambda Y(x) + \cos(x) - \lambda \sin(x), \quad Y(0) = 0$$

   is $Y(x) = \sin(x)$. Find the asymptotic error formula (9.43) in this case. Also compute the Euler solution for $0 \leq x \leq 6$, $h = 0.2, 0.1, 0.05$, and $\lambda = 1, -1$. Compare the true errors with those obtained from the asymptotic estimate

$$Y(x_n) - y_n \doteq hD(x_n)$$

6. Repeat problem 5 for the problem 1(d) of section 9.2. Compare for $1 \leq x \leq 6$, $h = 0.2, 0.1, 0.05$.

7. For the example (9.30) of section 9.2, with the numerical results in Table 9.2, use Richardson extrapolation to estimate the error $Y(x_n) - y_h(x_n)$ when $h = 0.05$. Also, produce the Richardson extrapolate $\tilde{y}_h(x_n)$ and compute its error. Do this for $x_n = 1, 2, 3, 4, 5, 6$.

8. Repeat problem 7 for the problems 1(a) to (d) of section 9.2.

9. Derive (9.49) from (9.43).

9.4 TAYLOR AND RUNGE–KUTTA METHODS

To improve on the speed of convergence of Euler's method, we look for approximations to $Y(x_{n+1})$ that are more accurate than the approximation

$$Y(x_{n+1}) \doteq Y(x_n) + hY'(x_n) \tag{9.50}$$

which led to Euler's method. Since this is a linear Taylor polynomial approximation, we first consider higher order Taylor approximations. Doing this will lead to a family of methods, depending on the order of the Taylor approximation being used.

To keep the initial explanations as intuitive as possible, we will develop a Taylor method for the problem

$$Y'(x) = -Y(x) + 2 \cos (x), \qquad Y(0) = 1 \tag{9.51}$$

whose true solution is $Y(x) = \sin (x) + \cos (x)$. To approximate $Y(x_{n+1})$ using information about Y at x_n, use the quadratic Taylor approximation

$$Y(x_{n+1}) \doteq Y(x_n) + hY'(x_n) + \frac{h^2}{2} Y''(x_n) \tag{9.52}$$

Its truncation error is

$$T_{n+1} = \frac{h^3}{6} Y'''(\xi_n), \qquad \text{some } x_n \leqq \xi_n \leqq x_{n+1} \tag{9.53}$$

To evaluate the right side of (9.52), $Y'(x_n)$ can be obtained directly from (9.51). For $Y''(x)$, differentiate (9.51) to get

$$Y''(x) = -Y'(x) - 2 \sin (x) = Y(x) - 2 \cos (x) - 2 \sin (x) \tag{9.54}$$

Then (9.52) becomes

$$Y(x_{n+1}) \doteq Y(x_n) + h[-Y(x_n) + 2 \cos (x_n)]$$
$$+ \frac{h^2}{2} [Y(x_n) - 2 \cos (x_n) - 2 \sin (x_n)].$$

By forcing equality, we are led to the numerical method

$$y_{n+1} = y_n + h[-y_n + 2 \cos (x_n)]$$
$$+ \frac{h^2}{2} [y_n - 2 \cos (x_n) - 2 \sin (x_n)], \qquad n \geqq 0 \tag{9.55}$$

with $y_0 = 1$. This should approximate the solution of the problem (9.51). And because the truncation error (9.53) contains a higher power of h than was true for Euler's method [see (9.32)], it is hoped that the method (9.55) will converge more rapidly.

Table 9.4 contains numerical results for (9.55) and for Euler's method, and it is clear that (9.55) is superior. In addition, if the results for stepsize $h = 0.1$ and 0.05 are compared, it can be seen that the errors are decreasing by a factor of about 4 when h is halved. This can be justified theoretically, as will be discussed later.

A Taylor method of order 3 for problem (9.51) can be obtained using the same procedure that led to (9.55). Differentiate (9.54) to obtain $Y'''(x)$,

$$Y'''(x) = Y'(x) + 2 \sin (x) - 2 \cos (x) = -Y(x) + 2 \sin (x) \quad (9.56)$$

Using this and the third order Taylor approximation

$$Y(x_{n+1}) \doteq Y(x_n) + hY'(x_n) + \frac{h^2}{2} Y''(x_n) + \frac{h^3}{6} Y'''(x_n)$$

we are led to the numerical method

$$y_{n+1} = y_n + h[-y_n + 2 \cos (x_n)] + \frac{h^2}{2} [y_n - 2 \cos (x_n) - 2 \sin (x_n)]$$

$$+ \frac{h^3}{6} [-y_n + 2 \sin (x_n)], \quad n \geq 0 \quad (9.57)$$

Table 9.4 Example of Second Order Taylor Method

		Taylor Method (9.55)		Euler
h	x	$y_h(x)$	Error	Error
0.1	2.0	0.492225829	9.25E − 4	− 4.64E − 2
	4.0	− 1.411659477	1.21E − 3	3.91E − 2
	6.0	0.682420081	− 1.67E − 3	1.39E − 2
	8.0	0.843648978	2.09E − 4	− 5.07E − 2
	10.0	− 1.384588757	1.50E − 3	2.83E − 2
0.05	2.0	0.492919943	2.31E − 4	− 2.30E − 2
	4.0	− 1.410737402	2.91E − 4	1.92E − 2
	6.0	0.681162413	− 4.08E − 4	6.97E − 3
	8.0	0.843801368	5.68E − 5	− 2.50E − 2
	10.0	− 1.383454154	3.62E − 4	1.39E − 2

It is left as problem 1 to repeat the calculations of Table 9.4 with this new method.

In general, to solve the initial value problem

$$Y'(x) = f(x, Y(x)), \qquad x_0 \leq x \leq b, \qquad Y(x_0) = Y_0 \qquad (9.58)$$

by the Taylor method, select an order of Taylor approximation and proceed as illustrated above. For order p, write

$$Y(x_{n+1}) \doteq Y(x_n) + hY'(x_n) + \ldots + \frac{h^p}{p!} Y^{(p)}(x_n) \qquad (9.59)$$

with the truncation error being

$$T_{n+1} = \frac{h^{p+1}}{(p+1)!} Y^{(p+1)}(\xi_n), \qquad x_n \leq \xi_n \leq x_{n+1} \qquad (9.60)$$

Find $Y''(x), \ldots, Y^{(p)}(x)$ by differentiating (9.58) successively, obtaining formulas implicitly involving only x_n and $Y(x_n)$. Substitute these formulas into (9.59) and then obtain a numerical method of the form

$$y_{n+1} = y_n + hy_n' + \frac{h^2}{2} y_n'' + \ldots + \frac{h^p}{p!} y_n^{(p)} \qquad (9.61)$$

by forcing (9.59) to be an equality.

If the solution $Y(x)$ and the function $f(x, z)$ are sufficiently differentiable, then it can be shown that the method (9.61) will satisfy

$$\max_{x_0 \leq x_n \leq b} |Y(x_n) - y_h(x_n)| \leq ch^p \max_{x_0 \leq x b} |Y^{(p+1)}(x)| \qquad (9.62)$$

The constant c is similar to that appearing in the error formula (9.38) for Euler's method. In addition, there is an asymptotic error formula

$$Y(x_n) - y_h(x_n) = h^p D(x_n) + O(h^{p+1}) \qquad (9.63)$$

with $D(x)$ satisfying a certain linear differential equation. The result (9.62) shows that higher order Taylor approximations lead to an equal, higher order of convergence. The asymptotic result (9.63) justifies the use of Richardson extrapolation to estimate the error and to accelerate the convergence.

Example

With $p = 2$, the formula (9.63) leads to

$$Y(x_n) - y_h(x_n) \doteq \tfrac{1}{3}[y_h(x_n) - y_{2h}(x_n)] \qquad (9.64)$$

This is left as problem 3 for the reader. To illustrate it, use the entries from Table 9.4 with $x_n = 10$.

$$y_{0.1}(10) = -1.384588757$$
$$y_{0.05}(10) = -1.383454154$$

From (9.64),

$$Y(10) - y_{0.05}(10) \doteq \tfrac{1}{3}[0.001134603] = 3.78\text{E-}4$$

And that is a good estimate of the true error 3.62E-4, given in Table 9.4.

Runge–Kutta Methods

The Taylor method is conceptually easy to work with, but it is bothersome and time-consuming to have to calculate the higher order derivatives. To avoid the need for the higher order derivatives, the Runge–Kutta methods evaluate $f(x, z)$ at more points, while attempting to equal the accuracy of the Taylor approximation. The methods obtained are fairly easy to program, and they are among the most popular methods for solving the initial value problem.

We will begin with Runge–Kutta methods of order 2, and later will consider some higher order methods. The Runge–Kutta methods have the general form

$$y_{n+1} = y_n + hF(x_n, y_n; h), \qquad n \geq 0, \qquad y_0 = Y_0 \qquad (9.65)$$

The quantity $F(x_n, y_n; h)$ can be thought of as some kind of average slope of the solution on the interval $[x_n, x_{n+1}]$. But its construction is based on making (9.65) act like a Taylor method. For methods of order two, we generally have

$$F(x, y; h) = \gamma_1 f(x, y) + \gamma_2 f(x + \alpha h, y + \beta h f(x, y)) \qquad (9.66)$$

The constants $\{\alpha, \beta, \gamma_1, \gamma_2\}$ are so chosen that when the true solution $Y(x)$ is substituted into (9.65), the truncation error

$$T_{n+1} \equiv Y(x_{n+1}) - [Y(x_n) + hF(x_n, Y(x_n); h)] \qquad (9.67)$$

will be $O(h^3)$, just as with the order 2 Taylor method.

It can be shown that this leads to

$$\gamma_1 = 1 - \gamma_2, \qquad \alpha = \beta = \frac{1}{2\gamma_2} \qquad (9.68)$$

with γ_2 arbitrary. Thus there is a family of Runge–Kutta methods of order 2, depending on the choice of γ_2. The three favorite choices are $\gamma_2 = 1/2, 3/4,$ and 1. With $\gamma_2 = 1/2$, we obtain the numerical method

$$y_{n+1} = y_n + \frac{h}{2} [f(x_n, y_n) + f(x_n + h, y_n + hf(x_n, y_n))], \qquad n \geq 0 \quad (9.69)$$

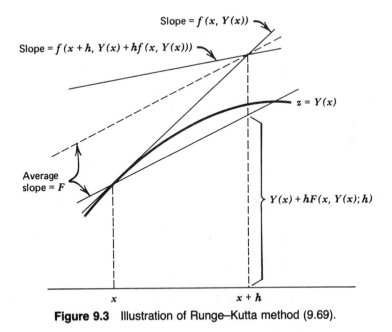

Figure 9.3 Illustration of Runge–Kutta method (9.69).

The number $y_n + hf(x_n, y_n)$ is the Euler solution at x_{n+1}. Using it, we obtain an approximation to the derivative at x_{n+1}, namely,

$$f(x_{n+1}, y_n + hf(x_n, y_n))$$

This and the slope $f(x_n, y_n)$ are then averaged to give an "average" slope of the solution on the interval $[x_n, x_{n+1}]$, giving

$$F(x_n, y_n; h) = \tfrac{1}{2}[f(x_n, y_n) + f(x_n + h, y_n + hf(x_n, y_n))]$$

This is then used to predict y_{n+1} from y_n, in (9.69). This definition is illustrated, in Figure 9.3, for $F(x, Y(x); h)$ as an average slope of Y' on $[x, x + h]$.

Example

Reconsider the earlier example (9.51),

$$Y'(x) = -Y(x) + 2 \cos(x), \qquad Y(0) = 1$$

Here

$$f(x, y) = -y + 2 \cos(x)$$

The numerical results from using (9.69) are given in Table 9.5. They show that the errors in this Runge–Kutta solution are comparable in accuracy to the results obtained with the Taylor method (9.55). In addition, the errors in Table 9.5 decrease by a factor of about 4 when h is halved.

Runge–Kutta methods of higher order can also be developed. A popular classical method is the following fourth order procedure.

$$v_1 = f(x_n, y_n)$$

$$v_2 = f(x_n + \frac{h}{2}, y_n + \frac{h}{2} v_1)$$

$$v_3 = f(x_n + \frac{h}{2}, y_n + \frac{h}{2} v_2) \tag{9.70}$$

$$v_4 = f(x_n + h, y_n + hv_3)$$

$$y_{n+1} = y_n + \frac{h}{6}[v_1 + 2v_2 + 2v_3 + v_4]$$

Table 9.5 Example of Second Order Runge–Kutta Method

| h | x | \multicolumn{2}{c}{Runge–Kutta (9.69)} |
		$y_h(x)$	Error
0.1	2.0	0.491215673	1.93E − 3
	4.0	− 1.407898629	− 2.55E − 3
	6.0	0.680696723	5.81E − 5
	8.0	0.841376339	2.48E − 3
	10.0	− 1.380966579	− 2.13E − 3
0.05	2.0	0.492682499	4.68E − 4
	4.0	− 1.409821234	− 6.25E − 4
	6.0	0.680734664	2.01E − 5
	8.0	0.843254396	6.04E − 4
	10.0	− 1.382569379	− 5.23E − 4

The truncation error in this method is $O(h^5)$. When the differential equation is simply

$$Y'(x) = f(x)$$

with no dependence of f on Y, this method reduces to Simpson's rule for numerical integration on $[x_n, x_{n+1}]$. The method (9.70) is easily programmed for a computer or hand calculator, and it is generally quite accurate. We leave its illustration as a problem for the reader.

If the solution $Y(x)$ of the initial value problem and the function $f(x, z)$ are sufficiently differentiable, and if the truncation error for the Runge–Kutta method is $O(h^{p+1})$, then it can be shown that the Runge–Kutta solution satisfies

$$\max_{x_0 \leq x_n \leq b} |Y(x_n) - y_h(x_n)| \leq ch^p \qquad (9.71)$$

The constant c depends on the derivatives of $Y(x)$ and the partial derivatives of $f(x, z)$. In addition, there is an asymptotic error formula of exactly the same general form (9.63) as was true for the Taylor method. Thus Richardson extrapolation can be justified for Runge–Kutta methods, and the error can be estimated. For the second order method (9.69), we obtain the error estimate

$$Y(x_n) - y_h(x_n) \doteq \tfrac{1}{3}[y_h(x_n) - y_{2h}(x_n)]$$

just as earlier for the second order Taylor method; see problem 3.

Example

Estimate the error for $h = 0.05$ and $x = 10$ in Table 9.5. Then

$$Y(10) - y_{0.05}(10) \doteq \tfrac{1}{3}[-0.0016028] = -5.34\text{E-}4$$

This compares closely with the actual error of $-5.23\text{E-}4$.

Error Prediction and Control

The easiest way to predict the error in a numerical solution $y_h(x)$ is to use Richardson extrapolation. Solve the initial value problem twice on the given interval $[x_0, b]$, with stepsizes $2h$ and h. Then use Richardson extrapolation to estimate $Y(x) - y_h(x)$ in terms of $y_h(x) - y_{2h}(x)$, as was done in (9.64) for a second order method. The cost of estimating the error in this way is an approximately 50% increase in the amount of computation, as compared to the cost of computing just $y_h(x)$. This may seem a large cost, but it is generally worth paying except for the most time-consuming of problems.

It would be desirable to have computer programs that would solve a differential equation on a given interval $[x_0, b]$ with an error less than a given $\epsilon > 0$. Unfortunately this is not possible with most types of numerical methods for the initial value problem. If at some point $\bar{x}$, we discover that $Y(\bar{x}) - y_h(\bar{x})$ is too large, then the error cannot be made smaller by merely decreasing h from thereon in the computation. The error $Y(\bar{x}) - y_h(\bar{x})$ depends on the cumulative effect of all preceding errors at points $x_n < \bar{x}$. Thus, to decrease the error at $\bar{x}$, it is necessary to repeat the solution of the equation from x_0, but with a smaller stepsize h. For this reason, most package programs for solving the initial value problem will not attempt to directly control the error, although they may try to monitor or bound it. Instead, they use indirect methods to affect the size of the error.

The error $Y(x_n) - y_h(x_n)$ is called the *global error* or total error at x_n. In contrast, the truncation error at x_n [see (9.60)] is called the *local error*, because it is the error introduced into the solution at step x_n. Most computer programs that contain error control are based on estimating the local error and then controlling it by varying h suitably; by so doing, they hope to keep the global error sufficiently small. If an error parameter $\epsilon > 0$ is given, the better programs choose the stepsize h to ensure that the local error T_{n+1} is much smaller, usually satisfying something like

$$|T_{n+1}| \leq \epsilon(x_{n+1} - x_n) \tag{9.72}$$

Then the global error is also kept small; for many differential equations, the global error will be less than $\epsilon(x_{n+1} - x_0)$.

To estimate the local error, various techniques can be used, including Richardson extrapolation. Recently another technique has been devised, and it has led to the currently most popular Runge–Kutta methods. Rather than computing a method of fixed order, one simultaneously computes using two methods of different orders. Then the higher order formula is used to estimate the error in the lower order formula. These methods are called Fehlberg methods; we will give one such pair of methods, of orders 4 and 5.

Define six intermediate slopes in $[x_n, x_{n+1}]$ by

$$v_0 = f(x_n, y_n)$$
$$v_i = f(x_n + a_i h, y_n + h \sum_{j=0}^{i-1} b_{ij} v_j), \qquad i = 1, 2, 3, 4, 5 \tag{9.73}$$

Then the fourth and fifth order formulas are given by

$$y_{n+1} = y_n + h \sum_{i=0}^{4} c_i v_i \tag{9.74}$$

$$\hat{y}_{n+1} = y_n + h \sum_{i=0}^{5} d_i v_i \tag{9.75}$$

The coefficients a_i, b_{ij}, c_i, d_i are given in Tables 9.6 and 9.7. The local error in the fourth order formula (9.74) is given by

$$T_{n+1} \doteq \hat{y}_{n+1} - y_{n+1} \tag{9.76}$$

It can be shown that this is a correct asymptotic result as $h \to 0$. Using this estimate, if T_{n+1} is too small or too large, then the stepsize can be varied so as to give a value for T_{n+1} of acceptable size.

The method (9.74) to (9.76) is easily programmed, and a simple-minded subroutine that implements it is given below, called FEHL45. To have a more completely automatic program requires a great deal more effort; for such a program, we recommend you check the library available from your computer center.

Table 9.6 Fehlberg Coefficients a_i, b_{ij}

i	a_i	b_{i0}	b_{i1}	b_{i2}	b_{i3}	b_{i4}
1	1/4	1/4				
2	3/8	3/32	9/32			
3	12/13	1932/2197	−7200/2197	7296/2197		
4	1	439/216	−8	3680/513	−845/4104	
5	1/2	−8/27	2	−3544/2565	1859/4104	−11/40

Table 9.7 Fehlberg Coefficients c_i, d_i

i	0	1	2	3	4	5
c_i	25/216	0	1408/2565	2197/4104	−1/5	
d_i	16/135	0	6656/12825	28561/56430	−9/50	2/55

Example

Solve

$$Y'(x) = -Y(x) + 2\cos(x), \qquad Y(0) = 1$$

whose true solution is $Y(x) = \sin(x) + \cos(x)$. Table 9.8 contains numerical results for $h = 0.25$ and 0.125. Compare the global errors with those in Tables 9.4 and 9.5, where second order methods were used. Also, it can be seen that the global errors in y_h decrease by factors of 17 to 21, which are fairly close to the theoretical value of 16 for a fourth order method. The truncation errors, estimated from (9.76), are included to show that they are quite different from the global error.

The method (9.73) to (9.76) uses $\hat{y}_{n+1}$ only for estimating the truncation error in the fourth order method. In practice, $\hat{y}_{n+1}$ is kept as the numerical solution rather than y_{n+1}; and thus $\hat{y}_n$ should replace y_n on the right sides of (9.73) to (9.75). The subroutine FEHL45 is easily converted by replacing the statement $Y = Y4$ with $Y = Y5$ near the end of the subroutine. The quantity in (9.76) will still be the truncation error in the fourth order method. Programs based on this will be fifth order, but they will vary their stepsize h to control the local error in the fourth order method. This tends to make these programs very accurate as regards global error.

Table 9.8 Example of Fourth Order Fehlberg Formula (9.74)

h	x	$y_h(x)$	$Y(x) - y_h(x)$	$\hat{y}_h(x) - y_h(x)$
0.25	2.0	0.493156301	$-5.71\text{E}-6$	$-9.49\text{E}-7$
	4.0	-1.410449823	$3.71\text{E}-6$	$1.62\text{E}-6$
	6.0	0.680752304	$2.48\text{E}-6$	$-3.97\text{E}-7$
	8.0	0.843864007	$-5.79\text{E}-6$	$-1.29\text{E}-6$
	10.0	-1.383094975	$2.34\text{E}-6$	$1.47\text{E}-6$
0.125	2.0	0.493150889	$-2.99\text{E}-7$	$-2.35\text{E}-8$
	4.0	-1.410446334	$2.17\text{E}-7$	$4.94\text{E}-8$
	6.0	0.680754675	$1.14\text{E}-7$	$-1.76\text{E}-8$
	8.0	0.843858525	$-3.12\text{E}-7$	$-3.47\text{E}-8$
	10.0	-1.383092786	$1.46\text{E}-7$	$4.65\text{E}-8$

Table 9.9 Example of Fifth Order Method (9.75)

h	x	$\hat{y}_n(x)$	$Y(x) - \hat{y}_n(x)$
0.25	2.0	0.493151148	$-5.58\text{E}-7$
	4.0	-1.410446359	$2.43\text{E}-7$
	6.0	0.680754463	$3.26\text{E}-7$
	8.0	0.843858731	$-5.18\text{E}-7$
	10.0	-1.383092745	$1.05\text{E}-7$
0.125	2.0	0.493150606	$-1.61\text{E}-8$
	4.0	-1.410446124	$8.03\text{E}-9$
	6.0	0.680754780	$8.65\text{E}-9$
	8.0	0.843858228	$1.53\text{E}-8$
	10.0	-1.383092644	$4.09\text{E}-9$

Example

Repeat the last example, but use the fifth order method described in the preceding paragraph. The results are given in Table 9.9. Note that the errors decrease by approximately 32 when h is halved, which is consistent with a fifth order method.

```
      SUBROUTINE FEHL45(F,Y,X,H,TRUN)
C
C     THIS COMPUTES A SINGLE STEP OF THE RUNGE-KUTTA-FEHLBERG METHOD
C     OF ORDER (4,5) FOR SOLVING THE DIFFERENTIAL EQUATION
C          Y'(X)=F(X,Y(X)).
C
```

```
C     INPUT:
C     (1) 'F' IS THE NAME OF A FUNCTION SUBPROGRAM FOR EVALUATING THE
C         DERIVATIVE F(X,Y) THAT DEFINES THE DIFFERENTIAL EQUATION.
C     (2) Y IS THE KNOWN VALUE OF THE SOLUTION AT THE POINT X.
C     (3) H IS THE STEPSIZE IN SOLVING THE DIFFERENTIAL EQUATION.
C         THE SOLUTION IS TO BE FOUND AT X+H.
C
C     OUTPUT:
C     (1) X WILL BE REPLACED BY X+H.
C     (2) Y WILL BE  REPLACED BY THE NUMERICAL SOLUTION BASED ON
C         USING THE FOURTH ORDER MEMBER OF THE FEHLBERG FORMULA
C         AT X+H.
C     (3) TRUN WILL BE SET TO THE ESTIMATED TRUNCATION ERROR IN THE
C         FOURTH ORDER SOLUTION Y; IT IS OBTAINED USING THE FIFTH
C         ORDER FEHLBERG SOLUTION.
C
C     TO SOLVE THE DIFFERENTIAL EQUATION ON A LONGER INTERVAL,
C     CALL THIS ROUTINE REPEATEDLY.
C

      DIMENSION A(5),B(5,0:4),C(0:4),D(0:5),V(0:5)
C     THE FOLLOWING CONSTANTS ARE NEEDED IN EVALUATING
C     THE FEHLBERG FORMULA.
      PARAMETER(B30=1932./2197.,B31=-7200./2197.,
     *    B32=7296./2197.,B40=439./216.,B42=3680./513.,
     *    B43=-845./4104.,B50=-8.0D0/27.0D0,B52=-3544./2565.,
     *    B53=1859./4104.,B54=-11./40.,A3=12./13.,
     *    C0=25./216.,C2=1408./2565.,C3=2197./4104.,
     *    D0=16./135.,D2=6656./12825.,D3=28561./56430.,
     *    D5=2./55.)
C     THE FOLLOWING ARRAYS ARE THE CONSTANTS IN THE FEHLBERG FORMULAS.
      DATA A/.25,.375,A3,1.0,0.5/,
     *    B/.25,.09375,B30,B40,B50,0.0,.28125,B31,-8.0,
     *    2.0,0.0,0.0,B32,B42,B52,0.0,0.0,0.0,B43,B53,
     *    0.0,0.0,0.0,0.0,B54/,
     *    C/C0,0.0,C2,C3,-.2/,
     *    D/D0,0.0,D2,D3,-.18,D5/
C
C     EVALUATE THE DERIVATIVES.
      V(0)=F(X,Y)
      DO 20 I=1,5
        SUM=0.0
        DO 10 J=0,I-1
10        SUM=SUM+B(I,J)*V(J)
20      V(I)=F(X+A(I)*H,Y+H*SUM)
C
C     EVALUATE THE FOURTH AND FIFTH ORDER FORMULAS.
      FSUM=0.0
      DO 30 I=0,4
30      FSUM=FSUM+C(I)*V(I)
      Y4=Y+H*FSUM
      FSUM=0.0
      DO 40 I=0,5
40      FSUM=FSUM+D(I)*V(I)
      Y5=Y+H*FSUM
C
```

```
C      SET THE OUTPUT VARIABLES.
       X=X+H
       Y=Y4
       TRUN=Y5-Y4
       RETURN
       END
```

PROBLEMS

1. Implement the numerical method (9.57) for solving the problem (9.51). Compute with stepsizes of $h = 0.1$, 0.05 for $0 \le x \le 10$. Compare to the values in Table 9.4, and also check the ratio by which the error decreases when h is halved.

   Hint: To make the programming easier, just modify the Euler program given in section 9.2.

2. Compute solutions to the following problems with a second order Taylor method. Use stepsizes $h = 0.2, 0.1, 0.05$.
   (a) $Y'(x) = [\cos (Y(x))]^2$, $0 \le x \le 10$, $Y(0) = 0$; $Y(x) = \tan^{-1}(x)$
   (b) $Y'(x) = 1/(1 + x^2) - 2 [Y(x)]^2$, $0 \le x \le 10$, $Y(0) = 0$;
        $Y(x) = x/(1 + x^2)$
   (c) $Y'(x) = \frac{1}{4} Y(x)[1 - \frac{1}{20} Y(x)]$, $0 \le x \le 20$, $Y(0) = 1$
        $Y(x) = 20/(1 + 19e^{-x/4})$
   (d) $Y'(x) = -[Y(x)]^2$, $1 \le x \le 10$, $Y(1) = 1$; $Y(x) = 1/x$
   These were solved previously in problem 1 of section 9.2. Compare your results with those earlier ones.

3. Recall the asymptotic error for Taylor methods, given in (9.63). For second order methods, this yields

   $$Y(x_n) - y_h(x_n) = h^2 D(x_n) + O(h^3)$$

   From this, derive the Richardson extrapolation formula

   $$Y(x_n) = \tfrac{1}{3}[4y_h(x_n) - y_{2h}(x_n)] + O(h^3)$$
   $$\doteq \tfrac{1}{3}[4y_h(x_n) - y_{2h}(x_n)] \equiv \tilde{y}_h(x_n)$$

   and the asymptotic error estimate

   $$Y(x_n) - y_h(x_n) = \tfrac{1}{3}[y_h(x_n) - y_{2h}(x_n)] + O(h^3)$$
   $$\doteq \tfrac{1}{3}[y_h(x_n) - y_{2h}(x_n)]$$

   Hint: Consider the formula

   $$Y(x_n) - y_{2h}(x_n) = 4h^2 D(x_n) + O(h^3)$$

   and combine it suitably with the earlier formula for $Y(x_n) - y_h(x_n)$.

4. Repeat problem 3 for methods of a general order $p \geq 1$. Derive the formulas

$$Y(x_n) \doteq \frac{1}{2^p - 1} [2^p y_h(x_n) - y_{2h}(x_n)] \equiv \tilde{y}_h(x_n)$$

with an error proportional to h^{p+1}, and

$$Y(x_n) - y_h(x_n) \doteq \frac{1}{2^p - 1} [y_h(x_n) - y_{2h}(x_n)]$$

5. Use problem 3 to estimate the errors in the results of Table 9.4, for $h = 0.05$. Also produce the Richardson extrapolate $\tilde{y}_h(x_n)$ and calculate its error. Compare its accuracy to that of $y_h(x_n)$.

6. Derive the second order Runge–Kutta methods (9.65) corresponding to $\gamma_2 = 3/4$ and 1 in (9.66). For $\gamma_2 = 1$, draw an illustrative graph analogous to Figure 9.3 for $\gamma_2 = 1/2$.

7. Solve the problem (9.51) with one of the formulas from problem 6. Compare your results to those in Table 9.5 for the formula (9.69) with $\gamma_2 = 1/2$.

8. Using (9.69), solve the equations in problem 2. Estimate the error using problem 3, and compare it to the true error.

9. Implement the classical procedure (9.70), and apply it to the equation (9.51). Solve it with stepsizes of $h = 0.25$ and 0.125. Compare with the results in Table 9.8, the fourth order Fehlberg example. *Hint:* Modify the Euler program of section 9.2.

10. Use the program of problem 9 to solve the equations in problem 2.

11. Solve the equations of problem 2 with the subroutine FEHL45 of this section. Use stepsizes of your choice. *Hint:* Combine FEHL45 with the Euler program of section 9.2, to have a driver program for the subroutine.

12. Consider the motion of a particle of mass in falling vertically under the earth's gravitational field, and suppose the downward motion is opposed by a frictional force $p(v)$ dependent on the velocity $v(t)$ of the particle. Then the velocity satisfies the equation

$$mv'(t) = -mg + p(v), \qquad t \geq 0, \; v(0) \text{ given}$$

Let $m = 1$ kg, $g = 9.8$ m/sec^2, and $v(0) = 0$. Solve the differential equation for $0 \leq t \leq 20$ and for the following choices of $p(v)$.
(a) $p(v) = -0.1v$, which is positive for a falling body
(b) $p(v) = 0.1v^2$

Find answers to at least three digits of accuracy. Graph the functions $v(t)$. Compare the solutions.

13. (a)   Using the Runge–Kutta method (9.69), solve

$$Y'(x) = -Y(x) + x^{-1}[1.1 + x], \quad Y(0) = 0$$

whose solution is $Y(x) = x^{1.1}$. Solve the equation on $[0, 5]$, printing the solution and the errors at $x = 1, 2, 3, 4, 5$. Use stepsizes $h = 0.1, 0.05, 0.025, 0.0125, 0.00625$. Calculate the ratios by which the errors decrease when h is halved. How does this compare with the theoretical rate of convergence of $O(h^2)$. Explain your results as best you can.

   (b)   What difficulty arises in attempting to use a Taylor method of order ≥ 2 to solve the equation of part (a)? What does it tell us about the solution?

9.5 MULTISTEP METHODS

Reformulate the differential equation

$$Y'(x) = .f(x, Y(X))$$

by integrating it over the interval $[x_n, x_{n+1}]$, obtaining

$$\int_{x_n}^{x_{n+1}} Y'(x)\, dx = \int_{x_n}^{x_{n+1}} f(x, Y(x))\, dx$$

$$Y(x_{n+1}) = Y(x_n) + \int_{x_n}^{x_{n+1}} f(x, Y(x))\, dx \tag{9.77}$$

We are going to develop numerical methods for finding the solution $Y(x)$ by approximating the integral in (9.77). There are many such methods, and we will consider only the most popular of them, the Adams–Bashforth (AB) and Adams–Moulton (AM) methods. These methods are the basis of some of the most widely used computer codes for solving the initial value problem. They are generally more efficient than the Runge–Kutta methods, especially if one wishes to find the solution with a high degree of accuracy.

To approximate the integral

$$\int_{x_n}^{x_{n+1}} g(x)\, dx, \qquad g(x) = f(x, Y(x)) = Y'(x) \tag{9.78}$$

approximate $g(x)$ by using polynomial interpolation and then integrate the interpolating polynomial. The AB methods use interpolation of degree q on the set of points $\{x_n, x_{n-1}, \ldots, x_{n-q}\}$, and AM methods use interpolation of degree q on the set of points $\{x_{n+1}, x_n, x_{n-1}, \ldots, x_{n-q+1}\}$. We will first consider AB methods, beginning with the AB method based on linear interpolation.

The linear polynomial interpolating $g(x)$ at $\{x_n, x_{n-1}\}$ is

$$p_1(x) = \frac{1}{h} [(x_n - x)g(x_{n-1}) + (x - x_{n-1})g(x_n)] \qquad (9.79)$$

Integrating over $[x_n, x_{n+1}]$, we obtain

$$\int_{x_n}^{x_{n+1}} g(x) \, dx \doteq \int_{x_n}^{x_{n+1}} p_1(x) \, dx = \frac{3h}{2} g(x_n) - \frac{h}{2} g(x_{n-1})$$

Using ideas similar to those used in sections 7.2 and 7.4 of Chapter 7, one can obtain the more complete result

$$\int_{x_n}^{x_{n+1}} g(x) \, dx = \frac{h}{2} [3g(x_n) - g(x_{n-1})] + \frac{5}{12} h^3 g''(\xi_n) \qquad (9.80)$$

for some $x_{n-1} \leq \xi_n \leq x_{n+1}$; also see problem 3. Applying this to the relation (9.77) gives us

$$Y(x_{n+1}) = Y(x_n) + \frac{h}{2} [3f(x_n, Y(x_n)) - f(x_{n-1}, Y(x_{n-1}))] + \frac{5}{12} h^3 Y'''(\xi_n)$$

Dropping the final term, the truncation error, we obtain the numerical method

$$y_{n+1} = y_n + \frac{h}{2} [3f(x_n, y_n) - f(x_{n-1}, y_{n-1})] \qquad (9.81)$$

This is an AB method of order 2, because its global error can be shown to be proportional to h^2.

With this method, note that it is necessary to have $n \geq 1$. Both y_0 and y_1 are needed in finding y_2; and y_1 cannot be found from (9.81). The value of y_1 must be obtained by another method. The method (9.81) is an example of a two step method, since values at x_{n-1} and x_n are needed in finding the value at x_{n+1}.

Example

Use (9.81) to solve

$$Y'(x) = -Y(x) + 2 \cos(x), \qquad Y(0) = 1 \qquad (9.82)$$

with the solution $Y(x) = \sin(x) + \cos(x)$. For illustrative purposes only, we take $y_1 = Y(x_1)$. The numerical results are given in Table 9.10. use $h = 0.05$. Note that the errors decrease by a factor of about 4 when h is halved, which is consistent with the numerical method being of order 2. The Richardson error estimate is also included in the table, using the estimate in (9.64) for second order methods. Where the error is decreasing like $O(h^2)$, the error estimate is quite accurate.

Adams methods are often considered to be less expensive than Runge–Kutta methods, and the main reason can be seen by comparing (9.81) with the second order Runge–Kutta method in (9.69). The main work of both methods is in the evaluations of the function $f(x, z)$. With second order Runge–Kutta methods, there are two evaluations of f for each step from x_n to x_{n+1}. In contrast, the AB formula (9.81) uses only one evaluation per step, provided past values of f are reused. There are other factors that affect the choice of a numerical method, but the AB and AM methods are generally more efficient in the number of evaluations of f that are needed for a given amount of accuracy.

A problem with multistep methods is in having to generate some of the initial values of the solution by using another method. For the second order AB method in (9.81), we must obtain y_1. And since the global error in $y_h(x_n)$ is to be $O(h^2)$, we must ensure that $Y(x_1) - y_h(x_1)$ is also $O(h^2)$. There are two immediate possibilities, using methods from preceding sections.

Case (i)   Use Euler's method:

$$y_1 = y_0 + hf(x_0, y_0) \qquad (9.83)$$

Table 9.10   An Example of the Second Order Adams–Bashforth Method

x	$y_h(x)$	$Y(x) - y_{2h}(x)$	$Y(x) - y_h(x)$	Ratio	$\frac{1}{3}[y_h(x) - y_{2h}(x)]$
2	0.49259722	2.13E−3	5.53E−4	3.9	5.26E−4
4	−1.41116963	2.98E−3	7.24E−4	4.1	7.52E−4
6	−0.67500371	−3.91E−3	−9.88E−4	4.0	−9.73E−4
8	0.84373678	3.68E−4	1.21E−4	3.0	8.21E−5
10	−1.38398254	3.61E−3	8.90E−4	4.1	9.08E−4

Assuming $y_0 = Y_0$, this has an error of

$$Y(x_1) - y_1 = \frac{h^2}{2} Y''(\xi_1)$$

based on (9.31). Thus (9.83) meets our error criteria for y_1. Globally, Euler's method is only $O(h)$, but a single step is $O(h^2)$.

Case (ii) Use a second order Runge–Kutta method, such as (9.69). Then $Y(x_1) - y_1$ will be $O(h^3)$, which is more than adequate.

Example

Use (9.83) with (9.81) to generate the last example (9.82). Then for $h = .05$ and $x = 10$, the error in the numerical solution turns out to be

$$Y(10) - y_h(10) = 9.90\text{E-}4$$

exactly the same as before for the results in Table 9.10.

Higher order Adams–Bashforth methods are obtained by using higher degree polynomial interpolation in the approximation of the integral in (9.78). For example, let $p_2(x)$ denote the quadratic polynomial that interpolates $g(x)$ at x_n, x_{n-1}, x_{n-2}. Then use

$$\int_{x_n}^{x_{n+1}} g(x)\, dx \doteq \int_{x_n}^{x_{n+1}} p_2(x)\, dx$$

Integrating this and including its error term (derived by other means), gives us

$$\int_{x_n}^{x_{n+1}} g(x)\, dx = \frac{h}{12} [23g(x_n) - 16g(x_{n-1}) + 5g(x_{n-2})] + \frac{3}{8} h^4 g'''(\xi_n)$$

for some $x_{n-2} \leq \xi_n \leq x_{n+1}$. Applying this to (9.77), the integral formulation of the differential equation, we obtain

$$Y(x_{n+1}) = Y(x_n) + \frac{h}{12} [23f(x_n, Y(x_n)) - 16f(x_{n-1}, Y(x_{n-1}))$$

$$+ 5f(x_{n-2}, Y(x_{n-2}))] + \frac{3}{8} h^4 Y^{(4)}(\xi_n)$$

By dropping the last term, the truncation error, we obtain the third order AB method

$$y_{n+1} = y_n + \frac{h}{12} [23y'_n - 16y'_{n-1} + 5y'_{n-2}], \qquad n \geq 2 \qquad (9.84)$$

where $y'_k \equiv f(x_k, y_k)$, $k \geq 0$. This is a three step method, requiring $n \geq 2$. Thus y_1, y_2 must be obtained separately by another method. We leave the illustration of (9.84) as problem 2 for the reader.

In general, it can be shown that the AB method based on interpolation of degree q will be a $(q + 1)$-step method and that its truncation error will be of the form

$$T_{n+1} = c_q h^{(q+2)} Y^{(q+2)}(\xi_n)$$

for some $x_{n-q} \leq \xi_n \leq x_{n+1}$. The initial values $y_1, \ldots, y_q$ will have to be generated by another method. If the error in these initial values satisfies

$$Y(x_n) - y_h(x_n) = O(h^{q+1}), \qquad n = 1, 2, \ldots, q \qquad (9.85)$$

then the global error in the $(q + 1)$-step AB method will also be $O(h^{q+1})$. In addition, the global error will satisfy an asymptotic error formula

$$Y(x_n) - y_h(x_n) = D(x_n) h^{q+1} + O(h^{q+2})$$

much as was true earlier for the Taylor and Runge–Kutta methods of section 9.4. Thus Richardson extrapolation can be used to accelerate the convergence of the method and to estimate the error.

To generate the initial values $y_1, \ldots, y_q$ for the $(q + 1)$-step AB method, and to have their errors satisfy the requirement (9.85), it is sufficient to use a Runge–Kutta method of order q. However, in many cases, people prefer to use a Runge–Kutta method of order $q + 1$, the same as the order of the $(q + 1)$-step AB method. There are other procedures used in the package programs for Adams methods, but we will not discuss them here.

The Adams–Bashforth methods of orders 1 through 4 are given in Table 9.11. The order 1 formula is simply Euler's method. In the table, $y'_k \equiv f(x_k, y_k)$.

Table 9.11   Adams–Bashforth Methods

q	Order	Method	Truncation Error
0	1	$y_{n+1} = y_n + hy'_n$	$\dfrac{1}{2}h^2 Y''(\xi_n)$
1	2	$y_{n+1} = y_n + \dfrac{h}{2}[3y'_n - y'_{n-1}]$	$\dfrac{5}{12}h^3 Y'''(\xi_n)$
2	3	$y_{n+1} = y_n + \dfrac{h}{12}[23y'_n - 16y'_{n-1} + 5y'_{n-2}]$	$\dfrac{3}{8}h^4 Y^{(4)}(\xi_n)$
3	4	$y_{n+1} = y_n + \dfrac{h}{24}[55y'_n - 59y'_{n-1} + 37y'_{n-2} - 9y'_{n-3}]$	$\dfrac{251}{720}h^5 Y^5(\xi_n)$

Example

Solve the problem (9.82) by using the fourth order AB method. Generate the initial values y_1, y_2, y_3 by using the true solution,

$$y_i = Y(x_i), \qquad i = 1, 2, 3$$

The results for $h = 0.125$ and $2h = 0.25$ are given in Table 9.12. Richardson's error estimate for a fourth order method is given in the last column. For a fourth order method, the error should decrease by a factor of about 16 when h is halved. In those cases where this is true, the Richardson error estimate is accurate. In no case is the error badly underestimated.

Comparing these results with those in Table 9.8 for the fourth order Fehlberg method, we see that the present errors appear to be very large. But note that the Fehlberg formula uses five evaluations of $f(x, z)$ for each step of x_n to x_{n+1}; and the fourth order AB method uses only one evaluation of f per step, assuming past evaluations are reused. If this AB method is used with an h that is only 1/5 as large (in order to have a comparable number of evaluations of f), then the present errors will decrease by a factor of about $5^4 = 625$. Then the AB errors will be much smaller than those of the Fehlberg method in Table 9.8, and the work will be comparable (measured by evaluations of f).

Table 9.12   Example of Fourth Order Adams–Bashforth Method

x	$y_h(x)$	$Y(x) - y_{2h}(x)$	$Y(x) - y_h(x)$	Ratio	$\frac{1}{15}[y_h(x) - y_{2h}(x)]$
2	0.49318680	$-3.96\mathrm{E}-4$	$-3.62\mathrm{E}-5$	10.9	$-2.25\mathrm{E}-5$
4	-1.41037698	$-1.25\mathrm{E}-3$	$-6.91\mathrm{E}-5$	18.1	$-7.37\mathrm{E}-5$
6	0.68067962	$1.05\mathrm{E}-3$	$7.52\mathrm{E}-5$	14.0	$6.12\mathrm{E}-5$
8	0.84385416	$3.26\mathrm{E}-4$	$4.06\mathrm{E}-6$	80.0	$2.01\mathrm{E}-5$
10	-1.38301376	$-1.33\mathrm{E}-3$	$7.89\mathrm{E}-5$	16.9	$-7.82\mathrm{E}-5$

Adams–Moulton Methods

As with the Adams–Bashforth methods, we begin our presentation of AM methods by considering the method based on linear interpolation. Let $p_1(x)$ be the linear polynomial that interpolates $g(x)$ at x_n and x_{n+1},

$$p_1(x) = \frac{1}{h}[(x_{n+1} - x)g(x_n) + (x - x_n)g(x_{n+1})]$$

Using it to approximate the integral in (9.78), we have

$$\int_{x_n}^{x_{n+1}} g(x)\, dx \doteq \int_{x_n}^{x_{n+1}} p_1(x)\, dx = \frac{h}{2}[g(x_n) + g(x_{n+1})]$$

which is just the simple trapezoidal rule of section 7.1, Chapter 7. Applying this to the integral formulation (9.77), and including the error term (7.26) for the trapezoidal rule, we obtain

$$Y(x_{n+1}) = Y(x_n) + \frac{h}{2}[f(x_n, Y(x_n)) + f(x_{n+1}, Y(x_{n+1}))]$$

$$- \frac{h^3}{12} Y'''(\xi_n) \tag{9.86}$$

Dropping the last term, the truncation error, we obtain the AM method

$$y_{n+1} = y_n + \frac{h}{2}[f(x_n, y_n) + f(x_{n+1}, y_{n+1})], \qquad n \geq 0 \tag{9.87}$$

It can be shown that this is a second order method, that its global error is $O(h^2)$.

Example

Solve the earlier example (9.82) using the AM method (9.87). The results are given in Table 9.13, for $h = 0.05$, $2h = 0.1$. The Richardson error estimate for second order methods is given in the last column of the table. In this case, the $O(h^2)$ error behavior is very apparent, and the error estimation is very accurate.

The most notably different feature of the AM method in (9.87) is that

Table 9.13 Example of Adams–Moulton Method of Order 2

x	$Y(x) - y_{2h}(x)$	$Y(x) - y_h(x)$	Ratio	$\frac{1}{3}[y_h(x) - y_{2h}(x)]$
2	$-4.59\mathrm{E}-4$	$-1.15\mathrm{E}-4$	4.0	$-1.15\mathrm{E}-4$
4	$-5.61\mathrm{E}-4$	$-1.40\mathrm{E}-4$	4.0	$-1.40\mathrm{E}-4$
6	$7.98\mathrm{E}-4$	$2.00\mathrm{E}-4$	4.0	$2.00\mathrm{E}-4$
8	$-1.21\mathrm{E}-4$	$-3.04\mathrm{E}-5$	4.0	$-3.03\mathrm{E}-4$
10	$-7.00\mathrm{E}-4$	$-1.75\mathrm{E}-4$	4.0	$-1.25\mathrm{E}-4$

y_{n+1} occurs on both sides of the equation. In general, the finding of y_{n+1} is a nonlinear rootfinding problem for the equation

$$z = y_n + \frac{h}{2}[f(x_n, y_n) + f(x_{n+1}, z)] \tag{9.88}$$

and it can be solved directly in only a small percentage of cases. Methods in which y_{n+1} must be found by solving a rootfinding problem are called *implicit methods*, since y_{n+1} is defined implicitly. In contrast, methods which give y_{n+1} directly are called *explicit methods*. All of the methods preceding (9.87) in this section and chapter were explicit methods.

The rootfinding methods of Chapter 4 can be applied to (9.88) to find its root y_{n+1}, but usually that will be a very time-consuming process. Instead, (9.88) is usually solved by a simple iteration technique. Given an initial guess $y_{n+1}^{(0)} \doteq y_{n+1}$, define $y_{n+1}^{(1)}, y_{n+1}^{(2)}$, etc. by

$$y_{n+1}^{(j+1)} = y_n + \frac{h}{2}[f(x_n, y_n) + f(x_{n+1}, y_{n+1}^{(j)})], \qquad j = 0, 1, 2, \ldots \tag{9.89}$$

It can be shown that if h is sufficiently small, then the iterates $y_{n+1}^{(j)}$ will converge to y_{n+1} as $j \to \infty$. Subtracting (9.89) from (9.87) gives us

$$y_{n+1} - y_{n+1}^{(j+1)} = \frac{h}{2}[f(x_{n+1}, y_{n+1}) - f(x_{n+1}, y_{n+1}^{(j)})]$$

$$y_{n+1} - y_{n+1}^{(j+1)} \doteq \frac{h}{2} \cdot \frac{\partial f(x_{n+1}, y_{n+1})}{\partial z}[y_{n+1} - y_{n+1}^{(j)}] \tag{9.90}$$

where the last formula is obtained by applying the mean-value theorem to $f(x_{n+1}, z)$, considered as a function of z. This formula gives a relation between the error in successive iterates. Therefore, if

$$\left| \frac{h}{2} \cdot \frac{\partial f(x_{n+1}, y_{n+1})}{\partial z} \right| < 1 \tag{9.91}$$

then the errors will converge to zero.

In practice one uses a good initial guess $y_{n+1}^{(0)}$, and one chooses an h that is so small that the quantity in (9.91) is much less than 1. Then the error $y_{n+1} - y_{n+1}^{(j)}$ decreases rapidly to a small quantity, and usually only one iterate needs to be computed.

The usual choice of the initial guess $y_{n+1}^{(0)}$ for (9.89) is based on the AB methods of orders 1 or 2. Use

$$y_{n+1}^{(0)} = y_n + hf(x_n, y_n) \tag{9.92}$$

or

$$y_{n+1}^{(0)} = y_n + \frac{h}{2}[3f(x_n, y_n) - f(x_{n-1}, y_{n-1})] \tag{9.93}$$

These are called *predictor formulas*, as they predict the root of the implicit method. In either of the two above cases, compute $y_{n+1}^{(1)}$ from (9.89) and accept it as the root y_{n+1}. With both methods of choosing $y_{n+1}^{(0)}$, the global error in the resulting solution $\{y_h(x_n)\}$ can be shown to still be $O(h^2)$. If the Euler predictor (9.92) is used to define $y_{n+1}^{(0)}$, then $y_{n+1}^{(1)}$ will be the Runge–Kutta method (9.69) from the last section. Thus we will illustrate the use of the predictor (9.93) and its effect on the numerical solution. For finding y_1, the first step, we use the Euler predictor formula.

Example

Repeat the last example, but using the procedure described in the last paragraph. Then the errors do not change by very much from those given in Table 9.10. For example, with $x = 10$ and $h = 0.05$, the error is

$$Y(10) - y_h(10) = -2.02\text{E-}4$$

This is not very different from the value of $-1.75\text{E-}4$ given in Table 9.10. The use of the iterate $y_{n+1}^{(1)}$ as the root y_{n+1} will not affect significantly the solution of most differential equations.

By integrating the polynomial of degree q that interpolates at the nodes $\{x_{n+1}, x_n, \ldots, x_{n-q+1}\}$ to the function $g(x)$ of (9.78), we obtain the AM method of order $q + 1$. It will be an implicit method, but in other regards the theory is the same as for the AB methods described earlier. The AM methods of orders 1 through 4 are given in Table 9.14. In the table, $y_k' \equiv f(x_k, y_k)$.

The effective cost of an AM method is two evaluations of the derivative $f(x, z)$ per step, assuming past values are reused. This includes one evaluation of f for the predictor, and then one evaluation of f in the iteration formula for the AM method, as in (9.89). When this is taken into consideration, then there is no significant gain in accuracy over the AB methods of the same order, for equal costs. Nonetheless, there are other properties of the AM methods that make them desirable to use for many types of differential equations. The desirable features relate to stability characteristics of numerical methods, and we will touch on this in the next section.

Some of the most popular computer codes for solving the initial value problem are based on using AM and AB methods in combination, as suggested above. These codes control the truncation error by varying the stepsize h and by varying the order of the method. The possible order is allowed to be as large as 12 or more; and this results in a very efficient numerical method when the solution $Y(x)$ has several continuous derivatives and is slowly varying. A thorough discussion of Adams methods and of one such computer code is given in Shampine and Gordon (1975).

An Illustrative Program

To aid in programming the methods of this section, we present a modification of the Euler program of section 9.2. The program implements the Adams–Bashforth formula of order 2, given in (9.81); and it uses Euler's method to generate the first value y_1, as in (9.83). We also give it in a double-precision version. This program can be easily modified

Table 9.14 Adams–Moulton Methods

q	Order	Method	Truncation Error
0	1	$y_{n+1} = y_n + hy_{n+1}'$	$-\dfrac{1}{2} h^2 Y''(\xi_n)$
1	2	$y_{n+1} = y_n + \dfrac{h}{2} [y_{n+1}' + y_n']$	$-\dfrac{1}{12} h^3 Y'''(\xi_n)$
2	3	$y_{n+1} = y_n + \dfrac{h}{12} [5y_{n+1}' + 8y_n' - y_{n-1}']$	$-\dfrac{1}{24} h^4 Y^{(4)}(\xi_n)$
3	4	$y_{n+1} = y_n + \dfrac{h}{24} [9y_{n+1}' + 19y_n' - 5y_{n-1}' + y_{n-2}']$	$\dfrac{-19}{720} h^5 Y^{(5)}(\xi_n)$

for other Adams methods, as is suggested in the problems. The function subprograms for $f(x, z)$ and $Y(x)$ should be given as with the program for Euler's method, but in double precision.

```
C       TITLE: DEMONSTRATION OF ADAMS-BASHFORTH METHOD OF ORDER TWO.
C
        IMPLICIT DOUBLE PRECISION(A-H,O-Z)
        COMMON/BLOCKF/NUMDE
C
C       INPUT PROBLEM PARAMETERS.
10      PRINT *, ' WHICH DIFFERENTIAL EQUATION?'
        PRINT *, ' GIVE ZERO TO STOP.'
        READ *, NUMDE
        IF(NUMDE .EQ. 0) STOP
        PRINT *, ' GIVE DOMAIN [X0,B] OF SOLUTION.'
        READ *, XZERO,XEND
        PRINT *, ' WHAT IS Y0=Y(X0)?'
        READ *, YZERO
C
20      PRINT *
        PRINT *, ' GIVE STEPSIZE H AND PRINT PARAMETER IPRINT.'
        PRINT *, ' LET H=0 TO TRY ANOTHER DIFFERENTIAL EQUATION.'
        READ *, H,IPRINT
        IF(H .EQ. 0.0D0) GO TO 10
C
        PRINT 1000,NUMDE,XZERO,XEND,YZERO,H,IPRINT
1000    FORMAT(//,' EQUATION',I2,5X,'XZERO=',1PD9.2,5X,'B=',D9.2,
       *           5X,'YZERO=',D12.5,/,' STEPSIZE=',D10.3,5X,
       *           'PRINT PARAMETER=',I3,/)
C
C       INITIALIZE.
        X0=XZERO
        Y0=YZERO
        F0=F(X0,Y0)
        X1=X0+H
        Y1=Y0+H*F0
        F1=F(X1,Y1)
        IF(IPRINT .GT. 1) THEN
            KBEG=2
          ELSE
            KBEG=1
            X2=X1
            Y2=Y1
            GO TO 50
        END IF
C
C       BEGIN MAIN LOOP FOR COMPUTING SOLUTION OF DIFFERENTIAL EQUATION.
30      DO 40 K=KBEG,IPRINT
          KBEG=1
          X2=X1+H
          IF(X2 .GT. XEND) GO TO 20
          Y2=Y1+H*(3.0D0*F1-F0)/2.0D0
          X0=X1
          X1=X2
```

```
       YO=Y1
       Y1=Y2
       FO=F1
       F1=F(X1,Y1)
40     CONTINUE
C
C      CALCULATE ERROR AND PRINT RESULTS.
50     TRUE=Y(X2)
       ERROR=TRUE-Y2
       PRINT 1001,X2,Y2,ERROR
1001   FORMAT(' X=',1PD10.3,5X,'Y(X)=',D17.10,5X,'ERROR=',D9.2)
       GO TO 30
       END
```

PROBLEMS

1. Implement the second order AB method and use Euler's method (9.83) to generate y_1. (The Fortran program at the end of the section would be the easiest way to do this.) Then solve the equations in problem 2 of section 9.4. Include the Richardson error estimate for $y_h(x)$ when $h = 0.1$ and 0.05.

2. Modify the Fortran program of this section to use the third order AB method. For y_1 and y_2, use one of the second order Runge–Kutta methods from section 9.4. Then repeat problem 1. Also solve the continuing example problem (9.82).

3. To make the error term in (9.80) a bit more believable, prove

$$\int_0^h g(x)\, dx - \frac{h}{2}[3g(h) - g(0)] = \frac{5}{12} h^3 g''(0) + O(h^4)$$

   Hint: Expand $g(x)$ as a quadratic Taylor polynomial about the origin, with an error term $R_3(x)$. Substitute that into the left side of the above equation, and obtain the right side.

4. Modify the Fortran program of this section to use the AM method of order 2. For the predictor, use the AB method of order 2; for the first step x_1, use the Euler predictor. Iterate the formula (9.89) only once. Apply this to the solution of the equations considered in problem 1, and produce the Richardson error estimate.

5. Check the differential equation programs available from your computer center, and see whether there is a multistep Adams code. If so, use it to solve the equations in problem 1 above. As error tolerances, use $\epsilon = 10^{-4}$ and $\epsilon = 10^{-8}$. Keep track of the number of evaluations of $f(x, z)$ that are used by the routine, and compare it to the number used in your own programs for Adams methods.

6. (a) Using the program of problem 1 for the AB method of order 2, solve

$$Y'(x) = -50Y(x) + 51 \cos(x) + 49 \sin(x), \qquad Y(0) = 1$$

for $0 \leq x \leq 10$. The solution is $Y(x) = \sin(x) + \cos(x)$. Use stepsizes of $h = 0.1, 0.02, 0.01$. In each case, print the errors as well as the answers.

(b) Using the program of problem 4 for the AM method of order 2, repeat part (a). Check the condition in (9.91).

(c) When the AM method of order 2 is applied to the equation in (a), the value of y_{n+1} can be found directly. Doing so, repeat part (a). Compare your results.

9.6 STABILITY OF NUMERICAL METHODS

The word *stability* has a multitude of meanings, both with respect to the differential equation under consideration and to the numerical method for its solution. For the stability of the initial value problem, we refer the reader back to the definitions in the latter part of section 9.1, in which stability referred to the effect on the solution $Y(x)$ of a perturbation in the initial value Y_0. The result (9.14) stated that if Y_0 was changed to $Y_0 + \epsilon$, then the solution $Y(x)$ was changed to $Y_\epsilon(x)$ with

$$\max_{x_0 \leq x \leq b} |Y(x) - Y_\epsilon(x)| \to 0 \text{ as } \epsilon \to 0 \qquad (9.94)$$

One of the meanings of numerical stability is that we would like the same type of property to hold for the numerical method being used to solve the initial value problem.

To aid in understanding what it would mean to lack numerical stability, we introduce and consider a special two step method. To begin, if $Y(x)$ is three times continuously differentiable, then it can be shown that

$$Y(x_{n+1}) = 3Y(x_n) - 2Y(x_{n-1})$$
$$+ \frac{h}{2}[Y'(x_n) - 3Y'(x_{n-1})] + \frac{7}{12}h^3 Y'''(\xi_n) \quad (9.95)$$

for some $x_{n-1} \leq \xi_n \leq x_{n+1}$; see problem 1. Assuming further that $Y(x)$ satisfies the differential equation

$$Y'(x) = f(x, Y(x))$$

and then dropping the truncation error term in (9.95), we obtain the numerical method

$$y_{n+1} = 3y_n - 2y_{n-1} + \frac{h}{2}[f(x_n, y_n) - 3f(x_{n-1}, y_{n-1})], \qquad n \geq 1 \quad (9.96)$$

This is a numerical method whose truncation error is $O(h^3)$, but in general its solution does not converge to $Y(x)$ as $h \to 0$.

To see that this is not a stable method, apply (9.96) to the very simple initial value problem

$$Y'(x) \equiv 0, \qquad Y(0) = 1 \tag{9.97}$$

whose solution is $Y(x) \equiv 1$. Then (9.96) becomes

$$y_{n+1} = 3y_n - 2y_{n-1}, \qquad n \geq 1 \tag{9.98}$$

since $f(x, z) \equiv 0$ in this case. If the initial values are

$$y_0 = y_1 = 1 \tag{9.99}$$

consistent with the true solution, then it is easy to verify that the solution of (9.98) is

$$y_n = 1, \qquad \text{all } n \geq 0 \tag{9.100}$$

However, change the initial values to

$$y_{\epsilon,0} = 1 + \epsilon, \qquad y_{\epsilon,1} = 1 + 2\epsilon \tag{9.101}$$

for some small ϵ. Then the solution of (9.98) can be shown to be

$$y_{\epsilon,n} = 1 + \epsilon 2^n, \qquad n \geq 0$$

To see what happens at a particular point, choose $x_n = 1$. Then $n = 1/h$, since $x_n = nh$. Thus at $x_n = 1$,

$$y_{\epsilon,n} = 1 + \epsilon 2^{1/h} \tag{9.102}$$
$$y_{\epsilon,n} - y_n = \epsilon 2^{1/h}$$

As $h \to 0$, the change in the solution at $x_n = 1$ becomes larger; and it goes to ∞ as $h \to 0$. The method (9.96) is unstable in its solution of (9.97) and, therefore, it is an unacceptable method in general.

All of the numerical methods developed in the preceding sections are stable in the general sense described above. For the Adams–Bashforth methods, we state this precisely, as a theorem.

Theorem 9.3

Let $q \geqq 0$ and let

$$y_{n+1} = y_n + h[c_0 y_n' + c_1 y_{n-1}' + \ldots + c_q y_{n-q}'], \qquad n \geqq q$$

denote the $(q + 1)$-step Adams–Bashforth method. Assume $f(x, z)$ and $\partial f(x, z)/\partial z$ are continuous for $-\infty < z < \infty$ and $x_0 \leqq x \leqq b$, and assume that

$$\max_{\substack{x_0 \leqq x \leqq b \\ -\infty < z < \infty}} |\partial f(x, z)/\partial z| < \infty$$

Let $\{y_h(x_n)\}$ denote the numerical solution obtained with the initial values

$$y_j = y_h(x_j) = Y(x_j), \qquad j = 0, 1, \ldots, q$$

and let $\{y_{\epsilon,h}(x_n)\}$ denote the solution obtained with initial values $y_{\epsilon,0}, \ldots, y_{\epsilon,q}$ satisfying

$$|y_j - y_{\epsilon,j}| \leqq \epsilon, \qquad j = 0, 1, \ldots, q \tag{9.103}$$

for some $\epsilon > 0$. Then for all sufficiently small ϵ, say $\epsilon \leqq \epsilon_0$, and all sufficiently small h, say $h \leqq h_0$, there is a constant c independent of h and ϵ for which

$$\max_{x_0 \leqq x_n \leqq b} |y_{\epsilon,h}(x_n) - y_h(x_n)| \leqq c\epsilon \tag{9.104}$$

Thus small perturbations in the initial data of the numerical method will result in small changes in the answer, independent of the grid size h of the numerical method.

We omit the proof of this theorem, although it is not a difficult result. The theorem can also be restated easily for Adams–Moulton methods, and the proof is much the same. In contrast, the example (9.96) to (9.102) shows that not all multistep methods are stable. The general theory of stability for multistep methods is quite involved and technically complicated, and we merely refer the reader to Atkinson (1978) for an introduction or to Isaacson and Keller (1966) for a fairly complete development. This general theory is also concerned with convergence of the numerical method, and it can be shown that a numerical method is stable if and only if it is a convergent method, subject to some minor restrictions. For Runge–Kutta methods, the theory is less involved, and it can be shown that all such methods are both stable and convergent.

Stability Regions

The stability result in Theorem 9.3 states that the method is well-behaved, provided the stepsize h is sufficiently small. We would like to further examine this restriction, as it has important practical consequences for the relative desirability of various numerical methods.

Examining the stability question for the general problem

$$Y'(x) = f(x, Y(x)), \qquad Y(x_0) = Y_0 \tag{9.105}$$

is too complicated. Instead, we examine the stability of numerical methods for the *model equation*

$$Y'(x) = \lambda Y(x) + g(x), \qquad Y(0) = Y_0 \tag{9.106}$$

whose exact solution can be found from (9.5). Questions regarding stability and convergence are more easily answered for this equation; the answers to such questions can be shown to usually be the answers to those same questions for the more general equation (9.105).

Let $Y(x)$ be the solution of (9.106), and let $Y_\epsilon(x)$ be the solution with the perturbed initial data $Y_0 + \epsilon$,

$$Y_\epsilon'(x) = \lambda Y_\epsilon(x) + g(x), \qquad Y_\epsilon(0) = Y_0 + \epsilon$$

Let $Z_\epsilon(x)$ denote the change in the solution,

$$Z_\epsilon(x) = Y_\epsilon(x) - Y(x)$$

Then subtracting the equation for $Y_\epsilon(x)$ from (9.106),

$$Z_\epsilon'(x) = \lambda Z_\epsilon(x), \qquad Z_\epsilon(0) = \epsilon \tag{9.107}$$

The solution is

$$Z_\epsilon(x) = \epsilon e^{\lambda x}$$

Typically in applications, we are interested in the case that $\lambda < 0$ or that λ is complex with a negative real part. In such a case, $Z_\epsilon(x)$ will go to zero as $x \to \infty$ and, thus, the effect of the ϵ perturbation dies out for large values of x. We would like the same behavior to hold for the numerical method that is being applied to (9.106). In our work, we will assume λ is real and negative.

If a Runge–Kutta method or multistep method is applied to the model equation (9.106), and if the initial values for it are perturbed in the manner of (9.103) of Theorem 9.3 or of (9.101) of the earlier example, then the perturbation in the solution $y_h(x)$ will lead to the numerical analog of (9.107). For example, consider the Euler method

$$y_{n+1} = y_n + h[\lambda y_n + g(x_n)], \qquad y_0 = Y_0$$

and its perturbation

$$y_{\epsilon,n+1} = y_{\epsilon,n} + h[\lambda y_{\epsilon,n} + g(x_n)], \qquad y_{\epsilon,0} = Y_0 + \epsilon$$

Let $z_{\epsilon,n} = y_{\epsilon,n} - y_n$, and then subtract the last two equations obtaining

$$z_{\epsilon,n+1} = z_{\epsilon,n} + h[\lambda z_{\epsilon,n}], \qquad n \ge 0, \qquad z_{\epsilon,0} = \epsilon$$

This is the Euler method applied to the solution of (9.107).

To examine the stability of a numerical method for the model equation with $\lambda = 0$, it is sufficient to apply the method to the perturbation equation (9.107). To simplify the notation, we drop the subscript ϵ, writing

$$Z'(x) = \lambda Z(x), \qquad x \ge 0, \qquad Z(0) = \epsilon \qquad (9.108)$$

To illustrate the analysis that can be carried out, continue the above analysis of Euler's method for (9.107). Then dropping the subscript ϵ,

$$z_{n+1} = (1 + h\lambda)z_n, \qquad z_0 = \epsilon$$

and it is easy to show that

$$z_n = (1 + h\lambda)^n z_0 = \epsilon(1 + h\lambda)^n, \qquad n \ge 0 \qquad (9.109)$$

From this, $z_n \to 0$ as $n \to \infty$ if and only if

$$|1 + h\lambda| < 1 \qquad (9.110)$$

$$-2 < h\lambda < 0 \qquad (9.111)$$

If h is sufficiently small, then this condition is satisfied and $z_n \to 0$ as $n \to \infty$. But for $h\lambda < -2$, the solution z_n will increase in size and, thus, the solutions $y_h(x_n)$ and $y_{\epsilon,h}(x_n)$ will grow apart as x_n increases.

If a numerical method is applied to the perturbation equation (9.108), then the set of values $h\lambda$ for which $z_n \to 0$ as $n \to \infty$ is called the *region of absolute stability* of the numerical method. For Euler's method, the interval $(-2, 0)$ is the real part of the region of absolute stability. One is also interested in values of λ that are complex with a negative real part, but that will not be considered here.

The larger a region of absolute stability, the less restrictive the condition on h to have numerical stability of the type just discussed. Thus a method with a large region of absolute stability is generally preferred over a method with a smaller region, provided that the accuracy of the two methods is similar. For the Adams methods of section 9.5, it can be shown that for AB and AM methods of equal order, the AM method will have the larger region of absolute stability. Consequently, Adams–Moulton methods are generally preferred over Adams–Bashforth methods.

Example

The AB method of order 2 has $-1 < h\lambda < 0$ as the real part of its region of stability; in contrast, the AM method of order 2 has $-\infty < h\lambda < 0$ as the real part of its region of stability. There is no stability restriction on h with this AM method.

The stability of a numerical method affects the convergence of the method. This can be shown theoretically, but we will just illustrate it numerically with an example. Solve the equation

$$Y'(x) = \lambda Y(x) + (1 - \lambda) \cos(x) - (1 + \lambda) \sin(x),$$

$$Y(0) = 1 \quad (9.112)$$

whose true solution is $Y(x) = \sin(x) + \cos(x)$. Euler's method is used for the numerical solution, and the results for several values of λ and h

are given in Table 9.15. Note that according to the formula (9.32) for the truncation error,

$$T_{n+1} = \frac{h^2}{2} Y''(\xi_n)$$

It will not depend on λ since $Y(x)$ does not depend on λ. But the actual global error depends strongly on λ, as illustrated in the table; and the behavior of the global error is directly linked to the size of λh and, thus, to the size of the stability region for Euler's method. The error is small, provided that λh is sufficiently small.

With a large region of absolute stability for a numerical method, the restriction on λh, and thus on h, is less stringent. Equations with λ negative but large in magnitude are called *stiff differential equations*. Their truncation error may be satisfactorily small with not too small a value of h, but the large size of $|\lambda|$ may force h to be smaller in order that λh be in the stability region. The second order Adams–Moulton method (9.87) is therefore very desirable because its stability region contains all λh where λ is negative or λ is complex with negative real part. For stiff differential equations, one must use a numerical method with a large region of absolute stability, or else h must be chosen very small.

Table 9.15 Euler's Method for (9.112)

λ	x	Error: $h = 0.5$	Error: $h = 0.1$	Error: $h = 0.01$
-1	1	$-2.46E-1$	$-4.32E-2$	$-4.22E-3$
	2	$-2.55E-1$	$-4.64E-2$	$-4.55E-3$
	3	$-2.66E-2$	$-6.78E-3$	$-7.22E-4$
	4	$2.27E-1$	$3.91E-2$	$3.78E-3$
	5	$2.72E-1$	$4.91E-2$	$4.81E-3$
-10	1	$3.98E-1$	$-6.99E-3$	$-6.99E-4$
	2	$6.90E+0$	$-2.90E-3$	$-3.08E-4$
	3	$1.11E+2$	$3.86E-3$	$3.64E-4$
	4	$1.77E+3$	$7.07E-3$	$7.04E-4$
	5	$2.83E+4$	$3.78E-3$	$3.97E-4$
-50	1	$3.26E+0$	$1.06E+3$	$-1.39E-4$
	2	$1.88E+3$	$1.11E+9$	$-5.16E-5$
	3	$1.08E+6$	$1.17E+15$	$8.25E-5$
	4	$6.24E+8$	$1.23E+21$	$1.41E-4$
	5	$3.59E+11$	$1.28E+27$	$7.00E-5$

Example

Apply the Adams–Moulton method of order 2 (9.87) to the solution of the preceding example (9.112). The results are shown in Table 9.16, with the stepsize $h = 0.5$. The error varies with λ, but there are no stability problems, in contrast to the Euler method. The solution of the AM method (9.87) for y_{n+1} was done exactly. There are difficulties with the iteration (9.89) when $|\lambda h|$ is large.

Rounding Error Accumulation

As with earlier topics in this text, the finite precision of the arithmetic will affect the numerical solution of a differential equation. To investigate this effect, consider Euler's method. Without rounding error, it is defined by

$$y_{n+1} = y_n + hf(x_n, y_n), \qquad n \geq 0, \ y_0 = Y_0$$

The simple arithmetic operations and the evaluation of $f(x_n, y_n)$ will usually contain errors due to rounding or chopping. Thus what is actually evaluated is

$$\hat{y}_{n+1} = \hat{y}_n + hf(x_n, \hat{y}_n) + \delta_n, \qquad n \geq 0, \qquad \hat{y}_0 = Y_0 \qquad (9.113)$$

The quantity δ_n will be based on the precision of the arithmetic, and its size is affected by that of $\hat{y}_n$. To simplify our work, we will simply assume

$$|\delta_n| \leq cu \max_{x_0 \leq x \leq x_n} |Y(x)| \qquad (9.114)$$

where u is the unit round of the computer (see problems 7 to 9 of chapter 2, section 9.2) and c is a constant of magnitude 1 or larger.

Table 9.16 Adams–Moulton Solution of Order 2 for (9.112): $h = 0.5$

x	Error: $\lambda = -1$	Error: $\lambda = -10$	Error: $\lambda = -50$
2	$-1.13E-2$	$-2.78E-3$	$-7.91E-4$
4	$-1.43E-2$	$-8.91E-5$	$-8.91E-5$
6	$2.02E-2$	$2.77E-3$	$4.72E-4$
8	$-2.86E-3$	$-2.22E-3$	$-5.11E-4$
10	$-1.79E-2$	$-9.23E-4$	$-1.56E-4$

To compare $\{\hat{y}_n\}$ to the true solution $Y(x)$, write

$$Y(x_{n+1}) = Y(x_n) + hf(x_n, Y(x_n)) + \frac{h^2}{2} Y''(\xi_n), \qquad (9.115)$$

which was obtained earlier in (9.31). Subtracting (9.113) from (9.115), we get

$$Y(x_{n+1}) - \hat{y}_{n+1} = Y(x_n) - \hat{y}_n + h[f(x_n, Y(x_n)) - f(x_n, \hat{y}_n)]$$
$$+ \frac{h^2}{2} Y''(\xi_n) - \delta_n, \qquad n \geq 0 \quad (9.116)$$

with $Y(x_0) - \hat{y}_0 = 0$. This equation is analogous to the error equation (9.33) of section 9.3, with the role of the truncation error in that equation replaced by the term

$$\frac{h^2}{2} Y''(\xi_n) - \delta_n = h\left[\frac{h}{2} Y''(\xi_n) - \frac{\delta_n}{h}\right] \qquad (9.117)$$

If the convergence theorem of Theorem 9.2 is derived from (9.116) rather than from (9.33) as before, then the error result (9.38) generalizes to

$$|Y(x_n) - \hat{y}_n| \leq c_1\left[\frac{h}{2} \max_{x_0 \leq x \leq b} |Y''(x)| + \frac{cu}{h} \max_{x_0 \leq x \leq b} |Y(x)|\right] \qquad (9.118)$$

where $x_0 \leq x_n \leq b$,

$$c_1 = \frac{e^{(b-x_0)K} - 1}{2K}$$

and K is the maximum of $\partial f(x, z)/\partial z$, defined in (9.37). The term in brackets on the right side of (9.118) is obtained by bounding the term in brackets on the right side of (9.117).

In essence, (9.118) says that

$$|Y(x_n) - \hat{y}_n| \leq \alpha_1 h + \alpha_2/h, \qquad x_0 \leq x_n \leq b$$

for appropriate choices of α_1, α_2. Note that α_2 is generally small because of u. Thus the error bound will initially decrease as h decreases; but at a critical value of h, call it h^*, the error bound will again increase, because of the term α_2/h. The same qualitative behavior is also true of the actual error $Y(x_n) - y_n$. Thus there is a limit on the attainable accuracy. This same analysis is valid for other numerical methods, with a term of the form

$$\frac{cu}{h} \max_{x_0 \leq x \leq 6} |Y(x)|$$

to be included as part of their global error. With rounded arithmetic, this behavior can usually be improved upon. But with chopped arithmetic, it is likely to be accurate in a qualitative sense; as h is halved, the contribution to the error due to the chopped arithmetic will double.

Example

Solve the standard problem

$$Y'(x) = -Y(x) + 2 \cos(x), \qquad Y(0) = 1$$

using Euler's method. Use a four digit decimal machine with chopped arithmetic, and then repeat the calculation with rounded arithmetic. Finally, give the results of Euler's method with exact arithmetic. The results with decreasing h are given in Table 9.17. The errors for the

Table 9.17 Example of Rounding Effects in Euler's Method

h	x	Chopped Decimal	Rounded Decimal	Exact Arithmetic
.04	1	$-1.00E-2$	$-1.70E-2$	$-1.70E-2$
	2	$-1.17E-2$	$-1.83E-2$	$-1.83E-2$
	3	$-1.20E-3$	$-2.80E-3$	$-2.78E-3$
	4	$1.00E-2$	$1.60E-2$	$1.53E-2$
	5	$1.13E-2$	$1.96E-2$	$1.94E-2$
.02	1	$7.00E-3$	$-9.00E-3$	$-8.46E-3$
	2	$4.00E-3$	$-9.10E-3$	$-9.13E-3$
	3	$2.30E-3$	$-1.40E-3$	$-1.40E-3$
	4	$-6.00E-3$	$8.00E-3$	$7.62E-3$
	5	$-6.00E-3$	$8.50E-3$	$9.63E-3$
.01	1	$2.80E-2$	$-3.00E-3$	$-4.22E-3$
	2	$2.28E-2$	$-4.30E-3$	$-4.56E-3$
	3	$7.40E-3$	$-4.00E-4$	$-7.03E-4$
	4	$-2.30E-2$	$3.00E-3$	$3.80E-3$
	5	$-2.41E-2$	$4.60E-3$	$4.81E-3$

answers obtained using decimal arithmetic are based on the true answers rounded to four digits.

Note that the errors with the chopped case are affected at $h = 0.02$, with the error at $x = 3$ larger than when $h = 0.04$ for that case. The increasing error is clear with the $h = 0.01$ case, at all points. In contrast the errors using rounded arithmetic are continuing to decrease, although the $h = 0.1$ case is being affected slightly, by comparison to the true errors when no rounding is present. The column with the errors for the case with exact arithmetic show that the use of the rounded decimal arithmetic has less effect on the error than does the use of chopped arithmetic.

PROBLEMS

1. To partially justify the formula (9.95), use a Taylor polynomial approximation of $Y(x)$ to show that

$$Y(2h) - \left\{ 3Y(h) - 2Y(0) + \frac{h}{2}[Y'(h) - 3Y'(0)] \right\}$$

$$= \frac{7}{12} h^3 Y'''(0) + O(h^4)$$

Hint: Expand $Y(x)$ about $x = 0$, and then evaluate the left side of the equation using the polynomial approximation.

2. Modify the Adams–Bashforth program of section 9.5 to use the numerical method (9.96). Use it to solve the continuing example problem

$$Y'(x) = -Y(x) + 2 \cos (x), \qquad Y(0) = 1$$

Compare the results with those using earlier methods. Use several stepsizes, and comment on your results.

3. Verify that $y_{\epsilon,n} = 1 + \epsilon 2^n$ is a solution of (9.98), (9.101).

4. Show that all points $h\lambda$ in $-\infty < h\lambda < 0$ are in the stability region of the Adams–Moulton method of order 2.
   Hint: Apply it to the numerical solution of (9.108), and check whether the numerical solution tends to zero as $n \to \infty$.

5. The Adams–Moulton method of order 1,

$$y_{n+1} = y_n + hf(x_{n+1}, y_{n+1}), \qquad n \geq 0$$

is called the *backward Euler's method*, and it is widely used in the numerical solution of time dependent partial differential equations. Show that the real part of its stability region is $-\infty < h\lambda < 0$.

6. Find the stability region for the Runge–Kutta method of order 2 that is given in (9.69) of section 9.4. Note that this method is the Adams–Moulton method of order 2, using an Euler predictor.
   Hint: Apply the method to numerically solve (9.108), and then investigate when the solution will tend to zero as $n \to \infty$.

7. Repeat the numerical example given in Table 9.15, but use the backward Euler method of problem 5. Solve the equation for y_{n+1} exactly, avoiding iteration.

8. Apply the backward Euler method (problem 5) to the numerical solution of

$$Y'(x) = \lambda Y(x) + g(x)$$

with $\lambda < 0$ and large in magnitude. Investigate how small h must be chosen in order that the iteration

$$y_{n+1}^{(j+1)} = y_n + hf(x_{n+1}, y_{n+1}^{(j)}), \qquad j = 0, 1, 2, \ldots$$

will converge to y_{n+1}. Is this iteration practical for large values of $|\lambda|$?

9. Solve the equation

$$Y'(x) = \lambda Y(x) + \frac{1}{1 + x^2} - \lambda \tan^{-1}(x), \qquad Y(0) = 0$$

with $Y(x) = \tan^{-1}(x)$ the true solution. Use Euler's method, the backward Euler method, and the Adams–Moulton method of order 2. Let $\lambda = -1, -10, -50$ and $h = 0.5, 0.1, 0.01$. Discuss the results.

9.7 SYSTEMS OF DIFFERENTIAL EQUATIONS

Although some applications of differential equations involve only a single first order equation, most applications involve a system of several such equations. In this section, we consider such systems and their numerical solution, showing how the methods of earlier sections apply to such systems.

To begin with a simple case, the general form of a system of two first order differential equations is

$$\begin{aligned} Y_1'(x) &= f_1(x, Y_1(x), Y_2(x)) \\ Y_2'(x) &= f_2(x, Y_1(x), Y_2(x)) \end{aligned} \qquad (9.119)$$

The functions $f_1(x, z_1, z_2)$ and $f_2(x, z_1, z_2)$ define the differential equations, and the unknown functions $Y_1(x)$ and $Y_2(x)$ are being sought. The initial value problem consists of solving (9.119), subject to the initial conditions

$$Y_1(x_0) = Y_{1,0}, \qquad Y_2(x_0) = Y_{2,0} \qquad (9.120)$$

Example

(a) The initial value problem

$$Y_1' = Y_1 - 2Y_2 + 4 \cos (x) - 2 \sin (x), \qquad Y_1(0) = 1 \qquad (9.121)$$
$$Y_2' = 3Y_1 - 4Y_2 + 5 \cos (x) - 5 \sin (x), \qquad Y_2(0) = 2$$

has the solution

$$Y_1(x) = \cos (x) + \sin (x), \qquad Y_2(x) = 2 \cos (x)$$

This example will be used later in a numerical example for Euler's method for systems.

(b) Consider the system

$$Y_1' = AY_1[1 - BY_2], \qquad Y_1(0) = Y_{1,0} \qquad (9.122)$$
$$Y_2' = CY_2[DY_1 - 1], \qquad Y_2(0) = Y_{2,0}$$

with $A, B, C, D > 0$. This is called the "Lotka–Volterra predator–prey model." The variable x denotes time, $Y_1(x)$ the number of prey (e.g., rabbits) at time x, and $Y_2(x)$ the number of predators (e.g., foxes). If there is only a single type of predator and a single type of prey, then this model is usually a good approximation of reality. The behavior of the solutions Y_1 and Y_2 is illustrated in problem 8.

The initial value problem for a system of m first order differential equations has the general form

$$Y_1' = f_1(x, Y_1, \ldots, Y_m), \qquad Y_1(x_0) = Y_{1,0}$$
$$\vdots \qquad\qquad\qquad (9.123)$$
$$Y_m' = f_m(x, Y_1, \ldots, Y_m), \qquad Y_m(x_0) = Y_{m,0}$$

We seek the functions $Y_1(x), \ldots, Y_m(x)$ on some interval $x_0 \leq x \leq b$. An example of a system of three equations is given below in (9.137).

The general form (9.123) is clumsy to work with, and it is not a convenient way to specify it when using a computer program for its solution. To simplify the form of (9.123), represent the solution and the differential equations using column matrices. Denote

$$Y(x) = \begin{bmatrix} Y_1(x) \\ \vdots \\ Y_m(x) \end{bmatrix},$$

$$f(x, z) = \begin{bmatrix} f_1(x, z_1, \ldots, z_m) \\ \vdots \\ f_m(x, z_1, \ldots, z_m) \end{bmatrix}, \qquad Y_0 = \begin{bmatrix} Y_{1,0} \\ \vdots \\ Y_{m,0} \end{bmatrix} \qquad (9.124)$$

Then (9.123) can be rewritten as

$$Y'(x) = f(x, Y(x)), \qquad Y(x_0) = Y_0 \qquad (9.125)$$

This looks like the earlier first order single equation, but it is general as to the number of equations. Computer programs for solving systems will almost always refer to the system in this manner.

Example

System (9.121) can be rewritten as

$$Y'(x) = AY(x) + G(x), \qquad Y(0) = Y_0$$

with

$$Y = \begin{bmatrix} Y_1 \\ Y_2 \end{bmatrix}, \qquad A = \begin{bmatrix} 1 & -2 \\ 3 & -4 \end{bmatrix},$$

$$G(x) = \begin{bmatrix} 4 \cos (x) - 2 \sin (x) \\ 5 \cos (x) - 5 \sin (x) \end{bmatrix}, \qquad Y_0 = \begin{bmatrix} 1 \\ 2 \end{bmatrix}$$

In the notation of (9.124),

$$f(x, z) = Az + G(x), \qquad z = [z_1, z_2]^T$$

Higher Order Differential Equations

In physics and engineering, the use of Newton's second law of motion leads to systems of second order differential equations, modeling some of the very important physical phenomena of nature. In addition, other applications lead to other higher order equations. To solve higher order equations, we reformulate them as equivalent systems of first order equations.

For simplicity, consider the second order equation

$$y''(x) = f(x, y(x), y'(x)), \tag{9.126}$$

where $f(x, z_1, z_2)$ is given. The initial value problem consists of solving (9.126) subject to the initial conditions

$$y(x_0) = y_0, \qquad y'(x_0) = y_0' \tag{9.127}$$

To reformulate this as a system of first order equations, denote

$$Y_1(x) = y(x), \qquad Y_2(x) = y'(x)$$

Then Y_1 and Y_2 satisfy

$$\begin{aligned} Y_1' &= Y_2, & Y_1(x_0) &= y_0 \\ Y_2' &= f(x, Y_1, Y_2), & Y_2(x_0) &= y_0' \end{aligned} \tag{9.128}$$

And starting from this system, it is straightforward to show that the solution Y_1 of (9.128) will also have to satisfy (9.126) to (9.127), thus demonstrating the equivalence of the two formulations.

Example

Consider the pendulum shown in Figure 9.4, of mass m and length l. The motion of this pendulum about its center line $\theta = 0$ is modeled by a second order differential equation derived from Newton's second law of motion. If the pendulum is assumed to move back and forth with negligible friction at its vertex, then the motion is modeled fairly accurately by

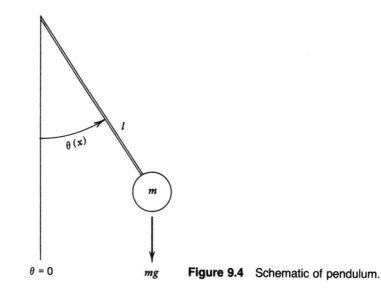

$\theta = 0$ mg **Figure 9.4**   Schematic of pendulum.

$$ml \frac{d^2\theta}{dx^2} = -mg \sin(\theta(x)) \tag{9.129}$$

with x being time and $\theta(x)$ the angle between the vertical center line and the pendulum. The description of the motion is completed by specifying the initial position $\theta(0)$ and initial angular velocity $\theta'(0)$.

To convert this to a system of two first order equations, write

$$Y_1(x) = \theta(x), \qquad Y_2(x) = \theta'(x)$$

Then (9.129) becomes

$$Y_1'(x) = Y_2(x) \qquad\qquad Y_1(0) = \theta(0) \tag{9.130}$$

$$Y_2'(x) = -\frac{g}{l} \sin(Y_1(x)) \qquad Y_2(0) = \theta'(0)$$

This is equivalent to the original second order equation.

A general differential equation of order m can be written as

$$\frac{d^m y}{dx^m} = f\left(x, y, \frac{dy}{dx}, \ldots, \frac{d^{m-1}y}{dx^{m-1}}\right) \tag{9.131}$$

and the initial conditions needed in solving it are given by

$$y(x_0) = y_0, \qquad y'(x_0) = y_0', \ldots, y^{(m-1)}(x_0) = y_0^{(m-1)} \qquad (9.132)$$

It is reformulated as a system of m first order equations by introducing

$$Y_1 = y, \qquad Y_2 = y', \ldots, Y_m = y^{(m-1)}$$

Then the equivalent initial value problem for a system of first order equations is

$$
\begin{aligned}
Y_1' &= Y_2 & Y_1(x_0) &= y_0 \\
&\;\vdots & &\;\vdots \\
Y_{m-1}' &= Y_m & Y_{m-1}(x_0) &= y_0^{(m-2)} \\
Y_m' &= f(x, Y_1, \ldots, Y_m) & Y_m(x_0) &= y_0^{(m-1)}
\end{aligned}
\qquad (9.133)
$$

A special case of (9.131) is the order m linear differential equation

$$\frac{d^m y}{dx^m} = a_0(x)y + a_1(x)\frac{dy}{dx} + \ldots + a_{m-1}(x)\frac{d^{m-1}}{dx^{m-1}} + b(x) \qquad (9.134)$$

This is reformulated as above, with

$$Y_m' = a_0(x)Y_1 + a_1(x)Y_2 + \ldots + a_{m-1}(x)Y_m + b(x) \qquad (9.135)$$

replacing the last equation in (9.133).

Example

The initial value problem

$$y''' + 3y'' + 3y' + y = -4 \sin (x),$$
$$y(0) = y'(0) = 1, \qquad y''(0) = -1 \qquad (9.136)$$

is reformulated as

$$
\begin{aligned}
Y_1' &= Y_2 & Y_1(0) &= 1 \\
Y_2' &= Y_3 & Y_2(0) &= 1 \\
Y_3' &= -Y_1 - 3Y_2 - 3Y_3 - 4\sin (x) & Y_3(0) &= -1
\end{aligned}
\qquad (9.137)
$$

The solution of (9.136) is $y(x) = \cos(x) + \sin(x)$, and the solution of (9.137) can be generated from it. This system will be solved numerically, later in this section.

Numerical Methods for Systems

The numerical methods of earlier sections can be applied without change to the solution of systems of first order differential equations. The numerical method should be applied to each equation in the system. Or more simply, apply the method in a straightforward way to the system written in the matrix form (9.125). The derivation of these numerical methods for the solution of systems is essentially the same as was done previously for a single equation. And the convergence and stability analyses are also done in the same manner.

To be more specific, we consider Euler's method for the general system of two first order equations that is given in (9.119). By following the derivation given for Euler's method in obtaining (9.31), Taylor's theorem gives

$$Y_1(x_{n+1}) = Y_1(x_n) + hf_1(x_n, Y_1(x_n), Y_2(x_n)) + \frac{h^2}{2} Y_1''(\xi_n)$$

$$Y_2(x_{n+1}) = Y_2(x_n) + hf_2(x_n, Y_1(x_n), Y_2(x_n)) + \frac{h^2}{2} Y_2''(\zeta_n)$$

for some ξ_n, ζ_n in $[x_n, x_{n+1}]$. Dropping the error terms, we obtain Euler's method for a system of two equations:

$$\begin{align} y_{1,n+1} &= y_{1,n} + hf_1(x_n, y_{1,n}, y_{2,n}) \\ y_{2,n+1} &= y_{2,n} + hf_2(x_n, y_{1,n}, y_{2,n}), \qquad n \geq 0 \end{align} \tag{9.138}$$

If $Y_1(x)$, $Y_2(x)$ are twice continuously differentiable, then it can be shown that

$$|Y_1(x_n) - y_{1,n}| \leq ch, \qquad |Y_2(x_n) - y_{2,n}| \leq ch$$

for all $x_0 \leq x_n \leq b$, for some constant c. In addition, the earlier asymptotic error formula (9.43) will still be valid; for $j = 1, 2$

$$Y_j(x_n) - y_{j,n} = D_j(x_n)h + O(h^2), \qquad x_0 \leq x_n \leq b$$

Thus Richardson's extrapolation and error estimation formulas will still be valid. The functions $D_1(x)$, $D_2(x)$ satisfy a particular linear system of differential equations, but we omit it here. Stability results for Euler's method generalize without any significant change, including the region of stability. Thus in summary, the earlier work for Euler's method generalizes without significant change to systems. The same is true of the other numerical methods given earlier, thus justifying our limitation to a single equation for introducing those methods.

Example

(a) Solve (9.121) using Euler's method. The numerical results are given in Table 9.18, along with Richardson's error estimate

$$Y_j(x_n) - y_{j,h}(x_n) \doteq y_{j,h}(x_n) - y_{j,2h}(x_n), \qquad j = 1, 2$$

In the table, $h = 0.05$, $2h = 0.1$. It can be seen that this error estimate is quite accurate, except for the one case $j = 2$, $x = 10$.

(b) Solve the third order equation in (9.136), by using Euler's method to solve the reformulated problem (9.137). The results for $y(x) = Y_1(x) = \sin(x) + \cos(x)$ are given in Table 9.19, for stepsizes $2h = 0.1$ and $h = 0.05$. The Richardson error estimate is again quite accurate.

Other numerical methods apply to systems in the same straightforward manner. And by using the matrix form (9.125) for a system, there is no

Table 9.18 Solution of (9.121) Using Euler's Method

j	x	$Y_j(x)$	$Y_j(x) - y_{j,2h}(x)$	$Y_j(x) - y_{j,h}(x)$	Ratio	$y_{j,h}(x) - y_{j,2h}(x)$
1	2	0.49315	$-5.65\mathrm{E}-2$	$-2.82\mathrm{E}-2$	2.0	$-2.83\mathrm{E}-2$
	4	-1.41045	$-5.64\mathrm{E}-3$	$-2.72\mathrm{E}-3$	2.1	$-2.92\mathrm{E}-3$
	6	0.68075	$4.81\mathrm{E}-2$	$2.36\mathrm{E}-2$	2.0	$2.44\mathrm{E}-2$
	8	0.84386	$-3.60\mathrm{E}-2$	$-1.79\mathrm{E}-2$	2.0	$-1.81\mathrm{E}-2$
	10	-1.38309	$-1.81\mathrm{E}-2$	$-8.87\mathrm{E}-3$	2.0	$-9.40\mathrm{E}-2$
2	2	-0.83229	$-3.36\mathrm{E}-2$	$-1.70\mathrm{E}-2$	2.0	$-1.66\mathrm{E}-2$
	4	-1.30729	$5.94\mathrm{E}-3$	$3.19\mathrm{E}-3$	1.9	$2.75\mathrm{E}-3$
	6	1.92034	$1.59\mathrm{E}-2$	$7.69\mathrm{E}-3$	2.1	$8.17\mathrm{E}-3$
	8	-0.29100	$-2.08\mathrm{E}-2$	$-1.05\mathrm{E}-2$	2.0	$-1.03\mathrm{E}-2$
	10	-1.67814	$1.26\mathrm{E}-3$	$9.44\mathrm{E}-4$	1.3	$3.12\mathrm{E}-4$

Table 9.19 Solution of (9.136) Using Euler's Method

x	$y(x)$	$y(x) - y_{2h}(x)$	$y(x) - y_h(x)$	Ratio	$y_h(x) - y_{2h}(x)$
2	0.49315	$-8.78\mathrm{E}-2$	$-4.25\mathrm{E}-2$	2.1	$-4.53\mathrm{E}-2$
4	-1.41045	$1.39\mathrm{E}-1$	$6.86\mathrm{E}-2$	2.0	$7.05\mathrm{E}-2$
6	0.68075	$5.19\mathrm{E}-2$	$2.49\mathrm{E}-2$	2.1	$2.70\mathrm{E}-2$
8	0.84386	$-1.56\mathrm{E}-1$	$-7.56\mathrm{E}-2$	2.1	$-7.99\mathrm{E}-2$
10	-1.38309	$8.39\mathrm{E}-2$	$4.14\mathrm{E}-2$	2.0	$4.25\mathrm{E}-2$

apparent change in the numerical method. For example, the Runge–Kutta method (9.69) is

$$y_{n+1} = y_n + \frac{h}{2}[f(x_n, y_n) + f(x_{n+1}, y_n + hf(x_n, y_n))], \qquad n \geq 0 \quad (9.139)$$

And interpreting this for a system of two equations with

$$y_n = \begin{bmatrix} y_{1,n} \\ y_{2,n} \end{bmatrix}, \qquad f(x_n, y_n) = \begin{bmatrix} f_1(x_n, y_{1,n}, y_{2,n}) \\ f_2(x_n, y_{1,n}, y_{2,n}) \end{bmatrix}$$

The method is

$$y_{j,n+1} = y_{j,n} + \frac{h}{2}[f_j(x_n, y_{1,n}, y_{2,n})$$

$$+ f_j(x_{n+1}, y_{1,n} + hf_1(x_n, y_{1,n}, y_{2,n}), y_{2,n} + hf_2(x_n, y_{1,n}, y_{2,n}))] \quad (9.140)$$

for $j = 1, 2$. It is easier to consider this in the form (9.139); for programming it on a computer, the matrix form is very convenient. We leave the illustration of (9.140) to the problems.

A Computer Program for Systems

Following is a program for Euler's method for a system of N first order differential equations. The program is written for ease of use, and it can be easily changed to another method, such as (9.140). Note the use of the PARAMETER statement in setting the dimensions of the arrays used in the program. The dimension statement is changed easily by changing the parameter statement, and the code is more transparent to read.

```
C       TITLE: DEMONSTRATION OF EULER'S METHOD FOR SYSTEMS.
C
C       THIS SOLVES THE INITIAL VALUE PROBLEM FOR THE FIRST ORDER
C       SYSTEM
C           Y'(X) = F(X,Y(X)) ,   XO .LE. X .LE. B ,  Y(XO)=YO.
C       THE FUNCTION F(X,Z) IS DEFINED BELOW, ALONG WITH THE TRUE
C       SOLUTION Y(X).  THE NUMBER OF THE PROBLEM TO BE SOLVED
C       IS SPECIFIED BY THE INPUT VARIABLE 'NUMDE', WHICH IS USED
C       IN THE SUBROUTINES 'FCN' AND 'TRUE'.  THE PROGRAM WILL
C       REQUEST THE PROBLEM PARAMETERS, ALONG WITH THE VALUES OF
C       'H' AND 'IPRINT'.  'H' IS THE STEPSIZE, AND 'IPRINT' IS
C       THE NUMBER OF STEPS BETWEEN EACH PRINTING OF THE SOLUTION.
C       USE H=0 AND NUMDE=0 TO STOP THE PROGRAM.
C
C       TO INCREASE THE ORDER OF THE SYSTEMS WHICH CAN BE
C       HANDLED, INCREASE THE SIZE OF 'MAXN' IN THE PARAMETER
C       STATEMENT GIVEN BELOW.
C
        IMPLICIT DOUBLE PRECISION(A-H,O-Z)
        PARAMETER(MAXN=3)
        DIMENSION YO(MAXN),Y1(MAXN),Y(MAXN),YZERO(MAXN),F(MAXN)
        COMMON/BLOCKF/NUMDE
C
C       INPUT PROBLEM PARAMETERS.
10      PRINT *, ' WHICH DIFFERENTIAL EQUATION?'
        PRINT *, ' GIVE ZERO TO STOP.'
        READ *, NUMDE
        IF(NUMDE .EQ. 0) CALL EXIT
        PRINT *, ' GIVE DOMAIN [XO,B] OF SOLUTION.'
        READ *, XZERO,XEND
        PRINT *, ' GIVE THE ORDER OF THE SYSTEM.  IT SHOULD BE'
        PRINT *, ' LESS THAN OR EQUAL TO MAXN=',MAXN
        READ *, N
        PRINT *, ' GIVE THE COMPONENTS OF YO, ONE AT A TIME.'
        DO 15 J=1,N
          PRINT *, ' GIVE COMPONENT',J
15        READ *, YZERO(J)
C
20      PRINT *, ' GIVE STEPSIZE H AND PRINT PARAMETER IPRINT.'
        PRINT *, ' LET H=0 TO TRY ANOTHER DIFFERENTIAL EQUATION.'
        READ *, H,IPRINT
        IF(H .EQ. 0.0) GO TO 10
C
C       INITIALIZE.
        XO=XZERO
        DO 25 J=1,N
25        YO(J)=YZERO(J)
        PRINT 1000, NUMDE,XZERO,XEND,H,IPRINT
1000    FORMAT(//,' EQUATION',I2,5X,'XZERO=',1PE9.2,5X,'B=',E9.2,
     *           /,' STEPSIZE=',E10.3,5X,'PRINT PARAMETER=',I3)
        PRINT 1001, N
1001    FORMAT(' ORDER=',I1,//,' J',4X,'YO(J)')
        PRINT 1002, (J,YZERO(J),J=1,N)
1002    FORMAT(I2,2X,1PE9.2)
        PRINT 1003
1003    FORMAT(/,5X,'X',7X,'J',6X,'APPROX Y(J)',7X,'ERROR')
C
C       BEGIN MAIN LOOP FOR COMPUTING SOLUTION OF DIFFERENTIAL EQUATION.
```

```
30      DO 40 K=1,IPRINT
          X1=XO+H
          IF(X1 .GT. XEND) GO TO 20
          CALL FCN(XO,YO,F)
          DO 35 J=1,N
            Y1(J)=YO(J)+H*F(J)
35          YO(J)=Y1(J)
          XO=X1
40        CONTINUE
C
C       CALCULATE ERROR AND PRINT RESULTS.
        CALL TRUE(X1,Y)
        DO 60 J=1,N
60        PRINT 1004, X1,J,Y1(J),Y(J)-Y1(J)
1004    FORMAT(F10.5,3X,I1,5X,1PE13.6,4X,E9.2)
        GO TO 30
        END


        SUBROUTINE FCN(X,Z,F)
C
C       THIS DEFINES THE RIGHT SIDE OF THE
C       DIFFERENTIAL EQUATION, F(X,Z).
C
        IMPLICIT DOUBLE PRECISION(A-H,O-Z)
        DIMENSION Z(*),F(*)
        COMMON/BLOCKF/NUMDE
C
        GO TO(10,20), NUMDE
10      CS=COS(X)
        SN=SIN(X)
        F(1)=Z(1)-2.0*Z(2)+4.0*CS-2.0*SN
        F(2)=3.0*Z(1)-4.0*Z(2)+5.0*(CS-SN)
        RETURN
20      F(1)=Z(2)
        F(2)=Z(3)
        F(3)=-3.0*(Z(3)+Z(2))-Z(1)-4.0*SIN(X)
        RETURN
        END


        SUBROUTINE TRUE(X,Y)
C
C       THIS GIVES THE TRUE SOLUTION OF
C       THE INITIAL VALUE PROBLEM, Y(X).
C
        IMPLICIT DOUBLE PRECISION(A-H,O-Z)
        DIMENSION Y(*)
        COMMON/BLOCKF/NUMDE
C
        GO TO(10,20), NUMDE
10      CS=COS(X)
        SN=SIN(X)
        Y(1)=SN+CS
        Y(2)=2.0*CS
        RETURN
```

```
20      SN=SIN(X)
        CS=COS(X)
        Y(1)=SN+CS
        Y(2)=CS-SN
        Y(3)=-SN-CS
        RETURN
        END
```

PROBLEMS

1. Let

$$A = \begin{bmatrix} -1 & -2 \\ 2 & -1 \end{bmatrix}, \qquad Y = \begin{bmatrix} Y_1 \\ Y_2 \end{bmatrix},$$

$$G(x) = \begin{bmatrix} -2e^{-x}+2 \\ -2e^{-x}+1 \end{bmatrix}, \qquad Y_0 = \begin{bmatrix} 1 \\ 1 \end{bmatrix}$$

Write out the two equations that make up the system

$$Y' = AY + G(x), \qquad Y(x_0) = Y_0$$

The true solution is $Y = [e^{-x}, 1]^T$.

2. Convert the system (9.137) to the general form of problem 1, giving the matrix A.

3. Convert the following higher order equations to systems of first order equations.
   (a) $y''' + 4y'' + 5y' + 2y = 2x^2 + 10x + 8$,
       $y(0) = +1$, $\quad y'(0) = -1$, $\quad y''(0) = 3$
       The true solution is $y(x) = e^{-x} + x^2$.
   (b) $y'' + 4y' + 13y = 40 \cos(x)$, $\quad y(0) = 3$, $\quad y'(0) = 4$
       The true solution is $y(x) = 3 \cos(x) + \sin(x) + e^{-2x} \sin(3x)$.

4. Convert the following system of second order equations to a larger system of first order equations. It arises from studying the gravitational attraction of one mass by another.

$$x''(t) = \frac{-cx(t)}{r(t)^3}, \qquad y''(t) = \frac{-cy(t)}{r(t)^3}, \qquad z''(t) = \frac{-cz(t)}{r(t)^3}$$

with c a positive constant and $r(t) = [x(t)^2 + y(t)^2 + z(t)^2]^{1/2}$, $t =$ time.

5. Using Euler's method, solve the system in problem 1. Use stepsizes of $h = 0.1, 0.05, 0.025$, and solve for $0 \le x \le 10$. Use Richardson's error formula to estimate the error for $h = 0.025$.

6. Repeat problem 5 for the systems in problem 3.

7. Modify the Euler program of this section to implement the Runge–

Kutta method given in (9.139). With this program, repeat problems 5 and 6.

8. Consider the predator–prey model of (9.122), with the particular constants $A = 4$, $B = 0.5$, $C = 3$, $D = 1/3$.

   (a) Show that there is a solution $Y_1(x) = C_1$, $Y_2(x) = C_2$, with C_1 and C_2 constants. What would be the physical interpretation of such a solution $Y(x)$?

   (b) Solve this system (9.122) with $Y_1(0) = 3$, $Y_2(0) = 5$, for $0 \leq x \leq 4$, and use the Runge–Kutta method of problem 7 with stepsizes of $h = 0.01$ and 0.005. Examine and plot the values of the output in steps of x of 0.1. In addition to these plots of $Y_1(x)$ vs. x and $Y_2(x)$ vs. x, also plot Y_1 vs. Y_2.

   (c) Repeat (b) for the initial values $Y_1(0) = 3$, $Y_2(0) = 1$, 1.5, 1.9 in succession. Comment on the relation of these solutions to one another and to the solution of part (a).

APPENDIX A

MEAN VALUE THEOREMS

This appendix discusses mean value theorems. These are often viewed as unnecessary for someone interested in applications; but, in fact, they are quite useful in deriving numerical methods, obtaining error estimates, and simplifying complicated expressions.

The mean value theorems are about functions $f(x)$ on an interval $[a, b]$. The theorems assert the existence of special points $x = c$ in $[a, b]$ for which some property of $f(x)$ is true. Since these theorems are discussed thoroughly in most calculus texts, we will just state the theorems and give illustrations of their use.

Theorem A.1 Intermediate Value Theorem

Let $f(x)$ be a continuous function on the interval $a \leq x \leq b$. Let

$$M = \max_{a \leq x \leq b} f(x), \qquad m = \min_{a \leq x \leq b} f(x)$$

Then for every value v satisfying $m \leq v \leq M$, there is at least one point c in $[a, b]$ for which $f(c) = v$.

This is an intuitive result. For example, consider the variation of temperature $T(t)$ throughout the day, where $T(t)$ denotes the temperature at time t. Generally, $T(t)$ is considered to be a continuous function of t. Thus, if T is 45°F at 7:30 A.M. and is 83°F at 3:00 P.M., then every temperature between 45°F and 83°F occurs at least once between 7:30 A.M. and 3:00 P.M. A possible graph of $T(t)$ is given in Figure A.1.

Example

Find the values of c in $0 \leqq x \leqq \pi$ for which

$$\sin(c) = 1/2$$

The graph of $\sin(x)$ is given in Figure A.2. Since the minimum of $\sin(x)$ on $[0, \pi]$ is $m = 0$ and the maximum is $M = 1$, we are guaranteed that there is a solution c to the equation. From a knowledge of the function $\sin(x)$, the critical values are easily shown to be $c = \pi/6$ and $5\pi/6$.

An important use of the intermediate value theorem is in simplifying complicated expressions. As a special case of a more general expression, consider

$$S = \tfrac{1}{3}[f(x_1) + f(x_2) + f(x_3)]$$

where $f(x)$ is continuous on an interval $[a, b]$ and x_1, x_2, and x_3 are points in that interval. Using the notation of Theorem A.1, we can show

$$S \leqq \tfrac{1}{3}[M + M + M] = M$$

and

$$S \geqq \tfrac{1}{3}[m + m + m] = m$$

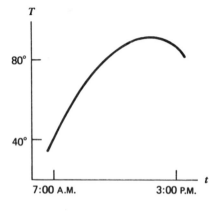

Figure A.1 Temperature vs. time.

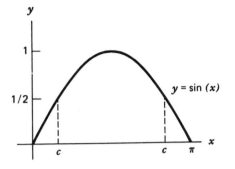

Figure A.2 Illustration of intermediate value theorem.

Thus S is a value in the interval $[m, M]$; and by the theorem, there is a point c in the interval for which $S = f(c)$, or

$$\tfrac{1}{3}[f(x_1) + f(x_2) + f(x_3)] = f(c) \tag{A.1}$$

for some c in $[a, b]$.

Let $f(x)$ be continuous on $[a, b]$ and let $x_1, x_2, \ldots, x_n$ be points in $[a, b]$. Let $w_1, w_2, \ldots, w_n$ be positive numbers for which

$$\sum_{j=1}^{n} w_j = 1$$

Form

$$S = \sum_{j=1}^{n} w_j f(x_j)$$

Then by an argument analogous to that of the last paragraph, there is a point c in $[a, b]$ for which $S = f(c)$; or

$$f(c) = \sum_{j=1}^{n} w_j f(x_j) \tag{A.2}$$

for some c in $[a, b]$.

Example

Let $f(x) = \sqrt{x}$, and consider

$$S = \frac{1}{5} [f(0) + f(0.25) + f(0.5) + f(0.75) + f(1)]$$

Here there are $n = 5$ points, and all $w_j = 1/5$. Find the value c of (A.2). First, $S \doteq 0.61463$. Thus we want

$$\sqrt{c} \doteq 0.61463$$
$$c \doteq (0.61463)^2 \doteq 0.3777$$

S is called an "average value" or "mean value" of the function $\sqrt{x}$ on $[0, 1]$, and it is attained at the above value of c.

Theorem A.2 Mean Value Theorem

Let $f(x)$ be continuous on the interval $a \leq x \leq b$, and also let it be differentiable for $a < x < b$. Then there is at least one point c in (a, b) for which

$$f(b) - f(a) = f'(c)(b - a) \tag{A.3}$$

Example

Find the values of c for which (A.3) is true for $f(x) = \sqrt{x}$ on $[0, 2]$. Figure A.3 is an illustration of the theorem in this case. The dotted line connects the points $(0, 0)$ and $(2, \sqrt{2})$. Solving for c in (A.3) gives us

$$\sqrt{2} - \sqrt{0} = \frac{1}{2\sqrt{c}}(2 - 0)$$

which has the solution $c = 1/2$. The graph shows the tangent line at $[\frac{1}{2}, f(\frac{1}{2})]$.

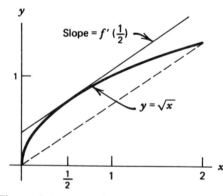

Figure A.3 Example of mean value theorem.

Note that by rewriting (A.3) as

$$\frac{f(b) - f(a)}{b - a} = f'(c)$$

we have a simple geometric interpretation of the theorem. It says that the straight line connecting the endpoints of the graph has the same slope as that of some tangent line to the curve at an intermediate point $(c, f(c))$. Check this with Figure A.3.

One of the uses of (A.3) is to estimate or bound the effect of an error in the argument of a function. Let x_T denote the true value of an argument, often unknown, and let x_A denote a known approximate value. Then

$$f(x_T) - f(x_A) = f'(c)(x_T - x_A)$$

with c some number between x_T and x_A. Since x_T and x_A are generally very close together, the value of $f'(x)$ will usually not vary a great deal for x between x_T and x_A and, thus,

$$f(x_T) - f(x_A) \doteq \begin{cases} f'(x_T)(x_T - x_A) \\ \quad\quad \text{or} \\ f'(x_A)(x_T - x_A) \end{cases} \tag{A.4}$$

Use the more convenient formula for your particular problem.

Example

Consider evaluating $\cos(\pi/6)$ using $\pi/6 \doteq 0.524$. To use (A.4), we let $x_T = \pi/6$, $x_A = 0.524$, and $f(x) = \cos(x)$. In general, $x_T - x_A$ will not be known and a bound for it will be used. But for illustrative purposes, we have $x_T - x_A \doteq -0.0004$. Using (A.4) gives us

$$\cos\left(\frac{\pi}{6}\right) - \cos(0.524) \doteq -\sin\left(\frac{\pi}{6}\right)\left(\frac{\pi}{6} - 0.524\right)$$

$$\doteq -\frac{1}{2}(-0.0004) = 0.0002$$

The actual error is 0.0002007.

The mean value theorem is also useful in obtaining bounds for the variation of a function based on the variation of the argument. If $|f'(x)|$ is bounded by a constant M everywhere on an interval $a \leqq x \leqq b$, then

$$|f(x) - f(z)| \leqq M|x - z| \qquad \text{(A.5)}$$

for all x, z in $[a, b]$.

Example

Take $f(x) = e^x$ for $0 \leqq x \leqq 1$. Then $f'(x)$ is bounded by e^1, and thus

$$|e^x - e^z| \leqq e|x - z|, \qquad 0 \leqq x, z \leqq 1$$

The mean value theorem is used in a number of ways other than those described here, and these occur in several places in the text. Also, the mean value theorem is a special case of Taylor's theorem, described in sections 1.1 and 1.2 of Chapter 1.

Theorem A.3   Integral Mean Value Theorem

Let $w(x)$ be a nonnegative integrable function on $[a, b]$, and let $f(x)$ be continuous on $[a, b]$. Then there is at least one point c in $[a, b]$ for which

$$\int_a^b f(x)w(x)\, dx = f(c) \int_a^b w(x)\, dx \qquad \text{(A.6)}$$

In particular, if we take $w(x) \equiv 1$, then

$$\int_a^b f(x)\, dx = f(c)(b - a) \qquad \text{(A.7)}$$

for some c in $[a, b]$.

A geometric interpretation of (A.7) is given in Figure A.4. For $f(x)$ nonnegative as well as continuous, the left side of (A.7) represents the area under the curve of $y = f(x)$ between $x = a$ and $x = b$. The right side of (A.7) represents the area of a rectangle with base $[a, b]$ and height $f(c)$. The theorem asserts that these two areas are equal for some choice of $f(c)$.

The result (A.6) is used in simplifying, estimating, and bounding integrals that occur in a number of situations. Many of these occur in

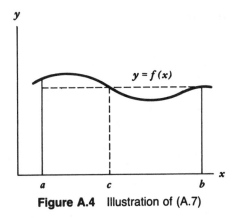

Figure A.4 Illustration of (A.7)

connection with Taylor's theorem, in section 1.2 of Chapter 1. As an example of the use of the integral mean value theorem, consider the integral

$$I = \int_a^b (b - x)(x - a)f(x) \, dx$$

Let $w(x) = (b - x)(x - a)$, and apply (A.6) to get

$$I = f(c) \int_a^b (b - x)(x - a) \, dx = \frac{(b - a)^3}{6} f(c) \qquad \text{(A.8)}$$

for some c in $[a, b]$. This integral occurs in the analysis of the trapezoidal numerical integration rule, in section 7.2 of Chapter 7.

Example

As a more concrete example, consider evaluating

$$g(t) = \int_0^t \frac{x^n}{1 + x^2} \, dx, \qquad n \geq 0$$

Take $w(x) = x^n$ and $f(x) = 1/(1 + x^2)$. Then from (A.6),

$$g(t) = \frac{1}{1 + c^2} \int_0^t x^n \, dx = \left(\frac{1}{1 + c^2}\right)\left(\frac{t^{n+1}}{n + 1}\right)$$

for some c between 0 and t. Thus for $t \geq 0$,

$$\left(\frac{1}{1 + t^2}\right)\frac{t^{n+1}}{n + 1} \leq g(t) \leq \frac{t^{n+1}}{n + 1}$$

PROBLEMS

1. Illustrate the intermediate value theorem by finding the values of c for which $f(c) = v$ for the given value of v.
   (a) $f(x) = 2x^2 - 5x + 2$, $v = -1$, $-5 \leq x \leq 5$
   (b) $f(x) = \cos(x)$, $v = 1/2$, $\pi \leq x \leq 3\pi$
   (c) $f(x) = \sin(x) + \cos(x)$, $v = \sqrt{2}$, $0 \leq x \leq \pi$

2. If $f'(x) > 0$ for $a \leq x \leq b$, what can be said about the number of solutions of $f(x) = v$?

3. Find the average value on $[a, b]$ of $f(x)$, based on (A.2) and the procedure given in the example following it. Also, find the point c in $[a, b]$ at which this average value is actually attained.
   (a) $f(x) = x^2$, $0 \leq x \leq 1$, $n = 3$, and $n = 6$
   (b) $f(x) = \sin(x)$, $0 \leq x \leq \pi$, $n = 3$, and $n = 6$

4. Illustrate the mean value theorem by finding the value of c for which (A.3) is satisfied with the following cases of $f(x)$ and $[a, b]$.
   (a) $f(x) = \cos(x)$, $0 \leq x \leq \pi/2$
   (b) $f(x) = x^3$, $0 \leq x \leq 1$
   (c) $f(x) = x^3 - x$, $-2 \leq x \leq 2$

5. Estimate the error in evaluating $f(x)$ with x_A rather than x_T.
   (a) $f(x) = \sqrt{x}$, $x_T = \pi$, $x_A = 22/7$
   (b) $f(x) = \log(x)$, $x_T = e$, $x_A = 2.72$
   (c) $f(x) = \cos(x)$, $x_T = \sqrt{2}$, $x_A = 1.414$

6. Derive the following inequalities.
   (a) $|\cos(x) - \cos(z)| \leq |x - z|$,   all x, z
   (b) $|e^x - e^z| \leq |x - z|$, for all x, $z \leq 0$
   (c) $|\tan(x) - \tan(z)| \geq |x - z|$, $-\pi/2 < x$, $z < \pi/2$
   (d) $py^{p-1}(x - y) \leq x^p - y^p \leq px^{p-1}(x - y)$, $0 \leq y \leq x$, $p \geq 1$.

7. Illustrate the integral mean value theorem by finding the points c for which the following results are true.

   (a) $\displaystyle\int_0^1 x^3\, dx = c^3$, c in $[0, 1]$

   (b) $\displaystyle\int_0^1 \frac{x - (1/2)}{\sqrt{x}}\, dx = \left(c - \frac{1}{2}\right)\int_0^1 \frac{dx}{\sqrt{x}} = 2\left(c - \frac{1}{2}\right)$, c in $[0, 1]$

(c) $\displaystyle\int_0^{\pi/2} x \sin(x)\, dx = \sin(c) \int_0^{\pi/2} x\, dx = \frac{\pi^2}{8} \sin(c)$, c in $[0, \pi/2]$.

8. Use the integral mean value theorem to obtain the following results.

(a) $\displaystyle\int_0^h x^2(h - x)^2 g(x)\, dx = \frac{h^5}{30} g(c)$, c in $[0, h]$

(b) $\displaystyle\int_a^x (t - a)^n f(t)\, dt = \frac{(x - a)^{n+1}}{n + 1} f(c)$, c in $[a, x]$.

9. An average value of $f(x)$ on $[a, b]$ can be found using (A.2), with

$$\text{Average value} = \frac{1}{n} \sum_{j=1}^{n} f(x_j)$$

Here
$x_1 = a$, $x_2 = a + h$, $x_3 = a + 2h$, ..., $x_n = b$, and $h = (b - a)/(n - 1)$.
If we write this as

$$\frac{1}{b - a} \sum_{j=1}^{n} f(x_j) \left(\frac{b - a}{n}\right)$$

the sum is recognizable as a Riemann sum or approximate integral. Letting $n \to \infty$, we have an increasingly accurate average value of $f(x)$ on $[a, b]$; in the limit, it is equal to

$$A = \frac{1}{b - a} \int_a^b f(x)\, dx$$

Using (A.7), this too can be written as $A = f(c)$ for some c in $[a, b]$. Use the integral formula for A to compute the average value of the functions in problem 3 and compare with the earlier results of that problem.

APPENDIX B

MATHEMATICAL FORMULAS

I. Algebra

(a) The quadratic formula: the solution of $ax^2 + bx + c = 0$, $a \neq 0$, is given by

$$x = \frac{-b \pm \sqrt{b^2 - 4ac}}{2a} = \frac{2c}{-b \mp \sqrt{b^2 - 4ac}}$$

(b) Logarithms and exponentials. For $a > 0$,

$$a^x = e^{x \log a}, \qquad \log_a x = \frac{\log x}{\log a}$$

with $\log(x) = \log_e x = \ln x$ the logarithm of x to the base e.

(c) Factorials. $0! = 1$

$$n! = n \cdot [(n-1)!]$$
$$= n(n-1)(n-2) \ldots (2)(1), \qquad n \geq 1$$

(d) Sums.

$$\sum_{j=1}^{n} j = 1 + 2 + 3 + \ldots + n = \frac{n(n + 1)}{2}$$

$$\sum_{j=1}^{n} j^2 = 1 + 4 + 9 + \ldots + n^2 = \frac{n(n + 1)(2n + 1)}{6}$$

$$\sum_{j=1}^{n} r^j = 1 + r + r^2 + \ldots + r^n = \frac{r^{n+1} - 1}{r - 1}, \qquad r \neq 1$$

$$(a + b)^n = \sum_{j=0}^{n} \binom{n}{j} a^{n-j} b^j = \binom{n}{0} a^n + \binom{n}{1} a^{n-1} b + \ldots + \binom{n}{n} b^n$$

$$\binom{n}{j} = \frac{n!}{j!(n - j)!}$$

(e) The fundamental theorem of algebra.
If $p(x)$ is a polynomial of degree n, there is a set of numbers r_1, $r_2, \ldots, r_n$ (possibly complex), for which

$$p(x) = c(x - r_1)(x - r_2) \ldots (x - r_n)$$

where c is the coefficient of x^n in $p(x)$. The numbers $\{r_j\}$ are unique, except for their order.

II. Geometry

(a) Areas.
　(i) Area of a triangle

$$A = \frac{1}{2} bh \qquad (b = \text{base}, \ h = \text{height})$$

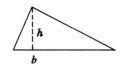

　(ii) Area of a trapezoid

$$A = h \left(\frac{b_1 + b_2}{2} \right)$$

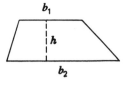

with b_1 and b_2 the lengths of the parallel sides.

(iii) Area of a circle

$$A = \pi r^2 \qquad (r = \text{radius})$$

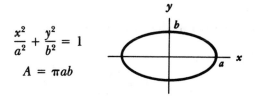

(iv) Area of a sector of a circle with central angle θ (in radians)

$$A = \frac{1}{2} r^2 \theta$$

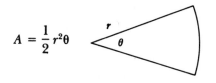

(v) Area of the ellipse with equation

$$\frac{x^2}{a^2} + \frac{y^2}{b^2} = 1$$

$$A = \pi ab$$

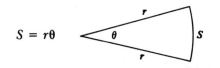

(vi) Length of a circular arc, with central angle θ (in radians)

$$S = r\theta$$

(b) Equations.
   (i) Equation of a straight line in the xy-plane

$$ax + by = c$$

(ii) Slope of a straight line,

$$m = \frac{y_2 - y_1}{x_2 - x_1}$$

with (x_1, y_1) and (x_2, y_2) points on the line

(iii) Equation of a plane in three dimensional space,

$$ax + by + cz = d$$

(iv) Equation of a circle in xy-plane

$$(x - x_0)^2 + (y - y_0)^2 = r^2$$

with (x_0, y_0) the center and r the radius.

(v) Equation of a parabola, with axis parallel to the y-axis.

$$y = ax^2 + bx + c, \qquad a \neq 0$$

III. Trigonometry

(a) Conversion between angles in degrees (θ_D) and angles in radians (θ_R).

$$\theta_R = \frac{\pi}{180} \theta_D$$

(b) Graphs.
    (i) sin (x) and cos (x)

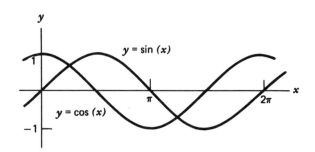

(ii)   tan (x)

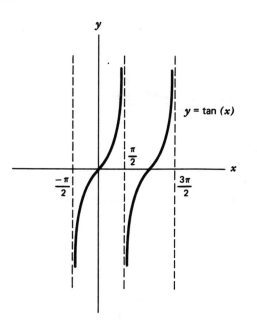

(iii) $\sin^{-1}(x) = \arcsin(x)$ and $\cos^{-1}(x) = \arccos(x)$

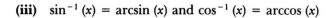

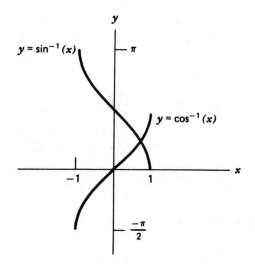

(iv) $\tan^{-1}(x) = \text{arctangent}(x)$

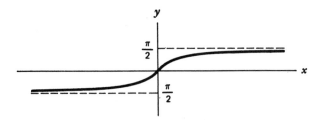

(c) Definitions.

$$\tan(x) = \frac{\sin(x)}{\cos(x)}, \qquad \cot(x) = \frac{1}{\tan(x)} = \frac{\cos(x)}{\sin(x)}$$

$$\sec(x) = \frac{1}{\cos(x)}, \qquad \csc(x) = \frac{1}{\sin(x)}$$

(d) Trigonometric identities.

$$\sin(-\alpha) = -\sin(\alpha), \qquad \cos(-\alpha) = \cos(\alpha)$$

$$\cos^2(\alpha) + \sin^2(\alpha) = 1, \qquad 1 + \tan^2(\alpha) = \sec^2(\alpha)$$

$$\sin(\alpha \pm \beta) = \sin(\alpha)\cos(\beta) \pm \cos(\alpha)\sin(\beta)$$

$$\cos(\alpha \pm \beta) = \cos(\alpha)\cos(\beta) \mp \sin(\alpha)\sin(\beta)$$

$$\tan(\alpha \pm \beta) = \frac{\tan(\alpha) \pm \tan(\beta)}{1 \mp \tan(\alpha)\tan(\beta)}$$

$$\sin(2\alpha) = 2\sin(\alpha)\cos(\alpha)$$

$$\cos(2\alpha) = \cos^2(\alpha) - \sin^2(\alpha) = 2\cos^2(\alpha) - 1 = 1 - 2\sin^2(\alpha)$$

$$\sin^2(\alpha) = \frac{1 - \cos(2\alpha)}{2}, \quad \cos^2(\alpha) = \frac{1 + \cos(2\alpha)}{2}$$

$$\sin(\alpha) \pm \sin(\beta) = 2\sin\left(\frac{\alpha \pm \beta}{2}\right)\cos\left(\frac{\alpha \mp \beta}{2}\right)$$

$$\cos(\alpha) + \cos(\beta) = 2\cos\left(\frac{\alpha + \beta}{2}\right)\cos\left(\frac{\alpha - \beta}{2}\right)$$

$$\cos(\alpha) - \cos(\beta) = -2\sin\left(\frac{\alpha + \beta}{2}\right)\sin\left(\frac{\alpha - \beta}{2}\right)$$

$$\sin^{-1}(x) + \cos^{-1}(x) = \frac{\pi}{2}$$

$$\tan^{-1}(x) = \frac{\pi}{2} - \tan^{-1}\left(\frac{1}{x}\right), x > 0$$

(e) Law of cosines.

$$c^2 = a^2 + b^2 - 2ab \cos(\gamma)$$

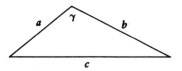

The angle γ is opposite to the side with length c.

IV. Calculus

(a) Definition of derivative.

$$f'(x) = \lim_{h \to 0} \frac{f(x + h) - f(x)}{h}$$

Other notations: $f'(x) = \dfrac{df(x)}{dx} = Df(x)$

(b) Properties of the derivative.

$$[f(x) + g(x)]' = f'(x) + g'(x)$$
$$[f(x)g(x)]' = f(x)g'(x) + f'(x)g(x)$$
$$[cf(x)]' = cf'(x), \ c = \text{constant}$$
$$[f(x)/g(x)]' = \frac{f'(x)g(x) - f(x)g'(x)}{[g(x)]^2}$$
$$[f(g(x))]' = f'(g(x)) \cdot g'(x) \qquad \text{(chain rule)}$$

(c) Derivatives of common functions.

$$\frac{d}{dx}(x^n) = nx^{n-1}, \qquad n \neq 0$$

$$\frac{d}{dx}[\cos(x)] = -\sin(x), \qquad \frac{d}{dx}[\sin(x)] = \cos(x)$$

$$\frac{d}{dx}[\tan(x)] = \sec^2(x), \qquad \frac{d}{dx}[\sec(x)] = \sec(x)\tan(x)$$

$$\frac{d}{dx}[\sin^{-1}(x)] = \frac{1}{\sqrt{1 - x^2}}, \qquad \frac{d}{dx}[\cos^{-1}(x)] = \frac{-1}{\sqrt{1 - x^2}}$$

$$\frac{d}{dx}[\tan^{-1}(x)] = \frac{1}{1+x^2}$$

$$\frac{d}{dx}[e^x] = e^x, \qquad \frac{d}{dx}[a^x] = [\log a]a^x$$

$$\frac{d}{dx}[\log(x)] = \frac{1}{x}, \qquad \frac{d}{dx}[\log_a x] = \frac{1}{x \log a}$$

(d) Definition of the definite integral.

$$\int_a^b f(x)\, dx = \lim_{n \to \infty} \sum_{j=1}^n f(p_j)(x_j - x_{j-1})$$

with $a = x_0 < x_1 < x_2 < \ \ldots \ < x_n = b$, p_j in $[x_{j-1}, x_j]$, and max $(x_j - x_{j-1}) \to 0$ as $n \to \infty$. The above sums are called Riemann sums.

(e) Properties of the definite integral.

$$\int_a^b [f(x) + g(x)]\, dx = \int_a^b f(x)\, dx + \int_a^b g(x)\, dx$$

$$\int_a^b cf(x)\, dx = c \int_a^b f(x)\, dx, \ c = \text{constant}$$

$$\int_a^b f(x)\, dx = \int_a^c f(x)\, dx + \int_c^b f(x)\, dx$$

$$\int_a^b f(x)\, dx = - \int_b^a f(x)\, dx$$

$$\int_a^b f(x)g'(x)\, dx = [f(x)g(x)]_a^b - \int_a^b f'(x)g(x)\, dx$$

$$\int_a^b f(g(x))g'(x)\, dx = \int_{g(a)}^{g(b)} f(u)\, du$$

(f) Fundamental theorem of the calculus.
Form 1. Let $F(x)$ be an antiderivative of $f(x)$. Then

$$\int_a^b f(x)\, dx = [F(x)]_{x=a}^{x=b} = F(b) - F(a)$$

Form 2.

$$\frac{d}{dx} \int_a^x f(t) \, dt = f(x)$$

(g) Table of antiderivatives.

$$\int x^n \, dx = \frac{x^{n+1}}{n+1} + c, \qquad n \neq -1$$

$$\int \frac{dx}{x} = \log |x| + c$$

$$\int e^x \, dx = e^x + c, \qquad \int a^x \, dx = \frac{a^x}{\log (a)} + c$$

$$\int x^\alpha \log (x) \, dx = \frac{x^{\alpha+1}}{\alpha + 1} \cdot \log (x) - \frac{x^{\alpha+1}}{(\alpha + 1)^2} + c, \qquad \alpha \neq -1$$

$$\int \frac{dx}{x^2 + a^2} = \frac{1}{a} \tan^{-1} \left(\frac{x}{a} \right) + c$$

$$\int \frac{dx}{\sqrt{a^2 - x^2}} = \sin^{-1} \left(\frac{x}{a} \right) + c$$

$$\int \sin (x) \, dx = -\cos (x) + c, \qquad \int \cos (x) \, dx = \sin (x) + c$$

$$\int \sec^2 (x) \, dx = \tan (x) + c, \qquad \int \sec (x) \tan (x) \, dx = \sec (x) + c$$

$$\int \tan (x) \, dx = \log |\sec (x)| + c$$

$$\int \sec (x) \, dx = \log |\sec (x) + \tan (x)| + c$$

$$\int \sin^2 (x) \, dx = \frac{x}{2} - \frac{\sin (2x)}{4} + c$$

$$\int \cos^2 (x) \, dx = \frac{x}{2} + \frac{\sin (2x)}{4} + c$$

APPENDIX C

```
┌─────────────────────────────────────────────┐
│                                             │
│                                             │
│                                             │
│                                             │
│          NUMERICAL ANALYSIS                 │
│          SOFTWARE PACKAGES                  │
│                                             │
└─────────────────────────────────────────────┘
```

The programs included in this text are designed to give insight into the numerical methods being studied. They are not intended to be used in a production setting that requires highly accurate, efficient, robust, and convenient codes. For such computer codes, one should take advantage of the many high-quality numerical analysis program packages that have been written in the past ten to fifteen years. In this appendix we list some of these packages, along with information on obtaining them. It should be noted that most of these packages are written in Fortran, and thus they are intended for use in a Fortran programming environment.

There are two widely used general libraries in numerical analysis, each containing programs for solving most standard numerical analysis problems. Both libraries are commercially available on most lines of computers and minicomputers, and both have subsets available on some microcomputers. In addition, both companies have additional numerical analysis packages that are not discussed in this appendix.

1. IMSL (International Mathematics and Statistics Library). This library contains over 500 subprograms for solving problems in numerical analysis and statistics. The library is organized into the following chapters:

A.   Analysis of variance

B.   Basic statistics

C. Categorized data analysis

D. Differential equations; quadrature; differentiation

E. Eigensystem analysis

F. Forecasting; econometrics; time series; transforms

G. Generation and testing of random numbers

I. Interpolation; approximation; smoothing

L. Linear algebraic equations

M. Mathematical and statistical special functions

N. Nonparametric statistics

O. Observation structure; multivariate statistics

R. Regression analysis

S. Samplings

U. Utility functions

V. Vector, matrix arithmetic

Z. Zeros and extrema; linear programming

For more information on the library, contact

IMSL, Inc.
NBC Building
7500 Bellaire Boulevard
Houston, Texas 77036-5085

2. NAG (Numerical Algorithms Group). This library also contains over 500 subprograms covering the basic problems in numerical analysis and statistics. The library is organized into chapters as follows. The chapter labels are based on a naming convention of the Association for Computing Machinery (ACM).

A02 Complex arithmetic

C02 Zeros of polynomials

C05 Roots of one or more transcendental functions

C06 Summation of series

D01 Quadrature

D02 Ordinary differential equations

D03 Partial differential equations

D04 Numerical differentiation

D05 Integral equations

E01 Interpolation

E02	Curve and surface fitting
E04	Minimizing or maximizing a function
F01	Matrix operations, including inversion
F02	Eigenvalues and eigenvectors
F03	Determinants
F04	Simultaneous linear equations
F05	Orthogonalization
G01	Simple calculations on statistical data
G02	Correlation and regression analysis
G04	Analysis of variance
G05	Random number generators
G08	Nonparametric statistics
G13	Time series analysis
H	Operations research
M01	Sorting
P01	Error trapping
S	Approximation of special functions
X01	Mathematical constants
X02	Machine constants
X03	Inner products
X04	Input/output utilities

For more information on the library, contact

Numerical Algorithms Group Ltd.
NAG Central Office
Mayfield House
256 Banbury Road
Oxford OX2 7DE, United Kingdom

Since 1972, a number of high-quality numerical analysis program packages have been written for specific problem areas. These packages are generally characterized by state-of-the-art algorithms, high accuracy and efficiency, robustness, and portability between different lines of computers. Many of the ideas of these packages have since been incorporated into the IMSL and NAG libraries, but it is still often advantageous to use these packages separately. They are usually documented more completely than the larger libraries, and having the actual Fortran source codes allows one to make modifications when needed. We list here only a few of these more specialized packages.

1. EISPACK. This solves eigenvalue-eigenvector problems for a wide variety of types of matrices. It was the first of the packages referred to in the preceding paragraph, and the currently available version is the third release of it. Like several of the other packages, it was developed jointly by researchers at Argonne National Laboratory and several universities, with financial support from the National Science Foundation. For a detailed description of the package and a listing of all of the programs in it, see the following two monographs:

B. T. Smith, J. M. Boyle, J. J. Dongarra, B. S. Garbow, Y. Ikebe, V. C. Klema, and C. B. Moler (1976), *Matrix Eigensystem Routines—EISPACK Guide*, 2nd ed., *Lectures Notes in Computer Science*, Volume 6 Springer-Verlag, 1976.

B. S. Garbow, J. M. Boyle, J. J. Dongarra, and C. B. Moler, *Matrix Eigensystem Routines—EISPACK Guide Extension Lecture Notes in Computer Science*, Volume 51 Springer-Verlag, 1977.

A machine-readable copy of the EISPACK code can be obtained from IMSL or from the NESC, whose address is

National Energy Software Center
Argonne National Laboratory
9700 South Cass Avenue
Argonne, Illinois 60439

2. LINPACK. This package is for the solution of simultaneous systems of linear algebraic equations. Special subprograms are included for many common types of coefficient matrices for such systems. Versions of the programs are available for four kinds of computer arithmetic: single and double precision real, and single and double precision complex. Error monitoring is available, along with checks on the conditioning of the matrix of coefficients for the linear system. The code is available through IMSL and NESC, just as for EISPACK. For documentation and a listing of all programs in LINPACK, see [J. J. Dongarra, J. R. Bunch, C. B. Moler, and G. W. Stewart, *LINPACK Users' Guide*, SIAM Publications, 1979.]

3. MINPACK. This contains programs for solving systems of nonlinear algebraic equations and nonlinear least squares problems. A second edition is to be released in the future and it will contain programs for unconstrained and constrained optimization problems. The code is available from IMSL and NESC, just as for EISPACK. A user's guide accompanies the package.

4. LLSQ.   This is a collection of programs to solve linear least squares problems. A complete discussion of the least squares problem together with a listing of the Fortran programs in LLSQ is given in [C. Lawson and R. Hanson, *Solving Least Squares Problems*, Prentice-Hall, 1974.] Machine-readable code is available from IMSL.

5. B-SPLINE.   This package contains subprograms for constructing, evaluating, differentiating, and integrating spline functions. The programs are a part of [C. deBoor, *A Practical Guide to Splines*, Springer-Verlag, 1978.] The code is available from IMSL.

Literature on Mathematical Software

There are several periodicals in which new programs or information about them are published. Several of these are published by the Association for Computing Machinery (ACM), and we will limit our references to those periodicals. Information on any of them can be obtained by writing to the following address:

ACM
11 West 42nd Street
New York, New York 10036

1. ACM Transactions on Mathematical Software.   This journal contains articles on the development and testing of numerical analysis software. It also contains new programs for a wide variety of numerical analysis problems. These programs are usually written very carefully, and they are in a portable form for use on different computers.

2. Collected Algorithms for the ACM.   This is a collection of all programs that have been published in ACM journals, and this collection is available as a separate subscription from the ACM. All new mathematics programs now appear in the *ACM Transactions on Mathematical Software*, listed above. Machine readable copies of the algorithms are available from IMSL.

3. ACM SIGNUM Newsletter.   The name SIGNUM stands for Special Interest Group in Numerical Mathematics, an organization devoted to numerical analysis within the ACM. Information on new software packages is often given first in this newsletter.

The discussion of this appendix is far from inclusive of all that is available in quality mathematical software. In addition, the choice of program packages is heavily skewed towards the United States, since the author is better acquainted with these programs.

CHAPTER 1

Section 1.1

2.(a) $p_1(x) = 1 + \frac{1}{2}(x - 1)$
$p_2(x) = 1 + \frac{1}{2}(x - 1) - \frac{1}{8}(x - 1)^2$

3.(c) $p_n(x) = 1 + \binom{1/2}{1}x + \binom{1/2}{2}x^2 + \ldots + \binom{1/2}{n}x^n$

5.(a) $p_4(x) = 1 - x + \frac{1}{2}x^2 - \frac{1}{6}x^3 + \frac{1}{24}x^4$

7. $g(x) \doteq 1 + \frac{1}{2}x + \frac{1}{8}x^2$, for small values of x

9.(a) $q(x) = f(a) + [(x - a)/(b - a)][f(b) - f(a)]$

Section 1.2

1. $e^x - p_3(x) = \dfrac{x^4}{4!} e^c$, c between 0 and x

$\displaystyle \max_{-1 \leq x \leq 1} |e^x - p_3(x)| \leq (1)^4 \frac{e^1}{24} \doteq 0.1133$

From Table 1.1, $e^1 - p_3(1) = 0.05161$

3.(b) $\left| \dfrac{\sin(x) - x}{x} \right| \leq 0.01$ for $|x| \leq 0.245$

4. Use degree $= 7$.

8.(b) $\tan^{-1}(x) = x - \dfrac{x^3}{3} + \ldots + \dfrac{(-1)^n x^{2n+1}}{2n+1} + R_{2n+1}(x)$

$R_{2n+1}(x) = (-1)^{n+1} \displaystyle\int_0^x \dfrac{t^{2n+2}\, dt}{1 + t^2}$

Section 1.3

2. Use degree $= 8$.
4. Form $p'(x)$ by direct differentiation in (1.25), and then let $x = z$.
5. Six multiplications are needed, if we assume that all coefficients have been calculated already.

CHAPTER 2.

Section 2.1

1.(b) $(5.8125)_{10}$
3.(b) $(100.01)_2$
3.(d) $\pi \doteq (11.001001)_2$
4. $(0.8)_{10} = (0.110011001100\ldots)_2$
5. $6/7$

Section 2.2

1. Thirty-two decimal digits
3. $n = 32$, $M \doteq 4.3 \times 10^9$, $-2^{-32} \leq \epsilon \leq 2^{-32}$
5. In single precision
  $2^{-129} \leq |x| < 2^{127}$
8. In single precision, $u = 16^{-5}$

CHAPTER 3

Section 3.1

1.(c) $x_T = 2.7182818\ldots$ (not repeating)
  $x_A = 2.71428571\ldots$ (repeating)

$x_T - x_A = 0.003996\ldots$
$\mathrm{Re}(x_A) \doteq 0.00147$
x_A has three significant digits.

3. $\lim_{t \to \infty} N(t) = N_c = 2N(0)$

5.(a) $\dfrac{1 - \cos(x)}{x^2} = \dfrac{2 \sin^2(x/2)}{x^2}$

(d) $\sqrt[3]{1 + x} - 1 = \dfrac{x}{(\sqrt[3]{x+1})^2 + \sqrt[3]{x+1} + 1}$

6.(b) Use degree 5 Taylor approximations to e^x and e^{-x}; then
$$\frac{e^x - e^{-x}}{2x} \doteq 1 + \frac{x^2}{3!} + \frac{x^4}{5!}$$
Error bounds can be calculated using the error terms for e^x and e^{-x}.

8. The roots are $x_1 = 20 + \sqrt{399}$, $x_2 = 20 - \sqrt{399}$.
To calculate x_2, use $\sqrt{399} = 19.975$ and the form
$$x_2 = \frac{1}{20 + \sqrt{399}} \doteq \frac{1}{39.975} \doteq 0.025016$$

Section 3.2

1.(a) $2.05265 \le x_T + y_T \le 2.05375$

(d) $\dfrac{8.4725}{0.0645} \le \dfrac{x_T}{y_T} \le \dfrac{8.4735}{0.0635}$

$131.356 \le \dfrac{x_T}{y_T} \le 133.441$

3.(a) Use (3.24), with $x_A = 1.473$, $|x_T - x_A| \le 0.0005$. Then
$\cos(x_T) - \cos(x_A) = -\sin(c)(x_T - x_A)$
$|\cos(x_T) - \cos(x_A)| \le [\sin(1.4735)](0.0005)$
< 0.0005

(c) $x_A = 1.4712$, $|x_T - x_A| \le 0.00005$
$$|\log(x_T) - \log(x_A)| = \frac{1}{c}|x_T - x_A| \le \frac{0.00005}{1.47115}$$
$$\doteq 0.000034$$
$$\doteq$$

4. $\mathrm{Rel}(\sqrt{x_A}) \doteq \frac{1}{2} \mathrm{Rel}(x_A)$ for x_A close to x_T.

9. $f(1 + 10^{-4}) \doteq (-1)^{n-1}(n - 1)! \cdot 10^{-4}$, $\quad f(1) = 0$
For $n = 8$, $f(1 + 10^{-4}) \doteq -0.5040$.

CHAPTER 4

Section 4.1

1.(a) 1.839287
  (c) 0.424031
3. $r = .09012$ or 9.012%
7. $n = 32$
8. $m - 1$

Section 4.2

2.(a) 1.839286755
  (c) 0.424031039
4. $x_{n+1} = \dfrac{1}{m}\left[(m-1)x_n + \dfrac{a}{x_n^{m-1}}\right]$

6.

B	Root
1	$-.5884017765$
10	$-.3264020101$
50	$-.1832913333$

Section 4.3

5.(b) $q_{n+1} = q_n + q_{n-1}, \qquad q_0 = q_1 = 1$

Section 4.4

2. For $\alpha(0) = 3$, $\alpha(\epsilon) \doteq 3 + 15.2\epsilon = 3.03$
4. $\alpha(\epsilon) - 5 \doteq 3125\epsilon$. An ill-conditioned problem.
5. The root has multiplicity 4.

CHAPTER 5

Section 5.1

1.(a) $P_1(x) = 2 - x$
  (b) $f(x) = \left(\dfrac{1 - 2e}{1 - e}\right) + \left(\dfrac{1}{1 - e}\right)e^x$

2. $p(x) = 3 - \cos(\pi x) + 2 \sin(\pi x)$

3.(a) 2.375036

7. $-x^2 + 4x - 3$

8.(a) $P_2(x) = P_1(x) = x + 1$

11. $L_0(x) = \dfrac{(x - x_1)(x - x_2)(x - x_3)}{(x_0 - x_1)(x_0 - x_2)(x_0 - x_3)}$

15. $q(x) = 4x^2 - 4x - 1$

Section 5.2

1. $f[x_0, x_1] = 2.363200$, $\qquad f'\left(\dfrac{x_0 + x_1}{2}\right) = 2.363161$

$f[x_0, x_1, x_2] = 1.193750$

3.(a) $p(x) = f(x) - f(x_0)$ is a polynomial of degree m, and $p(x_0) = 0$. Thus $(x - x_0)$ is a factor of $p(x)$, and $p(x)/(x - x_0)$ is a polynomial of degree $m - 1$.

5.(a) $\dfrac{f'(x_1) - 2f[x_0, x_1] + f'(x_0)}{(x_1 - x_0)^2}$

6.

n	x_n	y_n	Dy_n	D^2y_n
0	0.1	0.2		
			0.4	
1	0.2	0.24		1
			0.6	
2	0.3	0.30		

with Dy_n and D^2y_n denoting first and second order divided differences. Then

$P_1(0.15) = 0.22$, $\qquad P_2(0.15) = 0.2175$

11. Using (5.28), with $m = n$, we get

$$f[x_0, x_1, \ldots, x_n] = \frac{n!}{n!} = 1$$

Section 5.3

1.(a) 1.25×10^{-5} $\qquad$ **(b)** 6.42×10^{-8}

3.(a) 1.07×10^{-5} $\qquad$ **(b)** 1.28×10^{-7}

4. $h \leq 0.00632$, say $h = 0.006$ or 0.005.

6. $h^4 e/24$

8. For $x_0 \leq x \leq x_2$,

$$|\log_{10} x - P_2(x)| \leq \frac{h^3}{9\sqrt{3}} \cdot \frac{2 \log_{10} e}{x_0^3}$$

11. $\quad |e^x - P_n(x)| \leq \dfrac{e}{(n+1)n^{n+1}}, 0 \leq x \leq 1$

Section 5.4

1.(c) $s(x) = \begin{cases} x^3 + 1 - x, & 0 \leq x \leq 1 \\ (2-x)^3 + (4x-3) - (2-x), & 1 \leq x \leq 2 \end{cases}$

2. $\quad \frac{2}{3}M_2 + \frac{1}{6}M_3 = 2$
$\frac{1}{6}M_2 + \frac{2}{3}M_3 + \frac{1}{6}M_4 = 2$
$\frac{1}{6}M_3 + \frac{2}{3}M_4 = 2$
where $M_i = s''(x_i),\quad M_1 = M_5 = 0$

3.(c) The equations for $M_i = s''(x_i)$ are
$\frac{1}{3}M_2 + \frac{1}{12}M_3 = 1$
$\frac{1}{12}M_2 + \frac{1}{3}M_3 + \frac{1}{6}M_4 = -\frac{7}{2}$
$\frac{1}{6}M_3 + \frac{2}{3}M_4 = 2$
The solution is
$M_2 = \frac{38}{7},\quad M_3 = -\frac{68}{7},\quad M_4 = \frac{38}{7}$

6. The equations for $M_i = s''(x_i)$ are
$\frac{1}{3}M_1 + \frac{1}{6}M_2 = \frac{1}{2}$
$\frac{1}{6}M_1 + \frac{2}{3}M_2 + \frac{1}{6}M_3 = \frac{1}{3}$
$\frac{1}{6}M_2 + \frac{2}{3}M_3 + \frac{1}{6}M_4 = \frac{1}{12}$
$\frac{1}{6}M_3 + \frac{1}{3}M_4 = \frac{1}{48}$
The solution is
$M_1 = 173/120,\quad M_2 = 7/60,\quad M_3 = 11/120,\quad M_4 = 1/60$
Using the resulting value of $s(x)$, we obtain
$s(\frac{3}{2}) = 0.652604$ $\qquad$ True $f(\frac{3}{2}) = \frac{2}{3} = 0.666667$
$s(\frac{5}{2}) = 0.403646$ $\qquad$ True $f(\frac{5}{2}) = 0.4$
$s(\frac{7}{2}) = 0.284896$ $\qquad$ True $f(\frac{7}{2}) = \frac{2}{7} = 0.285714$

CHAPTER 6

Section 6.1

2.(a) $\quad \max_{-1 \leq x \leq 1} \left| \tan^{-1}(x) - t_1(x) \right| = 0.215$
$t_3(x) = x - \frac{1}{3}x^3$

(b) $\quad \max_{-1 \leq x \leq 1} \left| \tan^{-1}(x) - m_3(x) \right| = 0.00495$

4. Use trigonometric identities. For example, $\cos(-x) = \cos(x)$
$\cos(x + 2n\pi) = \cos(x)$, any integer n.

5.(a) From (6.8), $\rho_3(f) \leq 0.00198$

(b) $\max\limits_{0 \leq x \leq \pi/2} |\cos(x) - m_3(x)| = 0.00137$

7.(a) Let $\theta = \tan^{-1}(x), x > 0$. Then $0 < \theta < \pi/2$, and $\tan(\theta) = x$. From using a right triangle with angle θ, it is clear that $\tan[(\pi/2) - \theta] = 1/x$, or $(\pi/2) - \theta = \tan^{-1}(1/x)$, as desired.

Section 6.2

1. $T_5(x) = 16x^5 - 20x^3 + 5x$

3. $x_m = \cos(m\pi/n), m = 0, 1, 2, \ldots, n$

6. $1/2^{n-1}$

8.(c) $S_{n+1}(x) = 2xS_n(x) - S_{n-1}(x), n \geq 1$

Section 6.3

2. $x_1 = \cos(\pi/4) = 1/\sqrt{2}, \qquad x_2 = \cos(3\pi/4) = -1/\sqrt{2}$

$$P_1(x) = e^{x_1} + (x - x_1)\frac{e^{x_1} - e^{x_2}}{x_1 - x_2}$$

$$\doteq 1.2606 + 1.0854x$$

3.(b) $P_3(x) = 0.9670576x - 0.18712962x^3$

$\max\limits_{-1 \leq x \leq 1} |\tan^{-1}(x) - P_3(x)| = 0.00590$

4. Use $t = \frac{1}{2}(1 + x), -1 \leq x \leq 1$, and approximate $f(x) = e^{(1+x)/2}$ on $[-1, 1]$.

7. Look at the floating-point representation, $t = \hat{t} \cdot 2^k$, with k an integer and $1/2 \leq \hat{t} < 1$.

9. Evaluate $p(-x)$ and equate it to $p(x)$.

CHAPTER 7

Section 7.1

1. $T_4(f) = 0.697024 \qquad I - T_4(f) = -0.00388$

$S_4(f) = 0.693254 \qquad I - S_4(f) = -0.000107$

2.(a)

n	T_n	Error	Ratio
2	26.516336	-25.2	
4	3.2490505	-1.95	13.0
8	1.6245252	$-3.22\mathrm{E}-1$	6.04
16	1.3757225	$-7.33\mathrm{E}-2$	4.39
32	1.3203119	$-1.79\mathrm{E}-2$	4.09
64	1.3068479	$-4.45\mathrm{E}-3$	4.02
128	1.3035057	$-1.11\mathrm{E}-3$	4.01
256	1.3026716	$-2.78\mathrm{E}-4$	4.00
512	1.3024632	$-6.95\mathrm{E}-5$	4.00

3.(a)

n	S_n	Error	Ratio
2	22.715077371	-21.4	
4	-4.5067112930	5.81	-3.69
8	1.0830168315	$2.19\mathrm{E}-1$	26.5
16	1.2927882745	$9.61\mathrm{E}-3$	22.8
32	1.3018416653	$5.52\mathrm{E}-4$	17.4
64	1.3023598879	$3.38\mathrm{E}-5$	16.3
128	1.3023915828	$2.10\mathrm{E}-6$	16.1
256	1.3023935531	$1.31\mathrm{E}-7$	16.0
512	1.3023936761	$8.20\mathrm{E}-9$	16.0

6.(b) $B_4(f) = 0.69317460$, $I - B_4(f) = -0.0000274$

9. 2

Section 7.2

1.(a) $\pi h^2/24$ **(b)** $h^2/6$

3.(a) $\bar{E}_n^T (f) = \dfrac{-h^2}{12} [e^\pi - 1]$

     $\bar{E}_{32}^T(f) = -0.01778.$ True $E_{32}^T(f) = -0.0179$

5. $\bar{E}_n^S(f) = \dfrac{13h^4}{1215},$ $h = 2/n$

     Choose $h \leq 0.031$, or $n \geq 65$.

     The number of nodes $= n + 1 \geq 66$.

9.(b)   Use $\dfrac{I_{2n} - I_n}{I_{4n} - I_{2n}} = \dfrac{(I - I_n) - (I - I_{2n})}{(I - I_{2n}) - (I - I_{4n})}.$

     Apply (7.39) and simplify.

10.(a) $p = 1.5$

11.(a) From problem 2(a) in section 7.1, with $n = 32$,
$$I - T_{32} \doteq \tfrac{1}{3}(T_{32} - T_{16}) = -0.01847$$
True $I - T_{32} = -0.0179$

14. $n \geq 396$

Section 7.3

1. $I_3 = 2.3503369$, $I - I_3 = 0.0000655$

2.(a)

n	I_n	Error
2	-19.244871326	20.5
3	9.0897287084	-7.79
4	0.9220525215	$3.80\text{E}-1$
5	1.1462063447	$1.56\text{E}-1$
6	1.3298098510	$-2.74\text{E}-2$
7	1.3003438200	$2.05\text{E}-3$
8	1.3024781041	$-8.44\text{E}-5$
9	1.3023918529	$1.83\text{E}-6$
10	1.3023936884	$-4.10\text{E}-9$

5.(a) $\int_0^1 f(x) \log (x)\, dx \doteq -f(\tfrac{1}{4}) \equiv I_1$

5.(b) $w_1 + w_2 = -1$
$$w_1 x_1 + w_2 x_2 = -\tfrac{1}{4}$$
$$w_1 x_1^2 + w_2 x_2^2 = -\tfrac{1}{9}$$
$$w_1 x_1^3 + w_2 x_2^3 = -\tfrac{1}{16}$$

(c) $I_1 = -0.968912$,   error $= 0.0228$

Section 7.4

1.(a)

h	$D_h f(0)$	Error Estimate	Error	Ratio
0.1	1.051709	$-5.00\text{E}-2$	$-5.17\text{E}-2$	
0.05	1.025422	$-2.50\text{E}-2$	$-2.54\text{E}-2$	2.03
0.025	1.012605	$-1.25\text{E}-2$	$-1.26\text{E}-2$	2.02
0.0125	1.006276	$-6.25\text{E}-3$	$-6.28\text{E}-3$	2.01
0.00625	1.003132	$-3.12\text{E}-3$	$-3.13\text{E}-3$	2.00

	h	$D_h f(0)$	Error Estimate	Error	Ratio
2.(a)	0.1	1.00166775	$-1.67\mathrm{E}-3$	$-1.67\mathrm{E}-3$	
	0.05	1.00041672	$-4.17\mathrm{E}-4$	$-4.17\mathrm{E}-4$	4.00
	0.025	1.00010417	$-1.04\mathrm{E}-4$	$-1.04\mathrm{E}-4$	4.00
	0.0125	1.00002604	$-2.60\mathrm{E}-5$	$-2.60\mathrm{E}-5$	4.00
	0.00625	1.00000651	$-6.51\mathrm{E}-6$	$-6.51\mathrm{E}-6$	4.00

4. $f'(x_2) \doteq D_h f(x_2) \equiv (1/12h)\{f(x_0) - 8f(x_1) + 8f(x_3) - f(x_4)\}$
      $f'(x_2) - D_h f(x_2) = (h^4/30)f^{(5)}(c), \qquad x_0 \le c \le x_4$

6.   Denote $\hat{D}_h f(x) = 2D_h f(x) - D_{2h} f(x)$, with $D_h f(x)$ defined by (7.65).

h	$D_h f(x)$	$\hat{D}_h f(x)$	$f(x) - \hat{D}_h f(x)$
0.1	-0.54243		
0.05	-0.52144	-0.50045	0.00045
0.025	-0.51077	-0.50010	0.00010
0.0125	-0.50540	-0.50003	0.00003

9.(b) $f''(t) \doteq \dfrac{1}{h^2}[-f(t + 3h) + 4f(t + 2h) - 5f(t + h) + 2f(t)]$

11. $|f'(x) - \bar{D}_h f(x)| \le \dfrac{h^2}{6}|f'''(c)| + \dfrac{\delta}{h}, \; x - h \le c \le x + h$, with $\bar{D}_h f(x)$

     the version of $D_h f(x)$ with rounded entries, and $\delta = \max\{|\epsilon_0|, |\epsilon_1|\}$.

13.   With $h = 0.2$, $D_h^{(2)} f(0.5) = 0.0775$.
     The part of the error due to rounding in table entries is
     $\dfrac{4\delta}{h^2} = \dfrac{4(0.00005)}{(0.2)^2} = 0.005$

CHAPTER 8

Section 8.1

1.(a)   Write $p(x) = a_0 + a_1 x + a_2 x^2 + a_3 x^3$. Then a_0, a_1, a_2, a_3 satisfy
     $a_0 + a_1 x_i + a_2 x_i^2 + a_3 x_i^3 = y_i, \qquad i = 0, 1, 2, 3$

3. $2x_1 + x_2 = 3$
     $x_{i-1} + 2x_i + x_{i+1} = 4, \qquad i = 2, 3, \ldots, n - 1$
     $x_{n-1} + 2x_n = 3$

4.
```
SIGN = 1.0
DO 10 I=1,N
    B(I)=SIGN
    SIGN=-SIGN
    DO 10 J=1,I
        A(I,J)=FLOAT(I)/FLOAT(J)
10      A(J,I)=A(I,J)
```

Section 8.2

1.(a)

$$\left[\begin{array}{ccc|c} 2 & 1 & -1 & 6 \\ 4 & 0 & -1 & 6 \\ -8 & 2 & 2 & -8 \end{array}\right] \begin{array}{c} m_{21} = 2 \\ \longrightarrow \\ m_{31} = -4 \end{array} \left[\begin{array}{ccc|c} 2 & 1 & -1 & 6 \\ 0 & -2 & 1 & -6 \\ 0 & 6 & -2 & 16 \end{array}\right]$$

$$\begin{array}{c} \longrightarrow \\ m_{32} = -3 \end{array} \left[\begin{array}{ccc|c} 2 & 1 & -1 & 6 \\ 0 & -2 & 1 & -6 \\ 0 & 0 & 1 & -2 \end{array}\right]$$

$x_3 = -2, x_2 = 2, x_1 = 1$

4. If the third equation is a multiple of the second, then there are an infinity of solutions. Solve for x_3 and then substitute into the first equation, to determine the relation between x_1 and x_2. If the third equation is not a multiple of the second, then the system is inconsistent and there is no solution. To better understand the possible behavior, try particular systems of the given special form.

7. $MD(A \rightarrow U) = \frac{1}{2}n(n + 1)$
$MD(b \rightarrow g) = n - 1$
$MD(g \rightarrow x) = \frac{1}{2}n(n + 1)$

Section 8.3

1.(a) $\begin{bmatrix} 5 & 6 \\ -1 & 6 \end{bmatrix}$ **(b)** $\begin{bmatrix} a & b + 2c \\ 0 & 3c \end{bmatrix}$

(c) $A = \begin{bmatrix} \frac{7}{9} & -\frac{4}{9} & -\frac{4}{9} \\ -\frac{4}{9} & \frac{1}{9} & -\frac{8}{9} \\ -\frac{4}{9} & -\frac{8}{9} & \frac{1}{9} \end{bmatrix}$, $B = I$

2. Write $A = [A_{\ast 1}, A_{\ast 2}, A_{\ast 3}]$, with $A_{\ast j}$ denoting the jth column of A, a column matrix. Then $AD = [\lambda_1 A_{\ast 1}, \lambda_2 A_{\ast 2}, \lambda_3 A_{\ast 3}]$. Column j of A is multiplied by λ_j.

4. $A = ww^T$, $A^2 = (ww^T)(ww^T) = w(w^Tw)w^T$
$= w(1)w^T = ww^T = A$

7. mnp

10. $a = d, b = c$

12.(a)
$$\begin{bmatrix} 0.4 & -0.1 & -0.1 \\ -0.1 & 0.4 & -0.1 \\ -0.1 & -0.1 & 0.4 \end{bmatrix}$$

(c)
$$\begin{bmatrix} -0.5 & -0.5 & -0.5 & -0.5 \\ -0.5 & -1.5 & -1.5 & -1.5 \\ -0.5 & -1.5 & -2.5 & -2.5 \\ -0.5 & -1.5 & -2.5 & -3.5 \end{bmatrix}$$

16. $[1, 1, -2, -2]^T$

Section 8.4

1.
$$L = \begin{bmatrix} 1 & 0 & 0 \\ 2 & 1 & 0 \\ -1 & \frac{1}{2} & 1 \end{bmatrix}, \quad U = \begin{bmatrix} 1 & 2 & 1 \\ 0 & -2 & 1 \\ 0 & 0 & \frac{1}{2} \end{bmatrix}$$

2.(a)
$$L = \begin{bmatrix} 1 & 0 & 0 \\ 2 & 1 & 0 \\ -4 & -3 & 1 \end{bmatrix}, \quad U = \begin{bmatrix} 2 & 1 & -1 \\ 0 & -2 & 1 \\ 0 & 0 & 1 \end{bmatrix}$$

6.(a)
$$L = \begin{bmatrix} 1 & 0 \\ -1 & 2 \end{bmatrix}$$

This is not the only possible value for L.

8. $[\frac{5}{6}, -\frac{2}{3}, \frac{1}{2}, -\frac{1}{3}, \frac{1}{6}]^T$

11.(a) $MD(A \to LU) = 2n - 2$

(b) $MD(x) = 3n - 2$

Section 8.5

3.(a) cond $(A) = 289$

4. cond $(A) = \dfrac{(1 + |c|)^2}{|1 - c^2|}$

5. Use $I = AA^{-1}$ and property (P3) in (8.102).

7.(b) cond $(A_n) = n2^{n-1}$

Section 8.6

1. $f^*(x) = 2.0810909x - 1.0837273,\ E = 0.236$

5.
$$\begin{bmatrix} 1 & \frac{1}{2} & \frac{1}{3} \\ \frac{1}{2} & \frac{1}{3} & \frac{1}{4} \\ \frac{1}{3} & \frac{1}{4} & \frac{1}{5} \end{bmatrix} \begin{bmatrix} a_1 \\ a_2 \\ a_3 \end{bmatrix} = \begin{bmatrix} w_1 \\ w_2 \\ w_3 \end{bmatrix}$$

$w_i = \int_0^1 x^{i-1} f(x)\, dx, \qquad i = 1, 2, 3$

7. $na + b \sum_{j=1}^{n} e^{-x_j} = \sum_{j=1}^{n} y_j$

$a \sum_{j=1}^{n} e^{-x_j} + b \sum_{j=1}^{n} e^{-2x_j} = \sum_{j=1}^{n} y_j e^{-x_j}$

Section 8.7

1.(a) $f(\lambda) = \lambda^2 - 5\lambda - 6, \ \lambda_1 = 6, \ \lambda_2 = -1$
$v^{(1)} = [2, 5]^T, \qquad v^{(2)} = [1, -1]^T$

2. $v^{(2)} = [1, 0, -1]^T, \qquad v^{(3)} = [1, 2, 1]^T$

6. $U = \begin{bmatrix} \dfrac{1}{\sqrt{3}} & \dfrac{1}{\sqrt{2}} & \dfrac{1}{\sqrt{6}} \\[2mm] -\dfrac{1}{\sqrt{3}} & 0 & \dfrac{2}{\sqrt{6}} \\[2mm] \dfrac{1}{\sqrt{3}} & -\dfrac{1}{\sqrt{2}} & \dfrac{1}{\sqrt{6}} \end{bmatrix}$

8. In both cases, $f(\lambda) = \lambda^2 - 2\lambda + 1 = (\lambda - 1)^2$

11.(b) $\lambda_1 = -12, \ \lambda_2 = 9, \ \lambda_3 = 3, \ v^{(1)} = [1, -1, 1]^T$

   (d) $\lambda_1 = 3, \ \lambda_2 = 2, \ \lambda_1 = 1, \ v^{(1)} = [-4, 3, 1]^T$

   (f) $\lambda_1 = 15, \ \lambda_2 = 5, \ \lambda_3 = 5, \ \lambda_4 = -1, \ v^{(1)} = [1, 1, 1, 1]^T$

CHAPTER 9

Section 9.1

1.(a) $y(x) = \sin(x)$

   (b) $y(x) = x/(x - 0.5)$

2.(a) $y(x) = ce^{\lambda x} - (1/\lambda)$, general solution
$y(x) = [1 + (1/\lambda)]e^{\lambda x} - 1/\lambda,$

3.(a) $y = \sqrt{x^2 + 4}$

5.(a) $\partial f/\partial z = -1$, equation is well-conditioned.

Section 9.2

1.(a)

x	$Y(x)$	$Y(x) - y_{0.2}(x)$	$Y(x) - y_{0.1}(x)$	$Y(x) - y_{0.05}(x)$
2	1.10715	$-3.32E-2$	$-1.63E-2$	$-8.11E-3$
4	1.32582	$-1.67E-2$	$-8.34E-3$	$-4.17E-3$
6	1.40565	$-9.70E-3$	$-4.86E-3$	$-2.44E-3$
8	1.44644	$-6.38E-3$	$-3.20E-3$	$-1.60E-3$
10	1.47113	$-4.54E-3$	$-2.28E-3$	$-1.14E-3$

2.(c)

x	$Y(x)$	$Y(x) - y_{0.5}(x)$	$Y(x) - y_{0.0625}(x)$
2	0.49315	$2.23E-1$	$-4.44E-3$
4	-1.41045	$1.34E+0$	$8.67E-3$
6	0.68075	$6.39E+0$	$-2.78E-3$
8	0.84386	$3.25E+1$	$-6.36E-3$
10	-1.38309	$1.65E+2$	$8.07E-3$

Section 9.3

2. $K = \sup_{0 \le x \le b} (1/(x + 1)) = 1$, for any $b \ge 0$

$\max_{0 \le x \le b} |Y''(x)| = \max_{0 \le x \le b} (2x/(x + 1)) = 2b/(b + 1)$

$|Y(b) - y_h(b)| \le h \left[\dfrac{e^b - 1}{2} \right] \left[\dfrac{2b}{b + 1} \right]$

For $h = 0.05$,

b	$Y(b) - y_h(b)$	Error Bound
1	0.0187	0.0430
2	0.0634	0.213
3	0.125	0.716
4	0.199	2.14

4.(b) Use $K = \max_{0 \le x \le b} |4Y(x)| = 2$,   for $b \ge 1$.

$\max_{0 \le x \le b} |Y''(x)| = \max_{0 \le x \le b} \left| \dfrac{-6x + 2x^3}{(1 + x^2)^3} \right| \doteq 1.46$,   for $b \ge 0.41$

5. Solve $D'(x) = \lambda D(x) - \sin(x)$, $D(0) = 0$

The solution $D(x) = [1/2(\lambda^2 + 1)]\{[\lambda \sin(x) + \cos(x)] - e^{\lambda x}\}$ and

$Y(x_n) - y_h(x_n) \doteq D(x_n)h$.

For $\lambda = -1$ and $h = 0.1$, the answers are as follows.

x	$D(x)h$	$Y(x) - y_n(x)$
1	−0.0167	−0.0169
2	−0.0365	−0.0372
3	−0.0295	−0.0299
4	0.00212	0.00258
5	0.0309	0.0319
6	0.0309	0.0316

8.(a) For the case of $2h = 0.1$, $h = 0.05$,

x	$Y(x) - y_h(x)$	$y_h(x) - y_{2h}(x)$	$\bar{y}_h(x)$	$Y(x) - \bar{y}_h(x)$
2	−8.11E−3	−8.22E−3	1.1070296	1.19E−4
4	−4.17E−3	−4.16E−3	1.3258165	1.12E−6
6	−2.44E−3	−2.43E−3	1.4056549	−7.28E−6
8	−1.60E−3	−1.60E−3	1.4464472	−5.84E−6
10	−1.14E−3	−1.14E−3	1.4711317	−4.06E−6

10. Note that $y_{n+1} = (1 + h)y_n$, $n \geq 0$. Use this to obtain $y_n = (1 + h)^n$, and then modify it to the desired form.

Section 9.4

2.(a) Following are some results for $h = 0.05$ and $2h = 0.1$. The Richardson error estimate (9.64) is also included.

x	$Y(x) - y_{2h}(x)$	$Y(x) - y_h(x)$	Ratio	$\frac{1}{3}[y_h(x) - y_{2h}(x)]$
2	2.05E−4	5.12E−5	4.0	5.12E−5
4	1.75E−4	4.28E−5	4.1	4.40E−5
6	1.01E−4	2.48E−5	4.1	2.70E−5
8	6.40E−5	1.57E−5	4.1	1.61E−5
10	4.37E−5	1.07E−5	4.1	1.10E−5

5.

x	$Y(x) - y_h(x)$	$\frac{1}{3}[y_h(x) - y_{2h}(x)]$	$\tilde{y}_h(x)$	$Y(x) - \tilde{y}_h(x)$
2	2.31E−4	2.31E−4	0.4931513	−7.24E−7
4	2.91E−4	3.07E−4	−1.4104300	−1.61E−5
6	−4.08E−4	−4.19E−4	0.6807432	1.16E−5
8	5.68E−5	5.08E−5	0.8438522	6.05E−6
10	3.62E−4	3.78E−4	−1.3830760	−1.67E−5

6. With $\gamma_2 = 1$, the RK method is

$$y_{n+1} = y_n + hf(x_n + \frac{h}{2}, y_n + \frac{h}{2}f(x_n, y_n))$$

7. Using the formula for $\gamma_2 = 1$ in problem 6, the error results are given below for $h = 0.05$, $2h = 0.1$. Richardson's error estimate is included.

x	$Y(x) - y_{2h}(x)$	$Y(x) - y_h(x)$	Ratio	$\frac{1}{3}[y_h(x) - y_{2h}(x)]$
2	1.46E−3	3.53E−4	4.1	3.69E−4
4	−6.68E−4	−1.67E−4	4.0	−1.67E−4
6	−8.31E−4	−1.97E−4	4.2	−2.11E−4
8	1.37E−3	3.33E−4	4.1	3.46E−4
10	−3.08E−4	−8.01E−5	3.8	−7.60E−5

8.(a) For $h = 0.05$ and $2h = 0.1$

x	$Y(x) - y_{2h}(x)$	$Y(x) - y_h(x)$	Ratio	$\frac{1}{3}[y_h(x) - y_{2h}(x)]$
2	6.46E−4	1.56E−4	4.1	1.63E−4
4	2.51E−4	6.10E−5	4.1	6.34E−5
6	1.26E−4	3.07E−5	4.1	3.18E−5
8	7.49E−5	1.82E−5	4.1	1.89E−5
10	4.94E−5	1.20E−5	4.1	1.25E−5

9. The results for $h = 0.125$ and $2h = 0.25$ are given below, along with the Richardson error estimate for $y_h(x)$, taken from problem 4.

x	$Y(x) - y_{2h}(x)$	$Y(x) - y_h(x)$	Ratio	$\frac{1}{15}[y_h(x) - y_{2h}(x)]$
2	3.02E−5	1.69E−6	18	1.90E−6
4	−3.64E−5	−2.20E−6	17	−2.28E−6
6	−1.44E−6	3.41E−8	−42*	−9.85E−8
8	3.74E−5	2.16E−6	17	2.35E−6
10	−2.97E−5	−1.83E−6	16	−1.86E−6

*Richardson extrapolation not justified, since ratio < 0.

12.(a) The true solution is $v(t) = -98[1 - e^{-0.1t}]$, $t \geq 0$.

Section 9.5

1.(a) Let $h = 0.05$, $2h = 0.1$.

x	$Y(x) - y_{2h}(x)$	$Y(x) - y_h(x)$	Ratio	$\frac{1}{3}[y_h(x) - y_{2h}(x)]$
2	5.10E−4	1.28E−4	4.0	1.27E−4
4	4.37E−4	1.07E−4	4.1	1.10E−4
6	2.54E−4	6.23E−5	4.1	6.40E−5
8	1.61E−4	3.93E−5	4.1	4.04E−5
10	1.09E−4	2.68E−5	4.1	2.76E−5

2.(a) Let $h = 0.05$, $2h = 0.1$. The Richardson error formula is from problem 4 of section 9.4. The order of convergence is $p = 3$.

x	$Y(x) - y_{2h}(x)$	$Y(x) - y_h(x)$	Ratio	$\frac{1}{7}[y_h(x) - y_{2h}(x)]$
2	1.27E−4	1.49E−5	8.5	1.60E−5
4	5.17E−6	6.53E−7	7.9	6.45E−7
6	−1.55E−6	−1.65E−7	9.4	−1.98E−7
8	−1.72E−6	−1.94E−7	8.9	−2.17E−7
10	−1.36E−6	−1.56E−7	8.7	−1.72E−7

4.(a) Let $h = 0.05$, $2h = 0.1$.

x	$Y(x) - y_{2h}(x)$	$Y(x) - y_h(x)$	Ratio	$\frac{1}{3}[y_h(x) - y_{2h}(x)]$
2	−1.47E−4	−3.14E−5	4.7	−3.86E−5
4	−1.08E−4	−2.40E−5	4.5	−2.80E−5
6	−6.11E−5	−1.37E−5	4.5	−1.58E−5
8	−3.81E−5	−8.61E−6	4.4	−9.84E−6
10	−2.58E−5	−5.85E−6	4.4	−6.66E−6

6.(a) The errors for the three stepsizes are given below. An explanation for the sharply differing values of the errors must await the material on regions of stability given in section 9.6.

x	$Y(x) - y_{0.1}(x)$	$Y(x) - y_{0.02}(x)$	$Y(x) - y_{0.01}(x)$
2	3.82E+13	1.37E−4	1.09E−6
4	2.06E+30	1.33E−4	−5.54E−8
6	1.11E+47	1.29E−4	−1.05E−6
8	5.98E+63	1.37E−4	9.27E−7
10	3.23E+80	1.34E−4	2.76E−7

Section 9.6

2.

	Errors $= Y(x) - y_h(x)$		
x	$h = 0.5$	$h = 0.1$	$h = 0.01$
1	$-4.32E-1$	$-6.31E+0$	$-8.11E+25$
2	$-2.15E+0$	$-8.10E+3$	$-1.32E+56$
3	$-9.78E+0$	$-1.04E+7$	$-2.14E+86$
4	$-4.57E+1$	$-1.33E+10$	$-3.47E+116$
5	$-2.16E+2$	$-1.71E+13$	$-5.63E+146$

4. The numerical solution of (9.108) yields

$$z_n = \epsilon \left(\frac{1 + \dfrac{h\lambda}{2}}{1 - \dfrac{h\lambda}{2}} \right)^n, \quad n \geq 0$$

Examine its behavior for $\lambda < 0$ as $n \to \infty$.

6. $-2 < h\lambda < 0$

8. $y_{n+1} - y_{n+1}^{(j+1)} = h\lambda[y_{n+1} - y_{n+1}^{(j)}], j \geq 0.$
Convergence of $y_{n+1}^{(j)}$ to y_{n+1} occurs only if $|\lambda h| < 1$, or $h < 1/|\lambda|.$

9. We give the values obtained with the backward Euler method when $\lambda = -50$. Compare these with the values for $\lambda = -1, -10$ for the same values of h. Note that the errors decrease by a factor of about 10 when h decreases from 0.1 to 0.01; this is consistent with the method being order 1 in its convergence rate.

	Errors $Y(x) - y_h(x)$		
x	$h = 0.5$	$h = 0.1$	$h = 0.01$
2	$1.01E-3$	$1.70E-4$	$1.64E-5$
4	$1.59E-4$	$2.88E-5$	$2.81E-6$
6	$4.82E-5$	$9.00E-6$	$8.87E-7$
8	$2.03E-5$	$3.86E-6$	$3.82E-7$
10	$1.04E-5$	$1.99E-6$	$1.97E-7$

Section 9.7

1. $Y_1' = Y_1 - 2Y_2 - 2e^{-x} + 2, \quad Y_1(0) = 1$
$Y_2' = 2Y_1 - Y_2 - 2e^{-x} + 1, \quad Y_2(0) = 1$

3.(b) Let $Y_1 = y$, $Y_2 = y'$. Then

$$Y' = \begin{bmatrix} 0 & 1 \\ -13 & -4 \end{bmatrix} Y + \begin{bmatrix} 0 \\ 40 \cos{(x)} \end{bmatrix}, \qquad Y(0) = \begin{bmatrix} 3 \\ 4 \end{bmatrix}$$

6.(b) Use the notation of 3(b). And the error estimate is

$$y(x) - y_h(x) \doteq y_h(x) - y_{2h}(x)$$

	Errors in $Y_1(x) = y(x)$			Error estimate
x	$h = 0.1$	$h = 0.05$	$h = 0.025$	for $h = 0.025$
2	$-2.13E-2$	$-5.12E-3$	$-1.65E-3$	$-3.47E-3$
4	$4.03E-2$	$2.09E-2$	$1.06E-2$	$1.03E-2$
6	$-5.14E-2$	$-2.54E-2$	$-1.26E-2$	$-1.28E-2$
8	$4.17E-4$	$-3.74E-4$	$-3.31E-4$	$-4.35E-5$
10	$5.10E-2$	$2.57E-2$	$1.29E-2$	$1.28E-2$

7.(b) Use the notation of 3(b). And the error estimate is

$$y(x) - y_h(x) \doteq \tfrac{1}{3}[y_h(x) - y_{2h}(x)]$$

	Errors in $Y_1(x) = y(x)$			Error estimate
x	$h = 0.1$	$h = 0.05$	$h = 0.025$	for $h = 0.025$
2	$8.20E-4$	$1.74E-4$	$4.06E-5$	$4.44E-5$
4	$-7.97E-3$	$-2.02E-3$	$-5.04E-4$	$-5.06E-4$
6	$4.77E-3$	$1.32E-3$	$3.44E-4$	$3.27E-4$
8	$3.95E-3$	$9.06E-4$	$2.15E-4$	$2.30E-4$
10	$-8.05E-3$	$-2.08E-3$	$-5.23E-4$	$-5.19E-4$

8.(a) $c_1 = 3$, $c_2 = 2$

REFERENCES

M. Abramowitz and I. Stegun, Eds. (1964), *Handbook of Mathematical Functions*, Dover Press, New York.

K. Atkinson (1978), *An Introduction to Numerical Analysis*, Wiley, New York.

R. Burden, J. Faires, and A. Reynolds (1982), *Numerical Analysis*, 2nd ed., Prindle, Weber, and Schmidt, Boston.

W. Cheney and D. Kincaid (1980), *Numerical Mathematics and Computing*, Brooks/Cole, Monterey, California.

S. Conte and C. deBoor (1980), *Elementary Numerical Analysis*, 3rd ed., McGraw-Hill, New York.

G. Dahlquist and A. Bjorck (1974), *Numerical Methods*, Prentice-Hall, Englewood Cliffs, N.J.

P. Davis and P. Rabinowitz (1975), *Methods of Numerical Integration*, 2nd ed., Academic Press, New York.

C. deBoor (1978), *A Practical Guide to Splines*, Springer-Verlag, New York.

J. Dennis and R. Schnabel (1983), *Numerical Methods for Unconstrained Optimization and Nonlinear Equations*, Prentice-Hall, Englewood Cliffs, N.J.

J. Dongarra, J. Bunch, C. Moler, and G. Stewart (1979), *LINPACK User's Guide,* SIAM, Philadelphia.

G. Forsythe, M. Malcolm, and C. Moler (1977), *Computer Methods for Mathematical Computations,* Prentice-Hall, Englewood Cliffs, N.J.

G. Forsythe and C. Moler (1967), *Computer Solution of Linear Algebraic Systems,* Prentice-Hall, Englewood Cliffs, N.J.

B. Garbow, J. Boyle, J. Dongarra, and C. Moler (1977), *Matrix Eigensystem Routines–EISPACK Extension,* Lecture Notes in Computer Science, Springer-Verlag, New York.

C.W. Gear (1971), *Numerical Initial Value Problems in Ordinary Differential Equations,* Prentice-Hall, Englewood Cliffs, N.J.

G. Golub and C. Van Loan (1983), *Matrix Computations,* Johns Hopkins University Press, Baltimore, Md.

P. Henrici (1964), *Elements of Numerical Analysis,* Wiley, New York.

P. Henrici (1982), *Essentials of Numerical Analysis,* Wiley, New York.

A. Householder (1970), *The Numerical Treatment of a Single Nonlinear Equation,* McGraw-Hill, New York.

E. Isaacson and H. Keller (1966), *Analysis of Numerical Methods,* Wiley, New York.

R. Johnson, *Numerical Methods: A Software Approach* (1982), Wiley, New York.

L. Johnson and R. Riess (1982), *Numerical Analysis,* 2nd ed., Addison-Wesley, Reading, Mass.

C. Lawson and R. Hanson (1974), *Solving Least Square Problems,* Prentice-Hall, Englewood Cliffs, N.J.

B. Parlett (1980), *The Symmetric Eigenvalue Problem,* Prentice-Hall, Englewood Cliffs, N.J.

S. Pizer and V. Wallace (1983), *To Compute Numerically: Concepts and Strategies,* Little, Brown, & Co., Boston.

M. Powell (1981), *Approximation Theory and Methods,* Cambridge Press, Cambridge.

A. Ralston and P. Rabinowitz (1978), *A First Course in Numerical Analysis,* 2nd ed., McGraw-Hill, New York.

W. Rheinboldt (1974), *Methods for Solving Systems of Nonlinear Equations,* SIAM, Philadelphia.

J. Rice (1964, 1968), *The Approximation of Functions,* Vol. I: *Linear Theory,* Vol. II: *Advanced Topics,* Addison-Wesley, Reading, Mass.

J. Rice (1981), *Matrix Computations and Mathematical Software,* McGraw-Hill, New York.

J. Rice (1983), *Numerical Methods, Software, and Analysis,* McGraw-Hill, New York.

L. Shampine and M. Gordon (1975), *Computer Solution of Ordinary Differential Equations,* W. Freeman, San Francisco.

B. Smith, J. Boyle, J. Dongarra, B. Garbow, Y. Ikebe, V. Klema, and C. Moler (1976), *Matrix Eigensystem Routines—EISPACK Guide,* Lecture Notes in Computer Science, Springer-Verlag, New York.

J. Stoer and R. Bulirsch (1980), *Introduction to Numerical Analysis,* Springer-Verlag, New York.

A. Stroud (1971), *Approximate Calculation of Multiple Integrals,* Prentice-Hall, Englewood Cliffs, N.J.

A. Stroud and D. Secrest (1966), *Gaussian Quadrature Formulas,* Prentice-Hall, Englewood Cliffs, N.J.

J. Wilkinson (1965), *The Algebraic Eigenvalue Problem,* Oxford University Press, New York.

Index

An asterisk (*) following a subentry name means that name is also listed separately with additional subentries of its own. A page number followed by a number in parentheses, prefixed by 'P', refers to a problem on the given page. For example, 140(P9) refers to problem 9 on page 140.